WEAPONS IN SPACE

WEAPONS IN SPACE

TECHNOLOGY, POLITICS, AND THE RISE AND FALL OF THE STRATEGIC DEFENSE INITIATIVE

AARON BATEMAN

The MIT Press
Cambridge, Massachusetts
London, England

The MIT Press would like to thank the anonymous peer reviewers who provided comments on drafts of this book. The generous work of academic experts is essential for establishing the authority and quality of our publications. We acknowledge with gratitude the contributions of these otherwise uncredited readers.

This book was set in Stone Sans and Stone Serif by Westchester Publishing Services. Printed and bound in the United States of America.

Library of Congress Cataloging-in-Publication Data

Names: Bateman, Aaron, author.
Title: Weapons in space : technology, politics, and the rise and fall of the
    Strategic Defense Initiative / Aaron Bateman.
Other titles: Technology, politics, and the rise and fall of the Strategic Defense
    Initiative
Description: Cambridge, Massachusetts : The MIT Press, [2024] | Includes
    bibliographical references and index.
Identifiers: LCCN 2023024011 (print) | LCCN 2023024012 (ebook) |
    ISBN 9780262547369 (paperback) | ISBN 9780262377393 (epub) |
    ISBN 9780262377386 (pdf)
Subjects: LCSH: Space control (Military science)—United States. | Strategic Defense
    Initiative. | Astronautics, Military—United States. | Space weapons. | United
    States—Relations—Soviet Union. | Soviet Union—Relations—United States. |
    Cold War.
Classification: LCC UG1523 .B38 2024 (print) | LCC UG1523 (ebook) |
    DDC 358/.8070973—dc23/eng/20230912
LC record available at https://lccn.loc.gov/2023024011
LC ebook record available at https://lccn.loc.gov/2023024012

10 9 8 7 6 5 4 3 2 1

CONTENTS

ACKNOWLEDGMENTS   vii

ABBREVIATIONS AND ACRONYMS   ix

THE "OTHER NIGHT SKY"   1

1   THE RISE AND FALL OF DÉTENTE ON EARTH AND IN SPACE   11

2   CAMPAIGN FOR THE HIGH GROUND   41

3   OUT OF THE BLACK   73

4   "EUROPE MUST NOT LEAVE SPACE TO THE AMERICANS".   99

5   OUT OF THE LABORATORY AND INTO SPACE   133

6   SDI AND THE NEW WORLD ORDER   165

SDI RECONSIDERED   197

A SENSE OF DÉJÀ VU   209

NOTES   217

BIBLIOGRAPHY   299

INDEX   315

ACKNOWLEDGMENTS

In writing this book, I have benefited greatly from a global network of friends and colleagues who assisted me with my journey in so many ways. Asif Siddiqi's encouragement was a significant factor in pursuing my study of SDI and placing it within the context of US military space policy and strategy. I am especially indebted to Larry Principe who has been an unwavering source of support and friendship, and so generously devoted so much of his time to reading and commenting on drafts of this manuscript. Bleddyn Bowen deserves special thanks for many stimulating conversations on space security, providing feedback on key chapters, and encouraging me, especially in moments of doubt. I also want to thank Oliver Barton, James Cameron, Elizabeth Charles, Dwayne Day, Joseph DeSutter, Thomas Ellis, Yulia Frumer, Peter Hays, Matthew Jones, Bob Kargon, John Krige, Bill Leslie, Clay Moltz, Michael Neufeld, Scott Pace, Ori Rabinowitz, Daniel Salisbury, Emma Salisbury, Brian Weeden, Peter Westwick, and James Wilson for discussions (and email exchanges) that have sharpened my arguments.

As a former Air Force intelligence officer, I submitted this book to the Department of Defense for security review. I want to thank the staff members within the Department of Defense Office of Prepublication Security Review who cleared this manuscript for public release. I am deeply appreciative of their professionalism and close communication at every step of the process. The views expressed in this publication are those of the author and do not necessarily reflect the official policy or position of the Department of Defense or the US government, and the clearance for public release of this manuscript by the Department of Defense does not imply Department of Defense endorsement or factual accuracy of the material.

Throughout this journey, my wife Marisa has been my constant support and cheerleader every step of the way. For more than a decade, she has been able to pick me up when I was down and has always encouraged me to achieve my goals. Without her, the project would have never been possible. She is truly the perfect partner.

Last, but certainly not least, I especially want to thank Anthony James Joes. Since I was eighteen years old, he has been my teacher, mentor, friend, and confidant. He believed in me before I believed in myself and has always encouraged me to pursue my intellectual passions. All of my achievements as an adult I owe to him and his never-ending encouragement and support. To Tony, I dedicate this book.

ABBREVIATIONS AND ACRONYMS

ABM	Anti-ballistic missile
ACDA	Arms Control and Disarmament Agency
ASAT	Anti-satellite
BMD	Ballistic missile defense
CIA	Central Intelligence Agency
DIA	Defense Intelligence Agency
ESA	European Space Agency
FCO	Foreign and Commonwealth Office
HOE	Homing Overlay Experiment
ICBM	Intercontinental Ballistic Missile
INF	Intermediate-Range Nuclear Forces
MDA	Missile Defense Agency
MHV	Miniature Homing Vehicle
MoD	Ministry of Defence
MOU	Memorandum of Understanding
NASA	National Aeronautics and Space Administration
NATO	North Atlantic Treaty Organization
NIE	National Intelligence Estimate
NRO	National Reconnaissance Office
NSA	National Security Agency
NSC	National Security Council
NSDD	National Security Decision Directive
OSD	Office of the Secretary of Defense
OSTP	Office of Science and Technology Policy

SDI   Strategic Defense Initiative
SDIO   Strategic Defense Initiative Organization
START   Strategic Arms Reduction Treaty
WEU   Western European Union

THE "OTHER NIGHT SKY"

Space is just another place where wars will be fought.
—Senator Barry Goldwater, 1984

On November 14, 2021, Russia fired off a missile into outer space and eviscerated a defunct satellite launched by the Soviet Union.[1] The impact generated thousands of pieces of harmful debris, forcing astronauts and cosmonauts onboard the International Space Station to shelter in place as a safety precaution.[2] China, India, and the US, in addition to Russia, have demonstrated the ability to destroy satellites in orbit.[3] The fact that advanced missile defense interceptors can be used as anti-satellite (ASAT) weapons means that as missile defense systems proliferate, the small club of ASAT-capable nations is set to grow. In what could easily be mistaken as an excerpt from a Tom Clancy novel, US Space Command reported in 2020 that Russia had tested a satellite that released subsatellites with "characteristics of a weapon."[4] The US shone a spotlight on its security anxieties in space when it announced that space is a "warfighting domain."[5] Combined, these events, and many others like them, have intensified global concern about a space arms race.

War in space would have far-reaching and catastrophic consequences. Debris produced from strikes on military satellites would not discriminate in its collateral damage to other space systems. Despite the fact that satellites are so important for modern civilization, there are very few formal international constraints—that is, legally binding mechanisms—in space. Today, the primary international agreement that limits certain military activities in space is the 1967 Outer Space Treaty, which forbids the deployment of nuclear weapons in space (among other stipulations) but does not ban nonnuclear space weaponry.[6] Vice President Kamala Harris announced in a 2022 speech at Vandenberg Space Force Base in California that the US will lead the way in developing an international consensus concerning "what is right, what is wrong, and what is acceptable" in space.[7] This will be no easy task due to the tension between maintaining a sustainable space environment for all and the objective of a growing number of countries to develop technologies that can be used to degrade and destroy the space systems that have become the bedrock of modern warfare.

Tension over the military use of space is certainly not a new phenomenon. The lack of consensus today concerning limits on space militarization is directly connected to issues left unresolved at the end of the Cold War. The US and the Russian Federation emerged out of the Cold War with historically unprecedented nuclear arms control agreements but no new treaties constraining military actions in space. This outcome was not the result of military space issues having been a marginal topic in arms control negotiations—quite the contrary. Rather, in the 1980s, the US government decided to prioritize freedom of action in space due in large part to the pursuit of space-based missile defense through the Strategic Defense Initiative (SDI), more commonly known as "Star Wars." From the standpoint of key American officials, space arms control would have handicapped the US in an arena in which it had a distinct technological advantage.

Although SDI looms large in narratives about the last decade of the Cold War, it is oftentimes viewed in isolation from broader developments in the military use of space in the 1970s and 1980s.[8] This situation is due in large part to the fact that US national security activities in space were some of the most secret areas of American statecraft during the Cold War.[9] In 2008, artist and geographer Trevor Paglen created an exhibit dedicated to what he calls the "other night sky."[10] Streaks of light in his photographs of the heavens were highly classified military and intelligence satellites, intrinsic parts of a hidden world largely invisible to the human eye. Secrecy and invisibility have made the "black world" of American national security space activities, especially during the 1970s and 1980s, mostly absent from political histories of the Cold War.[11] Underlining the obscurity of national security space topics is the fact that there are very few political and diplomatic histories that even mention the National Reconnaissance Office (NRO), the agency responsible for developing and operating US intelligence satellites from 1961 to the present time.[12] The US government did not even acknowledge the NRO's existence until 1992.[13] During the Cold War, the NRO was the secret sibling of the more well-known National Aeronautics and Space Administration (NASA).

Scholars of the Cold War have correctly pointed out that space was militarized from the beginning of the Space Age.[14] Rockets that launched satellites could also be used as ballistic missiles, and satellites served important national security functions such as intelligence gathering. Using the term "militarization" to cover all national security space activities obfuscates the fact that during the Cold War, there were two American national security space programs. The Department of Defense oversaw the military space program, which included early warning, communications, navigation, space surveillance, and weather satellites. The NRO, responsive primarily to the needs of civilian policymakers rather than the military, managed the intelligence space program: that is, satellite reconnaissance systems. (The third element of American space activities was the civil program, which included human space exploration and scientific satellites.) Over time, the boundaries between the intelligence, military, and civil space programs would become increasingly porous as their infrastructural connections grew. Importantly, the civil, military, and intelligence space programs involved different interest groups, and, as we shall see, their agendas did not always neatly align.

Since the beginning of the US satellite reconnaissance program, officials feared that the Soviet Union might develop the means to destroy American satellites just as Soviet air defense had shot down U-2 pilot Francis Gary Powers in May 1960. In an effort to protect reconnaissance satellites, American presidents were very cautious concerning any action that might conceivably provoke the Soviet Union to act aggressively in space. Stability in space was critical, since satellites constituted the largest source of intelligence on the Soviet Union. Beginning in the 1960s, the US also developed nuclear early warning, communications, weather, and navigation satellites to support a wide range of military functions. Along with intelligence satellites, these space systems comprised what I term "American national security space infrastructure." Although it was mostly American built, the entirety of NATO, in addition to non-NATO partners in the Five Eyes intelligence alliance, depended on this infrastructure.[15]

The 1970s were a key time of transition in the nature of superpower competition in space. The moon race came to an end, leading to aspirations that the US and the Soviet Union would use space for cooperative measures such as the 1975 Apollo–Soyuz Test Project, a joint American–Soviet space mission. In the post-Vietnam era of financial belt tightening, the US forewent explorations deeper into space and focused instead on the space shuttle that was confined to operations in low Earth orbit. In 1972, the superpowers signed the Interim Agreement limiting strategic weapons and the Anti-Ballistic Missile (ABM) Treaty, whose boundaries SDI would later test, which became visible symbols of détente.[16] At this same time, there was growing attention on the emergence of new technologies, including lasers and more accurate sensors, that could enable the creation of sophisticated missile defense and ASAT systems that might alter the strategic balance.

With the advent of more advanced space technologies in the 1970s, the US and the Soviet Union began to more closely integrate satellites into their combat forces. This was the early era of satellites that could provide targeting information in near real time to deployed land and naval units. Due to this "tactical shift" in the use of space systems, President Gerald Ford concluded that "treating space as a sanctuary" was against the national interest and approved the development of an ASAT weapons program so that the US would have the ability to deny the Soviet Union use of space in wartime.[17] Fundamentally, by the mid-1970s, senior US officials had come to expect that superpower conflict would extend into space. This policy laid the foundation for the more militarized space strategy that SDI would come to embody. By the end of the 1970s, both the US and the Soviet Union were developing ASAT weapons. In this period, we find that the US military space program shifted away from its focus on "passive" satellites gathering data, and added the development of combat systems, including ASATs and later space-based missile defense, although the latter two would never actually be deployed.

President Ronald Reagan, whom the *New York Times* described as "the biggest space enthusiast to occupy the White House,"[18] embraced what Andrew Butrica calls the "conservative space agenda" that projected the ideology of the so-called New Right into space.[19] Thought leaders in this movement included New Gingrich and Barry Goldwater, who became vocal space promoters. Reagan's eight years in office witnessed the establishment of SDI, a space station project, the testing of an American ASAT in space, and policies that would contribute to the emergence of a commercial space market. A central element in the conservative space agenda was the intrinsic connection of military power and economic prosperity in space. This conception of

US space prowess was in many ways similar to nineteenth-century military theorist Alfred Thayer Mahan's observations concerning the inseparability of commercial-maritime and naval interests.[20]

More recent scholarship on Reagan has interpreted SDI through the lens of the president's abhorrence of nuclear weapons.[21] Nuclear abolitionism certainly was a significant factor in the genesis of SDI, but the program quickly became the centerpiece of Reagan's effort to use space technologies to advance US economic and national security interests. Throughout the remainder of his presidency, public diplomacy on SDI had, at times, diverse and contradictory themes. At various points, SDI was simultaneously an answer to the nuclear dilemma, a cooperative endeavor that would be shared with the Soviet Union, a form of leverage in arms control negotiations, a response to Soviet space and strategic defense programs, and a catalyst for civilian technological innovation and space exploration. Reagan sincerely believed in all of these elements and maintained that SDI was concurrently a vehicle for competition and cooperation—an approach not unlike that of John F. Kennedy to the Apollo program.[22]

Before going any further, it is necessary to explain briefly what exactly SDI was. Although SDI is oftentimes framed as one thing, or one system, it was in actuality an umbrella for research into multiple technologies with a wide variety of applications, including missile defense, space surveillance, and ASAT weapons. SDI became a formal program in 1984, but it was not until 1987 that the Pentagon began to solidify a concept for a specific strategic defense system that might be deployable in the following decade. SDI entailed a radical shifting in the size and scope of US national security space infrastructure from a focus on very few but highly sophisticated satellites to a massive number of space systems in orbit. This shift required space-launch technologies that in the 1980s existed only in the imagination of engineers and space exploration enthusiasts. At that time, existing space launch (e.g., the shuttle) could not cost-effectively deliver systems into space. The price tag of the infrastructure for space-based missile defense would become one of the central issues in shaping plans for a strategic defense system. From an infrastructural perspective, SDI further entailed the military space program becoming the dominant element in American national security space policy.

SDI's political significance has been the subject of much debate among historians of the Cold War, but its technological elements are generally the backdrop for political discussions concerning it.[23] Neglect of SDI's technological details is likely due to the fact that Reagan's expansive vision of comprehensive space-based defense did not come to fruition. This book moves SDI's technological dimensions to the center stage of the narrative and, in doing so, shows that technological choices were simultaneously political decisions with significant implications for US defense strategy and foreign relations, especially arms control negotiations.

Problematically for Reagan, who promoted SDI as a tool of peace, space-based missile defense was inextricably linked to the Pentagon's development of technologies and doctrine for executing "space control" operations in wartime: that is, destroying Soviet satellites, thereby denying the Soviet Union use of space. Although SDI is oftentimes associated with "exotic" technologies such as laser weapons, the concepts for strategic defense systems that would emerge by the end of the 1980s were built around off-the-shelf technologies, including hit-to-kill missiles that use kinetic force to destroy their targets. These hit-to-kill missiles could be used in both missile defense and ASAT roles. Throughout the 1980s, the Soviet Union urged the US to agree to ban what Moscow called "space-strike weapons," which included ASAT and space-based missile defense technologies.[24] The US rejected Moscow's characterization of space-based missile defense as an offensive space weapon. Definitions aside, the entanglement between missile defense and ASAT technologies precluded the US government from accepting *any* space arms control proposals out of fear that limits on missile defense and ASAT technologies would constrain the Pentagon's space plans. This is, therefore, an example of technological choices shaping political outcomes.[25]

In examining the rationale behind specific technological choices and considering their political consequences, this book moves beyond the characterization of SDI as a "fantasy" that would never "work."[26] Reagan's expansive vision that SDI "might one day enable us to put in space a shield that missiles could not penetrate, a shield that could protect us from nuclear missiles just as a roof protects a family from rain,"[27] certainly overlooked many technological obstacles.[28] US policy documents reveal, however, a disconnect between Reagan's rhetoric about replacing deterrence with strategic defense and the Pentagon's attempt to harmonize space-based missile defense with existing deterrence strategies.[29] Defense analysts were also considering whether ASATs could enhance the US deterrence posture, which was similarly a source of disagreement in the defense community. There were, moreover, divergent ideas about what kind of space-based missile defense systems might actually strengthen deterrence and thus dissuade Moscow from launching a first strike. Consequently, terms such as "system effectiveness" were shaped by various interpretations of deterrence theory and Soviet defense doctrine and were not purely based in technical analyses. Importantly, the emphasis on *space-based* missile defense was reflective of ideas about how space should be used and represented a rejection of any limits on military activities in the cosmos. Looking carefully at the reasoning behind SDI technological decisions shows that its trajectory was not in any way inevitable or determined by technical limits alone. Rather, military considerations, political conditions, and cultural factors also shaped its evolution.[30]

Although SDI would become a contentious and dominant issue in American–Soviet relations, it was controversial among NATO allies as well. Western European states pressed the US to consider space arms control that would have limited both space-based missile defense and ASATs. By the mid-1980s, the White House began making a more concerted effort to assuage allies' concerns about the arms race moving into space. The head of SDI, Lieutenant General James Abrahamson, presented SDI as a "new space renaissance" that would serve as a catalyst for dual-use technologies, and the Reagan administration sought to use the prospect of technology transfer as a way to reduce allied hostility toward the program. European politicians viewed SDI not only in strategy-military terms but also through an industrial and economic lens due to its implications for what they saw as a growing technology gap between the US and Europe. Multiple European states, as well as Japan and Israel, would become involved in SDI research and development with the hopes of influencing US foreign policy decisions and securing advanced technologies with civil and industrial applications. The predicament of Western European allies either to choose a collective response to SDI or to pursue bilateral arrangements guiding their integration into SDI research and development quickly became enmeshed in the politics of European integration.

Including the perspective of Western European states on space militarization in the late 1970s and 1980s expands our view of alliance dynamics and the politics of European integration in this period. We find that the military use of space was an invisible, although essential, part of transatlantic defense and intelligence cooperation, especially between the US and the UK.[31] With these factors in mind, this book frames space militarization in this period as a transnational phenomenon and thus contributes to a growing body of histories of twentieth-century spaceflight that move our perspective beyond a purely bipolar superpower dynamic.[32]

The fact that SDI captured the public imagination in both utopian and dystopian ways created significant challenges for the White House in its public diplomacy concerning space. For some, space-based missile defense entailed turning the cosmos into a discrete realm where wars would be fought without affecting life on Earth. Conversely, many opponents feared that moving the arms race into space would increase the likelihood of nuclear conflagration below and forestall further exploration of the cosmos for the betterment of humankind. The very idea of deploying weapons in space challenged more than two decades of US government messaging that emphasized its commitment to the peaceful use of outer space. Popular conceptions of SDI aside, discussing military space operations, a highly classified world, entailed moving restricted knowledge into the public domain, exacerbating existing tension between secrecy and openness in both military and intelligence space programs.

Two core questions in the late 1970s and throughout the1980s were what the nature of an arms race in space would be and whether it could be stopped with arms control. Within the US and British governments, for example, there was sincere anxiety that the Soviet Union could secure strategic advantages in space. British defense officials and diplomats, who were generally hostile to the further militarization of space with ASATs and space-based missile defenses, expressed worries in a classified study that Moscow could secure a "Pax Sovietica based on the domination of space"[33] if the US did not stay ahead of its main adversary in space technologies. The American and British intelligence records reveal more ambiguity than certainty about what the Soviet Union could do, and more importantly *wanted* to do, in space. The space arms race was not an action–reaction phenomenon, with the US and the Soviet Union attempting to mirror each other's actions directly. Soviet and American disagreement on what technologies could be considered "space weapons" only complicated discussions regarding the limits that could be imposed with arms control.

The history of SDI is often truncated to the years 1983–1987 (the latter being when Soviet General Secretary Mikhail Gorbachev de-linked SDI from negotiations over intermediate-range nuclear weapons). One must, however, follow SDI's evolution a little bit longer into the 1990s to uncover its impact on continued negotiations to limit space militarization, to understand why space-based missile defenses were not deployed, and to discover SDI's connections to ideas concerning spacepower in the post–Cold War era. The first Gulf War demonstrated the importance of space technologies for modern American combat power and led to a surge, however brief, in support for missile defense. These factors would prove critical for shaping US policy concerning military space strategy more broadly, and space-based missile defense in particular, in the immediate years after the dissolution of the Soviet Union. With the end of the Cold War, momentum for space arms control was quickly lost. Importantly, even though Reagan's space-based defense vision would not materialize, SDI led to a permanent US infrastructure for producing technologies with missile defense and space warfare applications in the form of the Missile Defense Agency (MDA), which currently oversees US missile defense research and development efforts.

THE ROAD AHEAD

The narrative is developed in chapters that proceed chronologically. Chapter 1 explores how key US officials adopted a more militarily competitive view of space beginning in the 1970s due to both emerging space technologies and the erosion of détente. These developments underscore the reality that military competition in space was already intensifying prior to Reagan's presidency. Chapter 2 situates SDI within Reagan's broader aim to use space technologies to enhance American leadership on the world stage. It moreover explores the tension that emerged between the president's characterization of SDI as a tool of peace and his pursuit of military space superiority. Chapter 3 analyzes the White House's strategy for promoting SDI on the world stage, its response to domestic and international criticisms and anxieties, and the inherent challenges of creating a public diplomacy campaign concerning military space technologies that were part of a highly classified area of US national security. Chapter 4 details how SDI made military space policy a source of contention in the transatlantic alliance and evaluates the strategic-military and industrial-economic motivations behind European attempts to shape the trajectory of US strategic defense research and development. Chapter 5 explains how specific technological choices concerning SDI affected US arms control strategy and convinced Reagan to eschew any limits on military space systems. Chapter 6 follows SDI through the transition into a post–Cold War world and investigates how the politics of this period shaped the evolution of space-based missile defense specifically and ideas about American spacepower more broadly. It further reveals why space arms control negotiations continued to be unsuccessful, even while Washington and Moscow agreed to unprecedented reductions in nuclear forces. "SDI Reconsidered" provides some key concluding ideas. And "A Sense of Déjà Vu" considers how issues left unresolved in the space arms control dialogue of the late Cold War, along with the entanglement of missile defense and ASAT technologies, continue to shape space security dynamics.

In telling the history of SDI, this book traces the emergence of intensified space militarization in the final two decades of the Cold War. It is the story of how controversy concerning the proper role of military space technologies in US statecraft shaped US relations with both the Soviet Union and transatlantic allies. The narrative takes place at the convergence of technology and politics, with the entanglement of missile defense and space warfare technologies disproportionately influencing the politics of space in this period. Tension is a central theme throughout this book—tension over what *should* the limits of military activities in space be. As such, it introduces the subject of space arms control into the mainstream history of the final two decades of the Cold War. By opening up the secret world of military space programs, we gain at least a partial window onto Paglen's "other night sky," and thereby discover a mostly hidden area of American foreign relations and defense policy. As space security once again emerges as a contentious subject on the world stage, this work takes on new relevance and urgency, providing important lessons and warnings for current diplomatic efforts to promote stability in the cosmos.

1   THE RISE AND FALL OF DÉTENTE ON EARTH AND IN SPACE

In the ornate St. Vladimir Hall of the Grand Kremlin Palace on May 26, 1972, President Richard Nixon and Soviet General Secretary Leonid Brezhnev signed the ABM Treaty and the Interim Agreement, both products of the Strategic Arms Limitation Talks (SALT).[1] In the midst of this historic occasion, Nixon expressed hope that the US and the Soviet Union would "cooperate in the exploration of outer space."[2] A joint space exploration mission would symbolize détente, but without military space technologies, this new phase in American–Soviet relations would not have been possible in the first place. In particular, reconnaissance satellites served as the primary mechanism for monitoring compliance with the SALT treaties—"verification" in the arms control lexicon. Media portrayals of satellites "policing the superpowers" created the perception that space technologies promoted superpower rapprochement.[3] By the time Nixon boarded Air Force One to return home, the arms race appeared to be decelerating with a new, cooperative era in space on the horizon.

Not even five years later, however, Moscow intensified ASAT testing in the context of a faltering détente. Soon thereafter, President Gerald Ford concluded that it went against the national interest to treat space as a "sanctuary," and in his final days in the White House, he approved the development of an ASAT weapon designed to attack Soviet satellites.[4] In response to Soviet military space developments, Central Intelligence Agency (CIA) Director Stansfield Turner warned the public in early 1978 that "the Russians can kill us in space."[5] Tilmann Siebeneichner notes that in this environment, "outer space was increasingly perceived as a sphere of permanent crisis and confrontation."[6] As détente crumbled, so too did the hope that space cooperation could reduce the prospect of conflict on Earth. The causes of this shift toward intensifying superpower military competition in space remains an understudied topic among Cold War historians. Now, with the declassification of relevant US national security documents, I show that the mid to late 1970s marked a transformation in the way key US officials viewed space; they concluded that the US needed tools for waging space warfare. Most importantly, it is in this transitionary period that we find the convergence of strategic ideas and technological developments that are core elements in the genealogy of SDI.

Since the dawn of the Space Age, US officials committed themselves to using space for peaceful purposes, although without excluding national security functions such as reconnaissance.[7] Since reconnaissance satellites constituted the largest source of intelligence on the Soviet Union, American space policy generally took a very cautious line with regard to any space programs (e.g., ASATs) that could provoke Moscow to act aggressively toward US reconnaissance satellites. The broader US arms control agenda leading to détente played a crucial role in restraining American ASAT development. But this situation began to change in the mid-1970s with the advent of new space technologies that US defense experts believed could "revolutionize conflict."[8] With satellites more directly enabling critical military functions such as precision targeting, Ford's advisors concluded that the Pentagon needed weapons capable of destroying Soviet satellites in wartime. These technological considerations, in the context of cooling relations with Moscow, were important factors in the adoption of a more militarily competitive US space policy.

As détente appeared increasingly moribund, tensions flared between American officials who sought strategic advantages in space and proponents of more stringent arms control measures constraining military space activities. As part of the SALT II framework, Jimmy Carter and Leonid Brezhnev commenced negotiations in 1978 to ban ASAT technologies, but their disagreements over the boundaries of space militarization proved insurmountable. As Carter left the White House, it appeared that a superpower arms competition in space was fast emerging. Yet, US–Soviet ASAT discussions in the late 1970s should not be dismissed as insignificant just because they did not produce a treaty. Rather, the space arms control dialogue of the 1980s was a continuation of talks in that decade. Fundamentally, the 1970s witnessed profound technological and political transformations that laid the foundations for the more expansive space militarization of the Reagan presidency.

PEERING THROUGH THE KEYHOLE IN THE 1950s AND 1960s

Even before the hysteria that followed in the wake of the Soviet Union's launch of Sputnik in November 1957, American-German rocket engineer Werner von Braun and US Air Force General Bernard Schriever were publicly advocating capabilities to obtain "space superiority."[9] At this time, several studies were underway concerning the US Air Force's future in space and the development of "anti-ICBM" (intercontinental ballistic missile) systems.[10] In October 1959, the US Air Force conducted the world's first ASAT test by air-launching a Bold Orion missile from a B-47 and aiming it at a US Explorer VI satellite.[11] Both ABM and ASAT weapons depended on nuclear warheads to destroy their targets; highly sophisticated electronics and sensors required for nonnuclear ABMs and ASATs did not yet exist. Problematically, nuclear ABMs and ASATs were not very accurate. High-altitude nuclear tests revealed that radiation blasting into space would indiscriminately destroy US satellites, underscoring the limited military utility of nuclear ASATs.[12] With the suspension of aerial overflights of the Soviet Union after the shooting down of U-2 pilot Francis Gary Powers in May 1960, US dependence on satellites for gathering intelligence on the Soviet Union would only further militate against pursuing nuclear ABMs and ASATs.

A little more than three months after the U-2 fiasco, the US successfully recovered film from Corona, its first photographic intelligence satellite.[13] The age of space reconnaissance had officially arrived and quickly became the largest source of intelligence on Soviet military strength.[14] To ensure the viability of space reconnaissance, Eisenhower sought to construct "an international political framework to place US reconnaissance satellites in both a political and psychological context favorable to protecting them from interference."[15] Consequently, he rejected any military proposal that might provoke the Soviet Union to act aggressively in space.[16] To this end, the president sided with his science advisor, George Kistiakowsky, who opposed an Air Force satellite inspector system (i.e., a satellite designed to approach and "inspect" Soviet satellites) on the grounds that it could "conceivably elicit a violent Soviet response against the vulnerable [US] intelligence satellites."[17]

National intelligence needs, rather than military agendas, overwhelmingly shaped Eisenhower's space strategy. The US government wanted to establish international protections for satellite observation, which it maintained was consistent with the peaceful use of outer space. Walter McDougall writes that "the American formula of space for 'peaceful' rather than for explicitly 'non-military' purposes" was a core tenet of US space policy.[18] And US officials would downplay military space activities. Early in John F. Kennedy's presidency, Secretary of Defense Robert McNamara established the NRO as a covert organization to oversee all satellite reconnaissance operations.[19] The head of the NRO answered to both the director of central intelligence and the secretary of defense, firmly placing satellite reconnaissance in the hands of civilian leaders rather than in those of the military.[20]

Protecting the viability of satellite reconnaissance played a central, although largely invisible, role in Kennedy's space policy, just as it had in Eisenhower's. In an attempt to avoid international political difficulties, the US government kept details concerning the NRO and its intelligence satellites secret, not even formally acknowledging the existence of either.[21] Details about NRO satellites were placed in a special security compartment called "Talent Keyhole," in reference to the idea that space reconnaissance systems allowed the US to peek through the keyhole of a closed Soviet society.[22] The NSAM 156 Committee, an interagency body that formulated recommendations concerning satellite reconnaissance, implored Kennedy to take whatever action necessary, "in word and deed," to legitimize reconnaissance from space.[23]

The US Air Force did not abandon hope that the US would turn its sights toward achieving space superiority, especially as anxiety about the Kremlin's ballistic missile and military space programs intensified. As the Soviet Union built more missiles, the US Army pushed for an ABM system.[24] Long-range ABMs designed to intercept their targets outside of the earth's atmosphere could of course be used in an ASAT role too. *The complexities that stemmed from the entanglement of missile defense and ASAT technologies are a recurring theme in this book.* In the early 1960s, the Department of Defense examined space-based missile defense as part of Project BAMBI (short for "Ballistic Missile Boost Intercept"), but determined that the requisite technology was not feasible.[25] McNamara introduced Nike-X, an ABM concept that involved both long-range and short-range interceptors, but opposed deployment of an ABM, fearing that the Soviets would just build more ICBMs to overwhelm US defenses.[26]

Senior officials pointed to Nikita Khrushchev's claim that the Soviet Union could deploy "loads [in space] that can be directed to any place on earth" as evidence that Moscow was contemplating the development of space-based nuclear weapons.[27] Notably, in the early 1960s, US intelligence lacked evidence of Soviet intent to deploy orbital nuclear delivery vehicles.[28] Nevertheless, in an effort to "meet the [potential] threat of Soviet orbital bombardment systems," in 1962, McNamara approved a ground-based nuclear ASAT project called "Program 437."[29] Sean Kalic observes that countering "the psychological threat posed by [Soviet orbital nuclear] systems" was an especially important factor in Program 437's creation.[30] McNamara doubted that an ASAT was indeed necessary, but Edward R. Murrow, the head of the US Information Agency, had hyperbolically warned him that if the Soviets did indeed deploy orbital nuclear weapons and the US did not have a defense against them, "you will see the first impeachment proceeding of an American President since Andrew Johnson."[31] Roswell Gilpatric, McNamara's deputy, believed that the US needed an ASAT program as "technological insurance" in case of expanded Soviet space militarization.[32]

Many senior officials at the Pentagon and the State Department were, nevertheless, "very nervous" about having a research effort for an ASAT program, let alone operationally deploying such a system.[33] To this end, US officials downplayed ASAT research and development and "propagate[d] the idea that interference with or attacks on any space vehicle of another country in peacetime are inadmissible and illegal."[34] Members of the NSAM 156 Committee advised the president that securing a ban on nuclear weapons in space would project a peaceful image of the US military space program, thereby contributing to the formation of an international political atmosphere favorable toward satellite reconnaissance.[35] A prohibition on nuclear weapons in space would also head off a new dimension in the arms race that could be both costly and destabilizing.[36]

After Kennedy's assassination, Lyndon Johnson carried the orbital nuclear ban agenda through to completion with the signing of the Outer Space Treaty in 1967.[37] Johnson affirmed that with the Outer Space Treaty, the US, Soviet Union, and other signatories were "taking the first firm step toward keeping outer space free forever from the implements of war."[38] Even though the treaty banned nuclear weapons in space, the Soviet Union deployed a Fractional Orbital Bombardment System (FOBS) that was designed to evade US early warning radars by having warheads approach the US along a polar orbit from the south.[39] McNamara tried to assuage public concerns about FOBS, saying that it remained "a system in which the disadvantages far outweigh the advantages as far as the attacker is concerned."[40] US intelligence analysts believed that a FOBS deployment was "unlikely in the foreseeable future to affect the basic strategic balance."[41] The Soviet FOBS was certainly less accurate than ICBMs. Publicly, the US government maintained that FOBS did not constitute a breach of the treaty, even though some officials at NASA and the State Department thought otherwise.[42] Despite this situation, the White House determined that it was better not to confront the Soviets about FOBS. According to Stephen Buono, "the Johnson administration was careful not to rattle cages over a comparatively disadvantageous weapon already in the testing phase."[43]

Since the Outer Space Treaty did not ban "conventional space weapons," its limited protections would become especially apparent when Moscow and Washington later pursued more sophisticated ASATs.[44] In the mid to late 1960s, Johnson continued to reject the military's push for new space programs, including a satellite inspector designed to rendezvous with its target, ascertain its function, and destroy it if needed.[45] Concurrently, the escalating war in Vietnam forced the Department of Defense to prioritize defense programs carefully, and more traditional needs such as bombers and ships quickly took precedence over new space endeavors. In this fiscally constrained environment, the Pentagon moved Program 437 into a limited operating capacity and terminated it in 1975 due to its lack of military utility.[46]

The Outer Space Treaty and the 1968 Treaty on the Non-Proliferation of Nuclear Weapons were only the initial parts of a more expansive arms control agenda aimed at progress in what Johnson termed "Strategic Arms Limitation Talks" (SALT). Although Richard Nixon would see SALT through to completion, Johnson laid much of the groundwork.[47] Johnson believed that securing limits on ABMs was especially urgent; he wanted to prevent an ABM race, which would have taken resources from his Great Society initiative. But since ABMs could be used as ASATs too, a treaty limiting them would also have significant implications for security in space.[48]

Toward the end of the 1960s, American officials had concluded that "space reconnaissance [was] an instrument of peace which [had] assisted the president and the free world . . . to [minimize] the threat of surprise attack that could plunge the world into another major war."[49] By the time Johnson left office in 1969, the US had built a national security space infrastructure composed of satellites used for navigation, intelligence, weather monitoring, early warning, and communications. This infrastructure formed a central part of the nation's—and, by extension, NATO's—strategic warning trip wire. These kinds of space programs were in harmony with the stated US commitment to use outer space for peaceful purposes. Formulating policies that protected these space systems was imperative because the US had much to lose from interference with its satellites. Space weapons, at this time, were of limited military utility and could have threatened satellite reconnaissance whose value to US national security had been reaffirmed time and time again. Moving the arms race into space would have undermined détente as well. As space technologies evolved and geopolitical conditions changed, however, the boundaries of American military space activities would expand.

SATELLITES POLICING THE SUPERPOWERS

Nixon came into office amid significant foreign policy challenges; he had to find a politically acceptable resolution to Vietnam and address the looming prospect of Soviet nuclear parity with the US. Hal Brands observes that Nixon inherited a situation in which the US was no longer the predominant nuclear power. The "global balance of power had shifted fundamentally since the early Cold War," which required a new approach to foreign relations.[50] At the same time, domestic political considerations prevented Nixon from obtaining support for a new wave of nuclear expansion. Instead, he pursued the Interim Agreement and the ABM Treaty (both products of SALT) to cope with this difficult political climate. Consequently, "Nixon, the nuclear hawk, became the arch proponent of agreements that would be the bane of other hawks until the US's withdrawal from the ABM Treaty in 2002."[51]

My purpose here is not to provide a history of the SALT-era treaties; scholars of the Cold War have already done this.[52] Rather, I will focus on a largely unexplored subject: the role of military space technologies in détente's emergence and how this new phase in superpower relations simultaneously shaped the field of American military space policy and strategy. In the 1960s, US officials recognized that on-site inspection of American and Soviet nuclear facilities would be impossible; it was reconnaissance satellites that provided a crucial "unilateral" verification mechanism.[53] Such satellite reconnaissance fundamentally undergirded the Interim Agreement and the ABM Treaty—the era's visible symbols of détente.[54]

Nixon's time in office witnessed a significant transition in American space activities. The president wanted to reap the political rewards of the Apollo 11 moon landing that happened on his watch, even though the foundational work took place under Kennedy and Johnson. Nixon described the Apollo 11 mission as "the greatest week in the history of the world since the Creation."[55] Yet, he soon made it clear that a follow-on to Apollo would not be forthcoming. Fiscal constraints and waning domestic political support for space exploration made NASA's more expansive space ambitions unrealistic. Rather than pursuing large-scale space "firsts," Nixon and Henry Kissinger, the president's national security advisor, brought the objectives of civil space endeavors down to earth and focused instead on using space cooperation to promote foreign policy objectives such as fostering European integration and improving relations with Moscow.[56] NASA hoped to secure a budget large enough for a space station and a reusable spacecraft called the "shuttle." Funding both would be an expensive proposition, and NASA needed to find a justification that would be acceptable to a president focused on cost saving.[57]

With its uncertain future in mind, NASA's leaders used national security and economic arguments to secure presidential support for the shuttle. A joint Department of Defense/NASA report noted that "in times of crisis our national leadership requires accurate information for decisions" and that a "mission-equipped" shuttle could deliver vital intelligence "on a crisis located anywhere in the world."[58] NASA's pledge that the shuttle would cost-effectively launch "all satellites foreign and domestic" won the support of the White House.[59] The fact that engineers designed the spacecraft's payload bay to fit the dimensions of Hexagon, the largest reconnaissance satellite in the nation's inventory, underlines the influence of national security considerations for the shuttle project.[60] Vance Mitchell notes that "NASA had previously limited its association with military and intelligence activities to protect its good name and public image," but now embraced national security applications for institutional survival.[61] Using the shuttle for civil and national security roles contributed to further blurring the lines between civil and national security space infrastructure.

In early January 1969, CIA Director Richard Helms wasted no time in sending then-president-elect Nixon a memorandum outlining the need for "a satellite-borne photographic reconnaissance system, capable of practically instantaneous transmission of pictures for interpretation in Washington."[62] Into the mid-1970s, imagery satellites still used only film-return canisters, meaning that it could take days or weeks to process the film before intelligence analysts could evaluate photographs, rendering them of little utility in fast-changing crisis situations. The Soviet invasion of Czechoslovakia in 1968 underscored that NRO satellites were not responsive during crises because the film containing images of Soviet military movements arrived too late.[63]

Although a digital "quick reaction" satellite was the most advanced concept, not everyone in the intelligence community believed it should be pursued over other simpler and less expensive imagery satellites using proven technologies.[64] Ray Cline, the intelligence chief at the State Department, described a digital imagery satellite as the "Cadillac," and cautioned Helms that "a relatively inexpensive quick reaction system" should not be overlooked.[65] But as with NASA, the White House would not spare NRO from the budgetary knife; some programs would have to be eliminated.[66] The president forewent an interim near real-time option based on simpler technologies, opting instead to move ahead with Zaman, soon redesignated Kennen, the more expensive digital near real-time concept.[67] This new satellite would lead to greater integration between space systems and military forces because it could provide timely intelligence to deployed combat units. This tactical shift in space technologies would have significant implications for US thinking about the role of space systems in warfare.

In the midst of making decisions about the future of US civil and national security space programs, Nixon received disconcerting intelligence reports about the Soviet Union testing a nonnuclear co-orbital ASAT designed to be launched on a ballistic missile, maneuver close to its target, and then destroy it. In the late 1960s, US intelligence had learned that the Soviets were working on a maneuverable satellite "more applicable to an anti-satellite role than any other mission objective," but the precise details of the program were unclear.[68] In 1971, new information surfaced indicating that the "Soviet program to develop and test an orbital interceptor system has progressed significantly" and that "the scope of the program [was] much broader than previously estimated."[69] Despite this, the CIA maintained that the Soviet Union was unlikely to use its co-orbital ASAT short of full-scale war.[70] Helms wrote to Kissinger that "our assumption that the Soviets rely heavily on their own satellites for intelligence about us and the Chinese" made actions that might "jeopardize their own capability" very remote.[71] Despite the CIA's assurances, the Soviet ASAT could not be ignored. Consequently, Kissinger directed a study of potential responses to the Soviet ASAT program.[72]

The US Air Force proposed a *nonnuclear* aircraft-launched ASAT, called Project Spike, "capable of intercepting and negating a Soviet satellite prior to its first overflight of [the US]."[73] Notably, electronics and sensors for homing in on a target had now advanced to a stage where a nonnuclear ASAT could indeed be contemplated, but the requisite technologies had not been integrated into a system and tested.[74] A nonnuclear ASAT would eliminate the problems associated with nuclear ASATs propagating radiation into space and damaging satellites indiscriminately. By this point, Program 437 had been mothballed, which meant that the US did not have a fully operational ASAT. Project Spike provided an opportunity to expand the Air Force's reach into space further. Problematically for the program's advocates, a new ASAT created political complications for ongoing arms control negotiations, and many senior officials believed that American ASATs would increase the vulnerability of US satellites by inviting, in effect, the Soviet Union to develop more of its own ASATs.

Despite the fact that new technologies made nonnuclear ASATs possible, Amrom Katz of the RAND Corporation vehemently opposed them. From his perspective, "the Soviets were stimulated, triggered, or catalyzed with their anti-satellite activity . . . by <u>our</u> (announced) 437 program (<u>our</u> anti-satellite program) started by President Kennedy" (emphasis in original). He argued that the Soviets did not necessarily have a cogent ASAT strategy and that the Soviet military-industrial complex pushed programs such as this due to the "kind of inter-service rivalry [that] we enjoy [in the US]," and that there was perhaps a government minister who said, "Goddamnit, we just gotta have an anti-satellite capability."[75] Overwhelming US dependence on satellites for intelligence on the Soviet Union certainly provided a compelling justification for a Soviet ASAT.

Katz and the panel that Kissinger convened to examine the ASAT problem rejected the idea that American ASATs had deterrent potential.[76] Paul Stares explains that "it was believed that the threat of retaliation in kind for an attack on US satellites was unlikely to deter the Soviet Union" because the US depended on satellites far more than the Soviet Union.[77] Kissinger's panel noted, however, that as the Soviets became more dependent on space systems, the ASAT–deterrence relationship might need to be revisited.[78] Katz tried to calm concerns about Moscow using ASATs as part of a first-strike strategy by pointing out that doing so would require eliminating nuclear early warning satellites located in geosynchronous orbit—a region far beyond the operating altitude of Soviet co-orbital ASATs.[79] When Moscow ceased ASAT testing in late 1971 to promote a more favorable climate for arms control negotiations, the ASAT issue became less urgent.[80]

In 1972, in addition to the Interim Agreement that constrained specific kinds of strategic weapons, Washington and Moscow signed the ABM Treaty. The latter limited each country's missile defense interceptors and specifically forbade the development and testing of sea-, air-, space-, or mobile land-based ABM systems.[81] According to James C. Moltz, "the signing of the ABM Treaty . . . had an impact on the space environment that exceeded the terms for earth . . . no ABM systems would be allowed in space."[82] Notably, Moscow and Washington did not include limits on ASATs. The interpretation of the ABM Treaty would become especially contentious during the Reagan administration after the president established SDI.

The provision in the ABM Treaty (and the Interim Agreement) that neither Moscow nor Washington would interfere with the other's national technical means of verification, NTM for short, had significant implications for space security. Moltz observes that "the agreements codified in legally binding terms constraints in space that had previously existed only as norms."[83] These constraints were, however, left ambiguous. Washington and Moscow never formally defined NTM, although Soviet negotiators alluded to using space systems for monitoring treaty compliance.[84] At that time, neither government publicly admitted to conducting satellite reconnaissance. Declassified US policy documents concerning verification show that NTM included reconnaissance satellites and clandestine technical intelligence sites in Iran, Ethiopia, and other places.[85] Keeping NTM vague circumvented the need for detailed discussions with Soviet representatives that might compromise certain sensitive intelligence sources and methods. In the public eye, however, NTM became a euphemism for reconnaissance satellites as media outlets reported that satellites would closely watch the superpowers to ensure treaty compliance.[86]

Since American and Soviet officials kept the nature of NTM ambiguous, they were similarly vague about the meaning of "interference." The NTM noninterference provisions did, nevertheless, signal a joint understanding that satellite operations in support of arms control verification were indeed legitimate. But when these same satellites were later increasingly used for military functions (e.g., providing targeting information in near real-time to land and naval forces), NTM protections became less certain. As the US and the Soviet Union developed new technologies such as laser weapons that could "blind" satellite optics, as well as jamming capabilities that could impede satellites without physically destroying them, the nature of interference would become even more complicated.

Even with the SALT agreements in place, national security officials remained concerned about the vulnerability of US satellites. In June 1973, Deputy Secretary of State Kenneth Rush informed acting Secretary of Defense William Clements that he had reviewed the results of a satellite vulnerability study and the potential development of ASAT technologies by the US. He agreed that Soviet ASATs were justifiably of concern but cautioned that "there are important political and other considerations . . . which should be taken into account" concerning an American ASAT. Such a program would be public knowledge and could become the subject of "domestic controversy" and "stir foreign concern." In light of the SALT treaties, Rush worried that doubts might be raised about US commitment to NTM noninterference but acknowledged that Moscow had proceeded with its ASAT efforts "despite our possible concerns in light of SALT agreements." Political conditions aside, he questioned the military utility of a nonnuclear ASAT.[87]

Rush's viewpoints reveal much continuity with prior US policy regarding ASATs. Yes, satellite vulnerability posed a serious problem for US national security, and the Soviet Union maintained an ASAT system. But even with new nonnuclear ASAT technologies, many senior officials still questioned their military utility for the US. Rush's observation that initiating ASAT development could create "domestic controversy" was justifiable because doing so might create the appearance that the US was no longer committed to the peaceful use of outer space. Most importantly, at this stage, the White House did not want to take any action that could undermine détente.

DÉTENTE UNRAVELS AND A NEW (LIMITED) SPACE COMPETITION BEGINS

In July 1975, more than 125 miles above the Earth, an American Apollo module docked with a Soviet Soyuz capsule. Millions of people around the world watched as smiling astronauts and cosmonauts shook hands. This was political theater at its finest; the coming together of representatives of the US and the Soviet Union in space symbolized the thaw in relations between the superpowers and the ending of the space race.[88] President Gerald Ford said that the joint mission opened the "door to useful cooperation in space . . . the day is not far off when space missions made possible by this first joint effort will be more or less commonplace."[89] In reality, this would be the first and last joint American–Soviet crewed space mission of the Cold War.

Only five months after the Apollo–Soyuz project, the spirit of hope and optimism about a new era in space started to dissipate.[90] In December 1975, the Soviets resumed ASAT testing due, at least in part, to "the apparent slow-down of SALT II negotiations."[91] According to Asif Siddiqi, late 1975 witnessed "the beginning of the most intensive series of tests in the IS [Soviet ASAT] program, one that would lead to a fully operational capability."[92] More tests in February and April of 1976 showed the Soviets performing rendezvous operations in space: that is, the Kremlin was practicing getting close enough to other satellites to be able destroy them.[93] As a consequence of Moscow's resumption of ASAT testing, the White House reexamined US ASAT policy.

In the summer of 1975, the intelligence community had completed a study on Soviet military space capabilities and claimed that the Soviet co-orbital ASAT had achieved operational status, demonstrated intercepts of targets up to 550 nautical miles in altitude, and would soon be capable of intercepts of up to 2,500 nautical miles.[94] Alarmingly, American intelligence personnel discovered Soviet "orbital operations required to intercept a satellite in geostationary orbit" where nuclear early warning satellites were located. Soviet forces had also improved their ability to use satellites to support military operations. The Soviet radar ocean surveillance satellite (RORSAT), underway since 1967, had the ability to determine some ship locations, allowing Soviet units to use RORSAT to find NATO vessels and destroy them. In 1975, the Soviet navy conducted its largest ever exercise, called *Okean* (Ocean), which involved more than two hundred vessels across the globe.[95] Expanding Soviet naval power, coupled with the Kremlin's new space-based tracking and targeting system, caught the president's attention.

Not long after *Okean*, intelligence reporting indicated that the Soviet military was becoming more reliant on space systems such as RORSAT. US defense planners wondered whether Moscow's growing reliance on satellites would encourage the Kremlin not to initiate hostilities in space. However, analysts quickly pointed out that Moscow's growing dependence on satellites would not necessarily "deter them from interfering with US satellites in the face of other compelling reason[s] to do so." But intelligence analysts allowed that increasing Soviet dependence on military satellites would "probably increase Soviet incentives not to interfere with US satellites" outside of wartime conditions.[96] New technologies exacerbated uncertainty about the potential for Soviet interference with American satellites. US intelligence reported that "in due course . . . [Moscow could] disable most low-altitude satellites with the large, probable laser system at Sary Shagan" in Kazakhstan.[97]

Acting on intelligence reporting about Soviet military space developments, in 1976, Ford appointed a special panel chaired by physicist and presidential science advisor Solomon Buchsbaum to examine the future of American military space strategy.[98] The National Security Council (NSC) asked Buchsbaum to study measures for decreasing satellite vulnerability, to conduct a reevaluation of ASAT policy, and to make predictions about the military uses of space over the next fifteen years or so. Since new technologies offered opportunities for using satellites in "direct support of tactical and strategic forces,"[99] the White House needed to understand better the political implications of more closely integrating intelligence satellites with military forces—a topic of growing interest since the early 1970s.[100] If satellites directly enabled combat actions such as precision strikes, then they could themselves become targets in wartime. Technological changes and the concurrent cooling in American–Soviet relations warranted a reappraisal of whether self-imposed boundaries governing US military space programs should be modified. Yet, any potential change in military space strategy had to be carefully weighed against the US commitment to use space for peaceful purposes, although the limits of what constituted "peaceful" space activities were open to interpretation.

The NSC expected the panel to finish its study in September 1976, but another test of the Soviet co-orbital ASAT system in April added even greater urgency for the report's completion.[101] Shortly after the Soviet test, National Security Advisor Brent Scowcroft met with Ford and Secretary of Defense Donald Rumsfeld to discuss the future of military space operations. Scowcroft described how advances in satellite technologies held the potential to "revolutionize conflict."[102] More specifically, satellites could enable long-range precision strikes on enemy locations, a fundamental aspect of what would later be called the revolution in military affairs.[103] Ford listened very carefully and described Buchsbaum's ASAT study as "really important."[104]

In an interim report, Buchsbaum detailed a menacing Soviet space threat. Moscow's ASATs could "completely deny US satellite photo reconnaissance missions for periods up to years [sic]." Due to this situation, Buchsbaum advocated increased US investment in ASAT countermeasures. Notably, he reaffirmed that an American ASAT would not deter the Soviet Union from using its ASATs to attack American satellites; consequently, a new US ASAT "would not contribute to the survivability of US space assets."[105] US satellites were few in number and therefore "juicy target[s]" for Soviet ASATs.[106] A few months after receiving this report, Ford signed a national security directive ordering the Pentagon to enhance satellite survivability measures.[107]

Whether to initiate a new ASAT program was an especially sensitive matter. Even though Buchsbaum maintained that ASATs did not have deterrent potential, he advised Ford that changing technological conditions provided a compelling military argument in their favor. In particular, the growing use of Soviet satellites in support of tactical military operations warranted a new ASAT system. According to Buchsbaum, "A major impact of new technology during the last five years has been the increasing (although not symmetric) use of space-based assets for direct support of military forces by both the US and the Soviets. Satellites already provide important support to the strategic and tactical forces of both sides—support that is greatly increasing the effectiveness of those combat forces. The panel is convinced that this trend toward effective integration of space assets into military combat operations will continue and that real-time space capabilities will become increasingly important—even essential to the effective use of military forces."[108] The panel underscored the threat posed by the Soviet Union's radar satellites that could track and target NATO vessels. The CIA assessed that these satellites could "detect large surface ships, such as aircraft carriers" and had "potential for providing targeting data to combatants at sea," anti-ship missiles in particular.[109]

Buchsbaum's final report emphasized that the US ASAT requirement "does not derive from a perceived military need to respond in kind to the appearance of the Soviet [ASAT] system." In other words, a mechanistic action–reaction was not driving the policy recommendation. Rather, countering "the growing military utilization of space by the USSR" for tactical-level operations made an American ASAT program vital.[110] Nullifying Soviet satellites used to support naval targeting alone provided "sufficient motivation to undertake an anti-satellite development effort."[111] The threat to NATO naval forces posed by RORSAT was, according to Buchsbaum, severe and "if not countered, brings the viability of the surface fleet into serious question."[112] Fundamentally, naval power and spacepower were closely linked.

Concern about the implications of a US ASAT initiative for American–Soviet arms control naturally arose as a consequence. Buchsbaum pointed out that since satellites used for treaty verification and those employed for tactical support were often one and the same, NTM noninterference provisions did *not* extend to satellites used for non-treaty verification functions. It must be emphasized that ASAT development did not preclude continued use of satellites for arms control verification; SALT prohibited "active interference, not the development and testing of means to interference."[113] This ASAT policy discussion took place in the context of preparation for wartime conditions when military needs would supersede arms control monitoring. To clear up ambiguity concerning interference, Buchsbaum pointed out that reaching an "understanding" with the Soviets about the destabilizing effects of attacks on the satellites linked with nuclear missions would be beneficial. Other diplomatic options included negotiating a ban on ASATs (not yet in existence) designed to attack satellites in higher orbits, many of which had nuclear command and control functions.[114]

A range of ASAT concepts could be pursued, all driven by the objective of being able to "destroy six to ten Soviet satellites in one week and to be able to carry out electronic jamming of Soviet radar ocean surveillance satellites."[115] Specialists examined several options, including a co-orbital concept similar to the Soviet ASAT, a direct-ascent capability (missile fired into space), and ground- and space-based lasers. They decided that a direct-ascent mode provided the greatest flexibility but required better sensors for homing in on a target.[116] Space-based lasers were ruled out on the grounds that "engineering problems associated with . . . such a system are formidable" and that the requisite technologies were at least ten to fifteen years away.[117] Notably, the report did not highlight any special sensitivities concerning the stationing of weapons in space, which would become a subject of great controversy in the 1980s.

The advent of laser and jamming weapons opened a whole new suite of capabilities for attacking satellites. Buchsbaum's recommendation to develop a satellite jamming system warrants special attention; major details concerning this option have only recently become available. Jamming provided the ability to interfere with satellites without physically destroying them. A jamming system combined with a nonnuclear ASAT interceptor created the "flexibility . . . necessary to deal with a reasonable range of crisis and conflict situations."[118] In other words, the US might need to be able to interfere with Soviet satellites in times of heightened tensions without physically destroying them.

In his final few days in office, Ford directed the Pentagon to develop a nonnuclear ASAT and an electronic warfare system (i.e., jamming capability) able to destroy Soviet military support satellites in wartime.[119] In the lead up to the final defense authorization request of Ford's presidency, Rumsfeld predicted that space systems could "materially influence the outcome of future conflicts" and that space would not remain a "relative sanctuary."[120] Above all else, establishing a new ASAT program signaled a dramatic shift in the relationship between space and American national security. Although treating space as a "sanctuary" had never been official US policy, neither had targeting Soviet intelligence and communications satellites. Further integration of space technologies into the American and Soviet war machines proved to be a critical factor in the US decision to develop new space warfare mechanisms.

These technological considerations cannot, of course, be divorced from the broader political atmosphere, especially changing views of détente, in the mid-1970s. When Ford became president, he promoted détente, saying that it was "in the best interest of the country."[121] In 1974, Ford and Brezhnev met in Vladivostok and laid the groundwork for a SALT II agreement. In its aftermath, both American and Soviet participants were euphoric; veteran Soviet diplomat Anatoly Dobrynin described it as the high point of détente.[122] Kissinger said that Soviets "had made almost all the [desired] concessions."[123] Nevertheless, upon Ford's return to the US, the media severely criticized the summit, and the president faced attacks from both sides of the political aisle.[124] Only one month prior to this, Ford had pardoned Nixon, and the Watergate scandal was once again thrust into the media limelight. Watergate had "tarnished everything associated with the Nixon administration, including détente."[125] According to Kissinger, it appeared that "the entire SALT process was floundering and might even collapse."[126]

In 1975, the same year that the Soviets resumed testing of their ASAT, Saigon fell, which "was a huge blow to the administration's standing among conservatives."[127] In 1976, the Soviet Union started deploying its SS-20 mobile intermediate-range ballistic missile that could rapidly strike anywhere in

Europe. This produced anxiety among European leaders about the credibility of NATO's deterrent and spawned the Euromissile Crisis.[128] Subsequently, Ford almost lost the 1976 contest for the Republican presidential nomination to Reagan who painted détente as a source of American weakness.[129] Reagan said that under Kissinger and Ford, the US had "become number two in military power in a world where it [was] dangerous—if not fatal—to be second best."[130] Reagan especially abhorred the ABM Treaty because he believed it constrained American technological advantages.

A key element of GOP criticism of détente was that the US had consistently underestimated Soviet military capabilities, and thus called into question the competence of the intelligence community. Members of the Committee on the Present Danger lobbied hard to have an independent review of CIA assessments, an effort resulting in the so-called Team B experiment.[131] Notable Team B members included détente skeptics such as Richard Pipes of Harvard, William Van Cleave of the University of Southern California, and Major General George Keagan, the head of Air Force intelligence. Team B received full access to classified intelligence and previous national intelligence estimates (NIEs). Its final report concluded that the intelligence community had "substantially misperceived the motivations behind Soviet strategic programs, and thereby tended consistently to underestimate their intensity, scope, and implicit threat." Team B, like the Buchsbaum panel, viewed space as a domain of military competition, saying that the Soviets sought "to deny the US the essential support of its space systems in potential future conflicts at all levels of the spectrum."[132] Several members of Team B and the Committee on the Present Danger would later go on to advise Reagan as president and shape his national security policies. Expanded space militarization thus became an intrinsic part of his "conservative space agenda."[133] From this point of view, space restraint held back American advantages in aerospace technologies. Space conservatives rejected anything associated with détente and were skeptical of arms control in general.

NEW NEGOTIATIONS ON SPACE MILITARIZATION

While many American conservative thinkers had largely abandoned détente, Jimmy Carter embraced it while also attempting to "change the domestic policies of the Soviet Union."[134] When Carter succeeded Ford in 1977, he inherited a very difficult global and domestic situation. The Soviets had deployed more SS-20 road-mobile intermediate-range ballistic missiles that could strike anywhere in Europe with very little warning. At the same time, anti-nuclear and anti-NATO sentiments were rising in Europe. In the US, the economy was suffering, and there was growing skepticism about arms control negotiations with the Soviet Union. In these complicated circumstances, Carter wanted to secure a SALT II agreement with the Soviet Union and to prevent the arms competition from expanding into space. Shortly after taking office, the president froze the Pentagon's ASAT program pending a policy review.[135]

At Carter's behest, during a March 1977 trip to Moscow, Secretary of State Cyrus Vance told Soviet Foreign Minister Andrei Gromyko that he "wanted to raise the issue of placing limits on the anti-satellite capabilities of both sides" and that these limits would help to "stabilize the strategic situation."[136] Gromyko acknowledged that ASATs were a problem and said that Moscow "would be prepared to examine any proposal the United States could submit."[137] Carter reiterated Vance's points in a letter to Brezhnev, writing that he sought an "agreement not to arm satellites nor to develop the ability to destroy or damage satellites."[138] Carter had concluded that since satellites were critical for arms control verification, the public perception that there was an emerging ASAT race could erode support for SALT II.

The Pentagon quickly pushed back on the White House's ASAT arms control objectives. The joint chiefs maintained that having the ability to destroy Soviet satellites in wartime remained vital. They argued that ASATs would also provide a credible deterrent against Soviet interference with US space systems during crises.[139] This deterrent argument contradicted the intelligence community's position that a US ASAT was unlikely to "deter [Moscow] from interfering with US satellites" in wartime.[140] Deterrence aside, the military service leaders reiterated that the Soviet radar ocean surveillance satellites seriously challenged the survivability of the US surface fleet. In reality, it is doubtful that these satellites actually threatened the "survivability" of the fleet due to their technical limitations, but improvements to these satellites over time were of justifiable concern.[141] If the US did not act promptly, the chiefs warned, Moscow could "gain superiority in space."[142]

By this point, the Pentagon had adopted an aircraft-launched ASAT concept called the "miniature homing vehicle" (MHV).[143] The Air Force would lead MHV research, development, and eventual operation. To determine the way forward on ASATs, the White House established a policy review committee focused on space issues and appointed Secretary of Defense Harold Brown as the lead. At the first meeting in August 1977, Brown proposed a "ban only on peacetime use of antisatellite systems" because "in wartime, arms control would not provide protection [for US satellites]." Zbigniew Brzezinski, Carter's National Security Advisor, disagreed with Brown, saying that "a comprehensive [ASAT] ban would serve [the US] security interest, reinforce stability, and support [US–Soviet] SALT efforts." He further argued that "just because the Soviets have something is no adequate reason for [the US] to acquire an ASAT." Lieutenant General William Smith, assistant to the chairman of the joint chiefs, however, opposed ASAT arms control on the grounds that adequate verification mechanisms were lacking.[144]

While the ASAT arms control debate raged, the Pentagon took steps to exploit the tactical potential of intelligence satellites more fully. Senior national security officials saw value in using NRO satellites to provide "responsive intelligence support" to military commanders. Yet, the Department of Defense, securing greater access to the nation's precious few intelligence satellites, raised concerns about the ability of the overhead reconnaissance inventory to meet all needs, for example treaty verification and national intelligence.[145] Such a move involved considerable political sensitivities as well; analysts predicted that "a visible military support role for reconnaissance satellites may increase the likelihood that [satellites] will become targets at certain levels of crises or conflicts."[146] From the standpoint of lawmakers, the benefits of expanding satellite usage in support of military objectives outweighed the risks. Consequently, in 1977, Congress directed each military service to create a Tactical Exploitation of National Capabilities Program (TENCAP) to integrate NRO satellites with military operations more closely.[147]

As the White House evaluated ASAT policy, space weapons captured the public's attention. George Keegan, the recently retired former head of US Air Force intelligence, stirred anxieties about "a fast-emerging beam weaponry 'gap' with the Soviet Union [in the lead]." He claimed that Soviet laser weapons would be able to "completely neutralize the American strategic deterrent."[148] Keegan's intelligence background gave him credibility, but unbeknownst to the public, these alarmist statements were not supported by intelligence estimates.[149]

The space-war hyperbole in the media only made limits on ASATs appear more urgent to the Carter administration. The interagency working group was frustrated that "exaggerated statements in the US press about lasers [had] raised public concerns about a possible arms competition in space."[150] There were already multiple international agreements that had direct bearing on the ASAT issue: the Limited Test Ban Treaty prohibited nuclear detonations in space; the Outer Space Treaty banned nuclear weapons in space; the International Telecommunications Convention outlawed harmful interference with radio services or communications; the 1971 "Agreement on Measures to Reduce the Outbreak of Nuclear War" required Moscow and Washington to notify each other in the event of interference with strategic warning systems; and the SALT agreements banned NTM interference.[151] The task at hand was to find ways to strengthen existing safeguards in the near term.

Members of the working group outlined four potential ASAT arms control approaches: (1) no limits at all; (2) a high threshold for use of ASATs against *any* satellites (this section is still mostly redacted); (3) selected limits such as a ban on laser and high-altitude ASATs but permission for low-altitude ASATs; and (4) a comprehensive ASAT arms limitation. Option 3 entailed creating a "partial sanctuary" for high-altitude satellites with nuclear early warning, command and control, and communications missions; all these functions were directly tied to crisis stability.[152] The State Department and the Arms Control and Disarmament Agency (ACDA) supported option 4, the Office of the Secretary of Defense (OSD) wanted option 2, and the Joint Chiefs of Staff endorsed option 1.[153] The Office of Science and Technology Policy (OSTP) recommended pushing forward with ASATs to deny the Soviets "a one-sided sanctuary in space for critical space systems that directly support their military forces."[154] Like Pentagon officials, OSTP believed that a verifiable ASAT arms control agreement with the Soviets was not really feasible.[155] Thus, at this stage, there was no consensus among national security officials about the ASAT arms control problem. It was therefore left to the Oval Office to make an executive decision regarding the path forward.

With public attention on space warfare and controversy over SALT II intensifying, Brzezinski believed that there would be no "better time to seek an ASAT arms control agreement" and advised the president to pursue a comprehensive ASAT ban. On September 23, 1977, Carter adopted Brzezinski's recommendation but directed the Pentagon to continue ASAT research, pending a treaty. He specifically prohibited any testing in space.[156] Electronic warfare systems (i.e., jamming capabilities) would not be included in an ASAT ban because verifying their elimination would be especially difficult and retaining them provided the US with a limited interference system should a crisis arise.

On the twentieth anniversary of Sputnik, Secretary of Defense Harold Brown held a news conference at the Pentagon in which he suggested the US and the Soviet Union were entering a new, and more militarized, phase in space competition. He described the "operational" Soviet ASAT as giving the Kremlin a distinct military advantage because "we rely a good deal on our space system[s]."[157] In contrast to the recent past, the possibility of war in space was publicly contemplated. Brown expressed hope that we "could keep space from becoming an area of active conflict" but then accused Moscow of "leaving [the US] with little choice" other than to "engage in a space weapons race."[158] Chairman of the Joint Chiefs General George Brown told Congress that due to growing American dependency on space, combined with the Soviet Union's operational ASAT capability, "the heretofore accepted sanctuary of space may be jeopardized."[159] The Pentagon endeavored to persuade the public and key lawmakers that the US could not risk falling behind in this high-technology arena.

Attention on conflict in space complicated the SALT II process by generating concerns about the dependability of satellites as verification tools. It is for this reason that Vance relayed to Soviet Ambassador Anatoly Dobrynin the urgency of the ASAT problem because of its direct "bearing on the ratification of the SALT [II] agreement."[160] Shortly thereafter, the *Washington Post* claimed that attacks on satellites "might take away one country's ability to police treaties like SALT and it could lead to a very cold resumption of the Cold War."[161] Due to news reports about Soviet "killer satellites," lawmakers raised the linkage of an ASAT ban with SALT II during a closed-door hearing with Paul Warnke, the director of ACDA. Senator John Glenn asked Warnke how ASAT deployments could affect SALT II. Warnke replied that deployment and *use* of ASATs meant that "you could forget about SALT agreements," but he quickly pointed out that "anti-satellite activity is barred in SALT II . . . [because] it would constitute interference with national technical means, which is prohibited,"[162] although, as stated above, interference had not actually been defined. Warnke reassured lawmakers that diplomats would commence ASAT arms control discussions with their Soviet counterparts in the near future.

While the White House continued to review its ASAT policy, new intelligence came to light suggesting that Moscow was further expanding its space warfare capabilities. Analysts predicted that "the Soviets [would] continue improving . . . their nonnuclear orbital interceptor, and possibly modify this system to permit intercepts of US satellites in synchronous and semi-synchronous orbits."[163] In other words, the Soviets might eventually be able to destroy higher-altitude satellites used for nuclear early warning and command and control. Laser weapons would permit Soviet interference with satellite operations short of their physical destruction. Not all concerns were centered on purely military considerations. Prior to this NIE, Brzezinski warned Carter that "the political consequences of a Soviet laser ASAT system in space might be substantial . . . shatter[ing] our sense of technical superiority as badly as it was when the first Sputnik was orbited."[164] The events of the 1980s would soon demonstrate the profound psychological power of space weapons, even when experts doubted their military utility.

The public discussion on space weapons soon moved beyond ASATs. The *New York Times* printed an article warning that Pentagon officials believed that "during the next decade the Soviet Union will be able to upgrade its combat capabilities in space, perhaps producing a new generation of laser-equipped spacecraft . . . if this happens, it will be 'Star Wars' for real."[165] *Aviation Week and Space Technology* published a piece about the potential applications for space-based laser weapons.[166] It described how lasers deployed in space could destroy ballistic missiles in their boost phase of flight. The author also used the term "space-based battle stations," which according to Donald Baucom, the official historian of the Strategic Defense Initiative Organization (SDIO), might have been the first appearance of this concept in US defense literature.[167] These battle stations would focus directed energy into a narrow beam in order to destroy incoming missiles. The primary source for this article was Maxwell Hunter, a senior aerospace engineer at Lockheed.[168] Hunter had concluded that "lasers in space could produce a revolution in warfare by ending the long-standing dominance of offensive weapons" and that it was therefore "a genocidal hoax" to treat space as a sanctuary.[169]

Hunter produced a paper in 1977 outlining his main ideas on strategic defense and caught the attention of Angelo Codevilla, a staffer for Senator Malcom Wallop. Codevilla and Wallop had become very interested in strategic defense and Soviet research into lasers that could be used for destroying ballistic missiles. Wallop was a member of the Senate Select Committee on Intelligence. So, both he and Codevilla had access to classified intelligence assessments of Soviet military research.[170] Codevilla introduced Hunter to Wallop, and Hunter convinced both of them that the US could develop "a constellation of laser-armed space-based battle stations" for missile defense.[171] Consequently, Wallop attempted to generate support for space-based missile defense in Congress. While he was not able to secure increased funding for laser weapons, his advocacy did attract thirty-nine senators who formed a "laser lobby" in the Senate.[172] Wallop detailed his arguments in favor of strategic defense in an article-length manuscript for *Strategic Review*. Reagan received a pre-circulated draft in the lead up to the 1980 election and responded with "supportive remarks."[173] Without knowing it, Wallop caught the attention of the most consequential person he could have hoped for.

Forestalling a space spiral in the arms race was no easy task. Warnke suggested that "all signs point to an uphill struggle to achieve [Soviet] consent to a comprehensive and verifiable [ASAT] agreement."[174] In an effort to place even more pressure on the Soviet Union to agree to ASAT limits, in March 1978, Carter removed the restriction on ASAT testing in space.[175] Brzezinski had convinced the president that approving testing would show Moscow that Washington "intend[ed] to seek equivalent capabilities as soon as possible unless the Soviets [were] willing to take positive steps to stop testing, dismantle, and agree to *substantive* verification techniques" (emphasis in original).[176] Carter's dual-track approach to ASATs foreshadowed NATO's decision in 1979 to move forward with deployment of intermediate-range nuclear forces in Europe while negotiating with the Soviet Union to limit these same systems.[177] The Pentagon maintained that verification would be extremely difficult because so many ASAT-related technologies were dual use. The Galosh anti-ballistic missile system around Moscow, for example, could have served as a ground-launched ASAT.

In May 1978, Carter approved a comprehensive space policy that addressed the nation's military, intelligence, and civil space programs.[178] The classified version established that "the United States shall seek a verifiable ban on anti-satellite capabilities, *excluding* electronic warfare [i.e., jamming]." Even if ASAT limits were secured, the administration maintained that at least "some R&D should be continued as a hedge against Soviet breakout." What is notably absent from this document is any mention of the military utility of an ASAT for negating the Soviet Union's radar satellites that were used to track US and allied naval vessels, which served as the original justification for the renewed American ASAT program. This omission was a further indication that, for Carter, the imperatives of arms control superseded any perceived military utility of having an operational ASAT. And Carter viewed ASATs as an impediment to his overarching arms control agenda.

In examining discussions among Carter and his advisors concerning objectives for ASAT arms control, we find multiple motivations at play. John Maurer writes that "historically, policymakers have embraced arms control pluralism, pursuing agreements that can advance multiple arms control objectives simultaneously."[179] These goals might include disarmament, stability, and advantage.[180] Cameron shows how domestic political considerations oftentimes disproportionately shaped US arms control initiatives during the Cold War.[181] Carter's ASAT strategy included elements of all of the above. As I already noted, Carter wanted an ASAT accord to bolster public confidence that satellites were indeed adequate arms control verification mechanisms. Concurrently, he sought to preserve US advantages in satellite technologies used for reconnaissance and military support. He further believed that strengthening satellite safeguards promoted stability. Even if he could not secure an ASAT ban, any limits on ASATs included in a SALT II agreement would have been a political victory for Carter.

In June 1978, American and Soviet diplomats met in Helsinki for exploratory talks on ASAT constraints. US representatives raised the prospect of a comprehensive agreement limiting the development and retention of ASAT systems, a prohibition on satellite attacks, and an end to ASAT testing. The Soviet side was reticent about agreeing to suspend ASAT testing, saying that "it was too early to consider such an understanding at [that] stage in the talks."[182] On the subject of prohibiting existing ASATs and acquiring new ones, the Soviets were quiet and requested more detailed information. Unexpectedly for the Americans, Soviet negotiators identified the space shuttle as a potential problem without going into any details.[183] The Soviets also wanted to discuss certain "unlawful" activities such as using satellites to broadcast television into another state without permission, clearly in reference to American propaganda. The parties reached a consensus that a hostile acts agreement could be beneficial. In summarizing these interactions, US diplomats described the atmosphere as "cordial" and judged that "the two sides have similar views of the main characteristics of the subject."[184]

The superpower ASAT talks drew greater international attention to the prospect of conflict in space. The Stockholm International Peace Research Institute "presented the future of global warfare as dependent on satellite technology."[185] The British weekly *Radio Times* advertised the BBC series "The Real War in Space" on the cover of an October edition and depicted Soviet ASATs destroying an American satellite with a laser weapon.[186] Siebeneichner notes that "rumors about the development of so-called killer-satellites and anti-satellite weapons (ASATs) made worldwide headlines."[187] Due to the prospect of superpower ASAT competition, in 1978, the United Nations Committee on Disarmament called for the "organization of negotiations on the prevention of an arms race in outer space."[188]

Amid growing public anxieties about a space arms race, Carter decided to declassify the existence of US photographic reconnaissance satellites. While it was known that both Washington and Moscow used reconnaissance satellites, the US government did not officially acknowledge them. A 1978 policy paper proposing declassification explained that "government spokesmen are prohibited from 'officially' stating that the US conducts satellite photography to monitor Soviet compliance with SALT. They are restricted to using the euphemism National Technical Means (NTM) . . . the term NTM, however, may be lost on less aware segments of the lay public."[189] Consequently, during a September 1978 speech at Cape Canaveral, Carter described how "photoreconnaissance satellites have become an important stabilizing factor in world affairs in the monitoring of arms control agreements."[190] Pointing to the essential role of reconnaissance satellites for SALT II was, however, a double-edged sword, making US space systems appear to be especially vulnerable due to the ongoing media uproar about Soviet killer satellites.

Moving into the next round of ASAT negotiations, the White House viewed the complete dismantlement of the Soviet ASAT system as an unlikely outcome. ASAT interceptors were very small and easily concealable, and thus it would be difficult to verify their elimination. All of the US agencies involved in deliberations over ASAT policy agreed that a hostile acts accord should prohibit only *physical* attacks on satellites, which had not been explicitly stated in SALT-era agreements concerning NTM. The Pentagon supported an indefinite test ban on high-altitude ASATs but opposed a low-altitude test ban due to its implications for the US ASAT program. The State Department and ACDA proposed a comprehensive test ban not to exceed three years.[191] The US ASAT was not expected to be ready for testing until after that latter ban expired, which meant that the American ASAT program would be unaffected by any of these interim constraints. Officials clearly agreed that unilateral restraint in ASAT development went against the US interest. Spurgeon Keeny, the deputy director of ACDA, argued that all potential means of attacking satellites should be prohibited, including lasers, but such a position was unrealistic due to the limits of verification capabilities.[192] Constraints on electronic means of attack (i.e., jamming) were still opposed. Fundamentally, Carter aimed to limit, but *not* eliminate, mechanisms for space warfare.

The White House remained focused on the linkage between an ASAT agreement and SALT II ratification. A one-year test ban, although very limited compared to Carter's original agenda for the talks, fulfilled the White House's desire "to reach [a] quick agreement with the Soviets in order to bolster SALT ratification prospects." This was not without risks; defense officials knew that an ASAT agreement "could backfire with Senators concerned about US–USSR military asymmetries."[193] Vance encouraged Carter to move ahead with an initial ASAT agreement, however limited, since it "would usefully complement the SALT Treaty by enhancing the security of our verification means."[194]

Since all of NATO depended on US satellites for vital defense functions, the outcome of the ASAT negotiations was of great consequence for the entirety of the transatlantic alliance. A UK Foreign and Commonwealth Office (FCO) study observed that "in view of NATO's increasing reliance on military satellites for intelligence gathering, navigation and C3 [command, control, and communications], the deployment of an effective Soviet system could pose a general threat to Western security."[195] There was also anxiety that an attack on military satellites could increase the likelihood of a nuclear confrontation. John Killick, the UK ambassador to NATO, worried that if ASATs were used to destroy either American or Soviet satellites, there could be strong reason "for the side whose satellites have been put out of action to have to contemplate first use of nuclear weapons for fear of themselves becoming the victim of a first pre-emptive strike."[196]

The British struggled to understand the American rationale for a new ASAT and wondered if a "general US unwillingness to concede inferiority in any area of military activity" was the primary explanation.[197] Despite its concerns, the FCO maintained that "we can rely on the Americans to protect our interests."[198] French President Valery Giscard d'Estaing strongly advocated limits on ASATs while addressing the UN General Assembly in 1978.[199] Despite this public concern about an arms race in space, French diplomats displayed little interest in US space arms control strategy.[200] In any case, French and broader

European concerns about ASATs would intensify during the Reagan administration, and European diplomats would take a much more activist role in space arms control in the 1980s.

In January and April 1979, the US and the Soviet Union commenced their second and third rounds of ASAT talks. Officials from both sides reached a consensus in favor of prohibiting "damaging or destroying each other's satellites [outside of wartime]" and displacement of each other's satellites. Talks hit a serious roadblock, however, when the Soviets demanded a halt to the shuttle test program as a precondition to an ASAT testing ban, which Washington rejected.[201] The Soviets viewed the shuttle as an ASAT-capable platform, since it could conceivably deploy weapons from its payload bay and had the ability to displace satellites using its grappling arm. A recent archival discovery reveals that some Soviet defense experts believed that the shuttle might be capable of carrying out nuclear bombardment from space.[202] Soviet objections to the space shuttle highlighted a complicated aspect of space arms control: even defining what constituted a space weapon was no easy task. This reality would continue to vex space arms control efforts up through the present writing.

The Soviets also insisted on banning "non-destructive interference with satellites," jamming for example. The ACDA and State Department countered that an ASAT test suspension should be confined to a ban only on means for "damaging or destroying" satellites. The Pentagon maintained that since electronic warfare systems did not *physically* damage satellites, they would be permissible under the US framework. Secretary of Defense Brown went out of his way to protect "some of our highly classified programs" for satellite interference.[203] Internally, members of the US delegation agreed that the US was "giving up nothing" in terms of military capabilities, since the Pentagon had no ASAT interceptor tests planned for the duration of the proposed US–Soviet ASAT testing suspension.[204]

Despite progress in key areas, Gromyko told Vance in June 1979 that "a substantial difference between the positions of the [two] sides existed" and that an ASAT agreement at the SALT II summit should not be expected.[205] The Soviets wanted a provision that satellites used for "hostile" or "illegal" acts—terms that were open to interpretation—would not be covered by a treaty. Moscow especially did not want Chinese satellites covered by an agreement.[206] These Soviet demands had the potential to "legitimize the retention and use of ASAT systems, thus undercutting the basic objective of the agreement."[207] Despite these differences, the State Department still held out hope that that an ASAT agreement might be reachable as a part of the SALT II agreement, but these hopes turned out to be more optimistic than realistic.[208]

The divergent US–Soviet agendas had reached an impasse. Consequently, the SALT II agreement signed by Brezhnev and Carter in Vienna on June 18, 1979, did not include any specific ASAT provisions. Talks on ASAT limits nevertheless continued until the Soviets invaded Afghanistan in December 1979. In response, Carter withdrew SALT II from the Senate ratification process, and the ASAT talks ground to a halt. The following year, Moscow indicated that it wanted to resume ASAT negotiations as part of an overall reinvigoration of the arms control dialogue. However, with Reagan's landslide electoral victory over Carter in November 1980, the Soviets would have to wait and see how the new administration would handle space arms control. Certainly, the lingering disagreements between Moscow and Washington concerning limits to military activities in space made progress in this arena appear to be an unlikely prospect.[209] And the tension between military freedom of action in space and the desire for arms control treaties would only intensify moving into the final decade of the Cold War.

CONCLUSIONS

At the beginning of the Space Age, satellites quickly became indispensable for US national security. The dependence of the US on space reconnaissance for intelligence concerning the Soviet Union disproportionately shaped US space strategy, which was, by and large, defined by taking a cautionary position on any action that could provoke the Soviet Union to act aggressively in space. From the 1960s to well into the 1970s, the notions that space weapons were of limited military utility and problematic due to their implications for broader arms control strategy were core tenets of American space policy. Since all of NATO depended on US space infrastructure, American officials strove to create a situation in which space was a free zone, in effect, for information functions (e.g., communications and reconnaissance). US and Soviet space weapons posed a threat to that framework.

Key US officials began to change their views on the utility of ASATs as American and Soviet defense planners began more closely integrating satellites with combat forces. Moreover, the emergence of nonnuclear ASATs allowed more precise strikes on enemy satellites. These technological changes, combined with cooling superpower relations, created the conditions for expanded space militarization beginning in the mid to late 1970s. Satellites that could be used for arms control monitoring increasingly had military support functions. Consequently, arguments that satellites were inherently "stabilizing" need to be qualified.[210] Certainly, satellites used for arms control monitoring opened the way for greater transparency, but some of those same satellites had the potential to enable direct attacks on enemy forces. With the latter in mind, we should not be surprised that American and Soviet officials sought capabilities for denying each other the use of critical military satellites in wartime conditions.

Less than a decade after the US won the moon race, media depictions of Soviet ASATs appeared to challenge American preeminence in the cosmos. Since many relevant details about national security space programs, especially those used for military support, were classified, the public perceptions concerning ASATs were shaped primarily by leaks and US government statements on superpower activities in space. Oftentimes, the psychological dimensions of the debate were at odds with the technical realities. The advent of ASAT negotiations raised the profile of space militarization in the diplomatic arena and forced US allies to begin paying attention to a subject that had been largely invisible in transatlantic relations.

A consistent theme in the 1970s—and one that continued to resurface— was lack of interagency agreement on military space matters. Most of the critical US decisions on ASATs were top down rather than the result of consensus in the national security establishment. Attempting to establish new boundaries on space militarization was especially challenging due to the growing tension between military space requirements and arms control agendas. Carter primarily viewed ASAT arms control as a means of bolstering support for SALT II. But the inability of Washington and Moscow to strengthen existing safeguards for satellites with a new arms control treaty only contributed further to the perception that superpower military competition would extend into the heavens. While events of the 1970s in no way made SDI inevitable, they did lay the groundwork for the intensified space militarization in the decade to come.

2 CAMPAIGN FOR THE HIGH GROUND

> Space capabilities may bring about the technological disarmament of nuclear weapons.
>
> —General Curtis LeMay (1962)

Approximately two months after Ronald Reagan's inauguration in January 1981, more than 250 people gathered at the US Air Force Academy for a symposium dedicated to examining the future of military activities in space. In a passionate speech, retired General Bernard Schriever, the father of the Air Force space and missile programs, described how he could visualize the day when weapons in space could "hold land, sea, and air systems hostage."[1] From his perspective, the US was at a crossroads regarding its space future. It could use military space technologies to secure a strategic advantage over the Soviet Union or pursue a restrained space agenda, which Schriever believed would be detrimental to US interests. Findings from this Air Force Academy conference would directly inform Reagan's first space policy, which was based on the premise that Moscow had initiated a campaign to seize the "high ground" of space.[2] Schriever's ideas served, moreover, as key elements in the agenda of a small but vocal group of "space zealots," primarily Air Force officers, who sought US military supremacy in space.[3]

Historians and political scientists have explained how Reagan's SDI emerged from a mixture of political and ideological forces.[4] Certainly, Reagan's abhorrence of mutually assured destruction was a very significant factor in his establishment of SDI.[5] However, SDI's connections to Reagan's broader space strategy remains an understudied topic.[6] In the first two years of Reagan's presidency, the White House was already signaling that space was going to play a prominent role in American grand strategy, in terms both of military and economic interests.[7] For Reagan, space militarization and commercialization were two sides of the same coin.[8]

SDI involved multiple, seemingly contradictory, elements of Reagan's broader space and foreign policy agendas. Technologies developed under the SDI aegis would ensure that the US remained the world leader in the military space arena. Concurrently, SDI would serve as a technological incubator with both defense and civilian applications.[9] In stark contrast with his competitive language, Reagan often described SDI as a tool of peace that he wanted to share with the Soviet Union.[10] Ideologically, the program was grounded in the astrofuturistic notion that the space frontier is a "site of renewal, a place where we can resolve the domestic and global battles that have paralyzed our progress on earth."[11] Importantly, Reagan sincerely believed that the US could indeed use SDI to promote peace while also pursuing military advantages in space.

Reagan established SDI without considering its many implications for US foreign policy. In terms of resources and political significance, it entailed the military space program taking the dominant role in US space policy. Although orbital weapons garnered the most attention, much of SDI's resources would be devoted to building a space infrastructure that could support the deployment and maintenance of vast numbers of systems in orbit. Despite the fact that the president characterized SDI as purely defensive, it involved technologies that could be used to attack other satellites.[12] Consequently, ASAT arms control emerged as an even more urgent issue on the world stage.

EMBRACING MILITARY COMPETITION IN SPACE

Reagan was elected on a political platform promising to reverse the trend of the US being on the defensive in the Cold War.[13] He had warned that the Soviet Union had exploited détente to get ahead in the arms race and to expand its influence around the globe, especially in the so-called Third World.[14] Since the US was still suffering from a sense of self-doubt after Vietnam, rebuilding national confidence was a central theme of Reagan's message.[15] The president believed it was possible to use the economic, political, and military tools of national power to moderate Soviet behavior and lower the overall tension between the superpowers.[16] Despite his belief that the US had the ability to turn the political tides in its favor, Reagan still feared that Moscow could surprise the US, especially in the strategic arms competition. Hal Brands notes that "Reagan's approach to grand strategy rested on a seemingly paradoxical idea—that the Soviet Union was simultaneously strong and weak."[17] The US military buildup in the 1980s was one of the most prominent symbols of Reagan's efforts to place pressure on the Soviet Union and to restore national confidence.

The White House identified space as an arena in which the US needed to expand its military power. Within six months of Reagan's January 1981 inauguration, the administration commenced a comprehensive evaluation of US space policy. This review aimed to establish the framework for ensuring that the US could maintain preeminent space leadership in the civil and national security arenas. Concurrently, the US military, the Air Force and the Navy in particular, and the intelligence community investigated ways for expanding the use of space systems for military objectives.[18]

In January 1980, Secretary of the Air Force and NRO Director Hans Mark suggested it was time for the Air Force to develop an official space doctrine to complement existing air doctrine, and he specifically called upon the US Air Force Academy to "apply [its] spectrum of academic expertise to the study of a doctrine for the military role in space."[19] In April 1981, the academy hosted a symposium dedicated to developing proposals for a comprehensive military space policy and doctrine. More than 250 people participated; most of them were active-duty and former military officers. During his keynote address, General Schriever criticized more than two decades of US restraint in space. He derided Eisenhower's policy of "space for peaceful purposes," saying that it "just haunted us. It haunted us constantly." The Reagan administration's interest in expanding US military space capabilities, including their combat potential, offered a chance for the rebirth of American military spacepower.[20]

Attendees resurrected astrodeterminist ideas and rhetoric from the 1960s that presented space as the decisive battlefield in the future. To this end, the symposium's final report declared that "the nation that controls space gains political leverage, if not control on the earth, and will again be in a position to alter the course of history."[21] From this perspective, space technologies provided a way for the US to break through the superpower nuclear stalemate. There was widespread agreement "that future systems will have this [terrestrial support] role as well as the *role of controlling space*" (emphasis added).[22] Specific recommendations included completing Air Force Manual (AFM) 1–6 ("military space doctrine") along with studying the establishment of a separate military service for space and a space command within the Air Force. (It would be another thirty-eight years before an independent space service came into existence.[23])

In assessing the impact of this conference, space policy expert Peter Hays observes that it "does seem to have been a landmark event in shaping general Air Force and military attitudes towards military space doctrine issues in the early 1980s."[24] Hays concludes that the "symposium helped to encourage the Air Force and the other services once again to think seriously about space control and high ground military space applications as they considered military space doctrine for the 1980s [and beyond]."[25] The final report, moreover, provided a framework for giving US military space elements more structure and an expanded mission: not only to support combat forces but also to *project* military force from and into space.

An expanded military role in space would require greater synergy between the national security and civil space programs.[26] April 1981, the same month as the Air Force Academy symposium, marked the first orbital flight of the space shuttle, officially called the "Space Transportation System" (STS). The potential of the shuttle for military missions was an important factor in reinvigorating military thought about space.[27] The US government planned for the shuttle to be the main vehicle for delivering military systems into orbit. It also represented a key technological bridge between the national security and civil space efforts. National Security Decision Directive (NSDD)-8, signed by Reagan on November 13, 1981, determined that "the STS [shuttle] will be the primary space launch system for both US military and civil government missions."[28] NASA's space shuttle would eventually be designated as the logistical workhorse for delivering SDI components into space and therefore contribute to a further blurring of the boundaries between civil and military space activities.[29]

With the shuttle as his backdrop, on July 4, 1982, Reagan delivered a speech at Edwards Air Force Base, describing how space technologies would play a central role in promoting American national security and economic agendas. Concurrently, the White House released a public version of its classified national space policy. The unclassified fact sheet stated that US space policy sought to maintain American space leadership, preserve the right of self-defense in space, and further its national interests through both civil and national security space endeavors. Other objectives of Reagan's space policy were fostering the growth of commercial space investment and transferring some government space missions to the private sector.[30] Military power and economic prosperity in space went hand in hand.

The classified version of Reagan's first space policy reflected key ideas presented at the Air Force Academy symposium in 1981. This was not by happenstance; Air Force Colonel Gil Rye, who was in charge of space policy on the NSC, attended the academy's space doctrine conference. In formulating a more expansive military space policy, Rye had an important ally in Air Force Chief of Staff General Lew Allen.[31] The general had a very unusual background for the Air Force's senior officer. He qualified as a pilot but never flew in combat. Allen possessed a doctorate in physics and spent a substantial portion of his career working on space programs, including time as a senior manager at the NRO and later as director of the National Security Agency (NSA).

Contrary to Carter's 1978 space policy that called for a comprehensive ASAT ban, Reagan's stated that "the United States will develop and deploy an ASAT capability" and that its purpose was to "deter threats to space systems of the United States and its allies, and within such limits imposed by international law [remainder redacted]."[32] The unclassified version affirmed the need to have an ASAT to "deny any adversary the use of space-based systems that provide support to hostile military forces."[33] Under the heading "Force Application," Reagan's policy further laid out the requirement to "be prepared to develop, acquire, and deploy space weapon systems and to counter adversary space activities, should national security conditions dictate."[34] According to media reports, the Pentagon planned "new areas of weaponry, particularly in space," including "prototype development of space-based weapons systems."[35] To prepare for "operationalizing" space, the Air Force activated a space command in Colorado Springs.[36]

Deterrence was a new, and controversial, justification for the ASAT program.[37] Thomas Schelling argued that a threat is more credible if the adversary perceives that it is "in the same currency, to respond in the same language, to make the punishment fit the character of the crime."[38] Applying Schelling's ideas to the superpower space balance in the early 1980s suggests that the Soviet Union would not have been deterred by a US ASAT because the Soviet Union was less dependent on space systems than the US. But as Soviet dependence on satellites grew, a US ASAT capability *might* have become a more effective deterrence mechanism in specific circumstances.[39] Whether the military utility of ASATs outweighed the potential consequences of a space arms competition would be a source of constant disagreement among US policymakers and defense experts.[40]

Soviet military space activities provided the Reagan administration with a ready justification for its even more militarized approach to space. Around the time that the Air Force announced its new command for space operations, the Soviet Union integrated ASATs into war games. After a Soviet military exercise in 1982 involving ASATs, General Lew Allen described how a Soviet ASAT "maneuver [in space] was 'remarkable' for its complexity and underscored the great determination of the Soviet Union [to expand its military might in space]."[41] This contrasted with his statement three years prior that "we give it [Soviet ASAT] a very questionable operational capability for a few launches. In other words it is a threat, but they have not had a test program that would cause us to believe it is a very credible threat."[42]

Allen's more alarmist characterization of the Soviet ASAT can be explained, at least in part, by newly acquired intelligence. An NIE on Soviet military space activities observed that in April 1980, the Soviets had renewed testing after a "standdown [sic] of nearly two years" and that they were evaluating new sensor homing technologies.[43] Even more disconcerting was intelligence suggesting that the Soviets could soon begin to test an ASAT prototype against satellites in higher orbits where especially important intelligence and nuclear early warning systems were located.[44] Analysts believed that the Soviet Union could develop space-based laser ASATs in the long term.[45] Space-based laser ASATs could be improved and eventually used in a missile defense role as well. At this same time, elements in the UK Ministry of Defence (MoD) warned that the threat posed by Soviet tracking and surveillance satellites was in danger of being underestimated.[46] A classified British space policy paper went so far as to suggest that the UK government should consider its own capability to "damage enemy satellites," although it would never be pursued.[47]

Reagan's near-term answer to the Soviet ASAT threat was the F-15 jet-launched MHV, born out of Ford's 1977 ASAT decision. Within a week of Reagan coming into office, the Pentagon awarded contracts to Boeing and Vought to develop the missile and homing vehicle.[48] The concept involved an F-15 launching a modified Boeing AGM-69 short-range attack missile that had an infrared homing device in the warhead, along with an infrared seeker that would enable it to track and engage its target more accurately.

Developing a homing-capable ASAT proved to be very difficult; the General Accounting Office (GAO) predicted that the MHV "will be [a] more complex and expensive task than originally envisioned, potentially costing in the tens of billions of dollars."[49] Consequently, GAO called for investigations into alternative ASAT weapons, especially ground-, air-, and space-based laser systems.[50] The Reagan White House nevertheless pushed forward with MHV and justified its cost by citing the Soviet Union's space weapons efforts.

Exactly one year before Reagan's "Star Wars" speech, laser weapons made the headlines. In March 1982, a member of Congress inadvertently read a transcript of a classified Department of Defense assessment about Soviet space lasers during an open hearing. The transcript quoted Richard DeLauer, the undersecretary of defense for research and engineering, as having said that "the Soviet military is well on its way to seizing the high ground of outer space, with the first big step the likely deployment of lasers there as early as next year [1983]." DeLauer predicted that the Soviets would deploy a "manned orbital space complex to be operational by about 1990 capable of effectively attacking ground, sea and air targets from space."[51] The *Washington Post* immediately thereafter published an article with the headline "Soviets Reported Ready to Orbit Laser Weapons" and included DeLauer's comments. In an attempt to curb hyperbole about an imminent Soviet space threat, DeLauer issued a clarification saying that 1983 was "the bottom range of possibilities which DoD considers likely." He later said that "the Soviets have about a 5 year lead in space based laser technology over the United States," and could possibly place such a weapon in orbit in five years.[52] This latter assessment was not supported by intelligence reporting.[53] But these sensationalist statements created the impression that a new space race had already begun and that the US was behind.

With military space issues becoming prominent in the national security discourse, the US Air Force devoted more attention to developing doctrine for conducting operations in space. To this end, the Air Force approved AFM 1–6 in October 1982. This was the first official US Cold War space doctrine and originated with Allen's directive in 1977 calling for the Air Force to create formal guidance for military space operations.[54] It identified spacepower as "a natural extension of the evolution of airpower" and argued for the need to "prevent space from being used as a sanctuary for aggressive systems by [US] adversaries." To protect US national security interests in space, AFM 1–6 listed three potential missions: (1) space-based weapons for deterrence,

(2) space-to-ground weapons, and (3) space control and superiority.[55] Space control entailed using military resources to be able to deny an adversary the use of space. According to Hays, despite its shortcomings, "AFM 1–6 [was] a major milestone in doctrinal thinking which was responsible to both top-down and bottom-up pressures.[56]

After having established his broad goals for military, intelligence, commercial, and civil space activities, in December 1982, Reagan approved a national security study to develop a more specific road map for achieving his space objectives.[57] According to Rye, the Soviet co-orbital ASAT was an important factor in initiating this directive.[58] The study was based on the two following premises: (1) "the Soviet Union [had] initiated a major campaign to capture the 'high ground' of space," and (2) "regardless of Soviet activities, the space medium offers significant potential for the enhancement of civil, commercial and national security capabilities."[59] Rye inserted requirements to examine the possibility of an independent military service for space, in addition to a unified space command. Even mentioning the idea of "a separate military department for space" ruffled many feathers in the Pentagon. According to Rye, this study sent Herb Reynolds, who worked on space policy in the Pentagon, "into orbit."[60] While the Air Force had already established a space command (and the Navy would create its own in 1983), this study called for an examination of a combatant command that would oversee *all* operational military space activities.

The space strategy study would not be completed until 1984, and it would have to contend with the announcement of SDI. By late 1982, US policy on space was more publicly militarized than it had been at any other time in the Space Age. As Stares notes, this more militarized approach to space did not originate in the Reagan era, but his presidency witnessed a departure from the established norm of downplaying national security activities in space.[61] Although SDI was not yet being formally considered, early in Reagan's presidency, the White House was already signaling its intent to greatly expand American military power in the cosmos.

THE LASER AND MISSILE DEFENSE ENTHUSIASTS

The more prominent role given to space systems in Reagan's defense strategy did not in any way make SDI inevitable or even likely. There was widespread enthusiasm within the defense and intelligence communities for perfecting space technologies designed to support American and allied combat forces, but little momentum existed for missile defense in space. Among senior military leaders, the US MHV ASAT program was not overwhelmingly popular, but it was not highly controversial either. Lack of concern about the MHV ASAT stemmed from the fact that it was a relatively small program that did not require a significant change in US defense policy and strategy. SDI, on the other hand, would come to represent a movement toward the most radical change in US strategy since the dawn of the Nuclear Age.

In explaining SDI's origins, historians have noted the significance of Reagan's nuclear abolitionism, his desire to place pressure on the Soviet Union, the Nuclear Freeze movement, and the MX missile basing debacle.[62] While all of these elements played a role in the emergence of SDI, so too did a network of space enthusiasts and laser advocates who convinced Reagan that embracing strategic defense was technologically feasible and strategically sound. Edoardo Andreoni argues that the importance of the "space defense enthusiasts" is overstated—that they "were not the driving force behind [SDI]" as FitzGerald suggested.[63] He maintains that the inability to find a consensus about the deployment of the new MX ICBM and the "rapid decline" in public support for the White House's modernization strategy were the crucial factors that led to SDI.[64] The latter certainly provided the occasion for SDI, but Reagan had indeed become convinced, due in large part to the "space defense enthusiasts," that advanced technologies had matured to a point where the US could begin to rely more on defensive systems. Simultaneously, anxiety about the Soviets getting ahead in this general area was not insignificant either and only reinforced the president's belief that the US needed to demonstrate its "space leadership" to the world through new space endeavors.

In the late 1970s, Senator Malcom Wallop and his senior aide Angelo Codevilla were waging a campaign to garner greater support for lasers and missile defense in the US Congress. Retired Air Force Major General George Keegan concurrently shared his alarmist views on Soviet research into laser-based missile defense with any media outlet that would listen to him; he also advocated greater investment in laser weapons.[65] A key figure who joined forces with Wallop and Codevilla to gain momentum for space-based missile defense was retired US Army Lieutenant General Daniel Graham; he had served most recently as the head of the Defense Intelligence Agency. Graham, like Keegan, tended to exaggerate Soviet military space and laser research programs. Graham had been a member of the controversial Team B experiment that accused CIA analysts of consistently underestimating Soviet military power. In 1979, Graham enlisted the assistance of Codevilla in writing *Shall America be Defended? SALT II and Beyond.*[66] In it, he advocated investment in missile defenses to alter the strategic balance in favor of the US. Graham went on to serve as an advisor to Reagan during his presidential campaign.[67]

Graham had the opportunity to brief Reagan on his ideas about missile defense during preparations for the Nashua, New Hampshire, presidential debate in February 1980. In a small motel the night before the debate, Graham walked Reagan through his concept for a missile defense system.[68] According to one account of this particular night, "Mr. Reagan listened attentively and took notes on some of the things Graham [had] said."[69] Graham's idea that advanced technologies could deliver the US out from under the dark cloud of mutually assured destruction deeply appealed to Reagan. Campaign advisor Martin Anderson said that Reagan "embraced the principle of a missile defense wholeheartedly."[70] Despite Reagan's enthusiasm, another senior campaign advisor, Michael Deaver, "vetoed the proposal that [missile defense] be made a campaign pledge."[71] He believed that a plan to change US defense strategy so radically would be treated with hostility by both sides of the political aisle. Graham's association with missile defense created a problem as well, since many people in national security circles considered him to be far from mainstream.[72] He and his space enthusiast colleagues were known as "space cadets" by their critics.[73] Deaver's advice prevailed during the campaign, but that did not stop Graham from continuing to push for missile defense in space both with Reagan's inner circle and in the public sphere.

Shortly after Reagan was elected in November 1980, Graham and Robert Richardson, a retired Air Force general, put together a study under the aegis of the conservative American Security Council Foundation to investigate the technologies associated with space-based missile defense. This project was eventually called "High Frontier"; the name was likely borrowed from Gerard O'Neil's book of the same name. Science fiction writer Jerry Pournelle also participated in the study. O'Neil was not pleased that the title of his book would become a "byword not for the peaceful humanization of space, but for a militaristic vision of space control."[74]

After several months of study, Graham had concluded that space-based missile defense was indeed feasible in the short term. Unlike Wallop and Codevilla, who sought laser missile defense in space, Graham believed that a near-term deployment was possible if the Department of Defense used existing kinetic kill technologies. He envisioned a missile defense system in space that "would most likely take the form of a space vehicle that would serve as a 'garage' for kinetic kill vehicles."[75] Notably, the initial phase of a planned US space-based missile defense system incorporated the idea of "garages" in orbit that housed kinetic weapons for destroying ballistic missiles.

Graham published an article in *Strategic Review* in which he called for the use of "off the shelf technologies" to build and deploy a strategic defense system as quickly as possible.[76] The multilayered concept embraced by Graham was originally advanced by SRI, a California-based research organization.[77] The price tag was listed in 1982 at $24 billion. By comparison, Apollo had cost approximately $28 billion (not adjusted for inflation), and the Manhattan Project had a price tag of $2 billion (also not adjusted for inflation). Officials in the US Army and Air Force were, however, very skeptical of High Frontier's plan.[78]

In 1981, Graham established the High Frontier organization with the sponsorship of the conservative Heritage Foundation in Washington. According to a US government history of SDI, "the founding axiom behind High Frontier was that with a technological 'end run,' concentrated on the deployment of defensive systems in outer space, the United States could regain the strategic superiority it had enjoyed over the Soviets in the 1950s and 1960s."[79] High Frontier sought not only to garner support for strategic defense in the US government, but also to generate enthusiasm for military space activities among the general population. To achieve the latter objective, it became a membership organization "using direct mail techniques to attract large numbers of average citizens."[80] By 1984, it had about forty thousand subscribers and published a regular newsletter.[81]

The brewer Joseph Coors, industrialist Karl Bendetsen, businessman William Wilson, and wealthy financier Jaquelin Hume—all members of Reagan's California Kitchen Cabinet—contributed to High Frontier and endorsed its findings.[82] The steering group for the organization included Edward Teller, presidential science advisor George Keyworth, and Lowell Wood of Lawrence Livermore National Lab.[83] Keyworth, a Teller protégé, was not yet a believer

in strategic defense. In a now-infamous January 1982 meeting, Teller, Bendetsen, Coors, and Wilson encouraged Reagan to push forward with strategic
defense and pledged that the public would welcome such a move.[84] The primary significance of the meeting was that it only served to reinforce Reagan's
optimism that advanced technologies could really lead to a decisive shift in
American defense strategy.

At the same time that Graham and his colleagues were promoting missile
defense in space, Guy Cook, a staffer for Congressman Robert Badham of
California, was leading a small group investigating the feasibility of space-
based missile defense. This was part of a larger national defense task force
in the US House Representatives that included Congressmen Dick Cheney
of Wyoming and Duncan Hunter of California.[85] According to Cook, both
Cheney and Hunter "thought that the idea of a space-based defense system
was something that should be investigated."[86] Cook identified Graham not
only as having been "a prevalent player" and an "influencer" but also as
someone who was viewed as "being a little on the edge."[87]

The conclusions contained in Cook's report were a mix of optimistic statements qualified by significant uncertainty. It said that "cost estimates are as
wide-spread as the deployment times for laser stations."[88] With a $24 bil-
lion investment, some experts believed that the US could put fifteen laser
battle stations in operation in about ten to twelve years' time. Despite the
unknowns, the report concluded that "there is enormous potential for space-
based lasers" and that a "space-laser program . . . may hold the future for the
security of the United States."[89] Notably, the report highlighted the infrastructural challenges with producing a strategic defense system, including
heavy launch capability, advanced software, and automation. The Pentagon
would have to confront these infrastructural issues as it moved forward with
various strategic defense concepts. It is not clear if Reagan was aware of the
report's findings.

Senator Wallop and his "laser lobby" in Congress pushed hard for greater
investment into "exotic" strategic defense technologies but were largely
unsuccessful.[90] After multiple defeats, Wallop went straight to the top and
requested a meeting with Secretary of Defense Caspar Weinberger and Undersecretary of Defense for Research and Engineering Richard DeLauer. The
senator made it clear that despite all the unknowns about strategic defense,
he sought a commitment from Reagan to develop and deploy laser space-
based defenses; he was not interested in pursuing only long-term research.

Weinberger was noncommittal, and DeLauer was openly skeptical.[91] Aside from the technical doubts about using laser weapons for missile defense, Wallop, like Graham, tended to overlook the complexities surrounding the requisite support technologies for tracking ballistic missiles, the creation of adequate command and control software, and the infrastructure for deploying and maintaining so many components in orbit.

In the first two years of Reagan's presidency, there was growing advocacy for strategic defense in space, but there was still significant skepticism among officials inside the national security establishment that it could (or should) be developed in the near term. There was support for continuing research into lasers that could be used for ASATs and eventually missile defense, but this was separate from the pursuit of an organized program to develop and deploy a missile defense capability on a set timeline. A 1982 GAO report, for example, urged greater US investment in laser research but stopped short of advocating a formal program for laser-based missile defense.[92] DeLauer said that "two weaknesses [for laser-based missile defense] . . . are the time it would take and the amount of money it would take. I think they are grossly underestimated." Lasers were "a military possibility," but he acknowledged that "major uncertainties still exist."[93] DeLauer supported research but cautioned against concluding that lasers would enable a comprehensive missile defense program anytime in the foreseeable future.[94]

INTELLIGENCE AND THE SOVIET SPACE THREAT

The role of intelligence in influencing Reagan's SDI decision has been largely reduced to the argument that senior officials in the White House greatly exaggerated the Soviet threat. FitzGerald writes that "to study this period [the Reagan era] is to reflect upon the extent to which our national discourse about foreign and defense policy is not about reality—or the best intelligence estimates about it—but instead a matter of domestic politics, history, and mythology."[95] John Prados argues that "Casey's CIA encouraged alarmist predictions about the imminent deployment of a Soviet laser [ballistic missile defense] BMD system that were then repeated in public diplomacy documents."[96]

CIA Director William Casey certainly believed that intelligence analysts had underestimated Soviet military power. He had been a member of the Committee on the Present Danger and embraced the findings of the Team B experiment. According to former CIA Director Robert Gates, Casey came to the CIA "primarily to wage war against the Soviet Union."[97] Despite these observations about Casey, there is a lack of evidence that he *altered* intelligence assessments. In evaluating the state of Soviet science and technology, there were often more questions than answers, which meant that intelligence predictions were rarely conclusive. What has emerged from declassified documents is that there was disagreement among intelligence analysts about the severity of the Soviet threat, especially with regard to space, missile defense, and directed-energy weapons research. Yet, there was also disconcerting intelligence reporting that the Soviet Union was planning a more expansive space weapons program.[98]

After the *Washington Post* reported DeLauer's comments about an imminent threat from Soviet lasers, an assistant to the CIA director sent a note to Larry Gershwin, who oversaw analysis concerning Soviet strategic programs, requesting that he set "the record straight" on the discrepancies between DeLauer's comments and a recent NIE. Gershwin wrote in response that there was "a good chance that [DeLauer's information was] derive[d] from a [DIA] briefing on Soviet military space [programs] that has been widely given through the [intelligence] community, including here at CIA."[99] Gershwin's memorandum was alluding to the disagreement in the intelligence community about the immediacy of the Soviet laser and space threats and the fact that the DIA tended to produce more aggressive assessments of Soviet military capabilities than the CIA.

Even if the CIA did not fully agree with DeLauer's statement about a substantial breakthrough in the Soviet laser and space weapons program being right around the corner, the Soviet military space threat was still a significant concern. In October 1982, the intelligence community presented Reagan with a report explaining how the Soviet space program was undergoing a "rapid expansion," reflecting the fact that Soviet leaders saw space as "an integral part of overall military economic, and political policy." It further predicted that Soviet space spending would double over the next four years to $12 billion per year.[100] According to Thomas Ellis, the briefing solidified Reagan's perception of a troubling Soviet space resurgence: Reagan's diary entry for August 8 notes that "there is no question but that they are working (twice as hard as us) to come up with military superiority in space."[101]

Reagan's National Security Advisor Richard Allen sent a memorandum to Casey in June 1981 requesting a comprehensive study of the Soviet missile defense program. Allen said that "Soviet ABM [missile defense] efforts are of critical concern to us" and that "some consider the analysis of the Intelligence Community [in this area] to be inadequate." He emphasized that "this review should involve a fresh examination of assumptions utilized within the Community in study [of] ABM-related issues, and should provide a thorough airing of inter-agency differences. It should give full weight to open source as well as classified material, and should consider in depth the question of Soviet strategic doctrine as it pertains to Soviet activities in the ABM area."[102] A key obstacle to providing a definitive assessment of the status of Soviet research into missile defense was all of the intelligence gaps: that is, lack of intelligence. There was very little human intelligence on sensitive Soviet military research programs because they were so difficult to penetrate. Consequently, the US primarily relied on communications intelligence (i.e., electronic intercepts) and satellite imagery.[103] Imagery intelligence could not, however, provide significant insight into what was happening inside Soviet military research institutes.

Approximately five months before Reagan's March 1983 "Star Wars" speech, the CIA submitted its special NIE on Soviet missile defense to the White House. It was just over thirty pages long and provided very technical descriptions of Soviet research into ground- and space-based missile defense technologies. The estimate concluded that it was unlikely that the Soviets would be able to "have a prototype space-based laser weapon system until after 1990 or an operational system until after the year 2000." The head of the DIA held the view that the Soviets could be able to deploy a ground-based laser weapons system by the early to mid-1990s.[104] Intelligence analysts pointed out the many "significant technological advances in large-aperture mirrors and in pointing and tracking accuracies" that were required to make laser-based missile defense a reality.

The executive summary, in isolation from the report's details, painted a somewhat disconcerting picture. It stated that "the great complexity and severe time constraints inherent in ballistic missile defense operation result in our having major uncertainties in any prediction of how well a Soviet ABM system would function."[105] The lack of certainty was only compounded by the "gaps in information and our analytical uncertainties [which have led to] understandably many differing conclusions and opinions [about Soviet ABM research]." Most significantly, the summary concluded that "the consequences of Soviet acquisition of a ballistic missile defense, despite uncertainties about its effectiveness, are so serious *that even a low probability of such an achievement is cause for concern"* (emphasis added).[106]

What specific influence this report had on Reagan's decision to pursue SDI is difficult to establish. Evaluating the impact of intelligence on policy is an especially challenging task for historians. This is due to the fact that the existence of an intelligence report rarely reveals anything about who read it. According to one intelligence scholar, "it is widely understood that many readers of intelligence reports rarely read past the executive summary."[107] It cannot be established who in the Reagan administration reviewed this 1982 report, but it is almost certain that members of the NSC read it. A memo stated that "NSC Staff members have emphasized the importance of this study to US strategic force planning, and regard it as one of the most important projects currently being undertaken by the Intelligence Community."[108]

Intelligence scholar Aleksandr Matovski argues that unknowns about foreign military research and development can lead to overestimation and then escalatory behavior.[109] Casey, in particular, was worried about the potential for Soviet technological surprises.[110] Reagan, Casey, and other members of the administration were products of the Sputnik generation, and they were concerned that the Soviet Union could surprise the US with a technological feat that could alter the strategic balance. Fear of Soviet technological surprise combined with intelligence about Moscow's space and missile defense research did not directly lead to SDI. These factors did, however, reinforce Reagan's belief that the US was in a new phase of a space competition with the Soviet Union that had significant implications for US national security.

RENDERING NUCLEAR WEAPONS "IMPOTENT AND OBSOLETE"

The political conditions in late 1982 presented Reagan with an opportunity to push forward with a large-scale strategic defense program. On June 12 of that same year, approximately one million protestors marched in New York City to "call for a freeze and reduction of all nuclear weapons."[111] The Nuclear Freeze movement was gaining momentum in the US at the same time that the Reagan administration was attempting to establish a deployment configuration for the MX, an upgraded ICBM with multiple independently targetable reentry vehicles. Ford had approved this program that was intended to redress the Soviet Union's numerical advantage in ICBM warheads. Efforts to create a survivable basing mode for MX turned into a political fiasco for Reagan. The White House proposed a concept known as "dense pack," in which the missiles were to be located in very close formation so that incoming enemy warheads would destroy one another without eliminating the majority of the MX missiles.[112]

Congress was largely skeptical of "dense pack" because it was predicated on unproven assumptions about the survivability of US missiles. Many legislators, moreover, in the Democratic-controlled House of Representatives "professed sympathy with the nuclear freeze . . . and saw in the MX an opportunity to defeat Reagan on a politically sensitive matter."[113] On December 7, 1982, the House of Representatives rejected "dense pack" and made MX funding contingent on the development of a suitable basing strategy. Problematically, the joint chiefs did not support the White House's basing strategy and were not making progress toward finding a suitable alternative. The MX debacle only further contributed to the president's sinking popularity. In December 1982, his approval rating declined to 41 percent; this was an all-time low for a postwar president in his second year.[114] A January poll showed that 57 percent of Americans believed that Reagan might involve the US in a nuclear conflict.[115] What many people did not realize was that Reagan shared the nuclear abolitionist convictions of the Nuclear Freeze movement, but the president also believed that the US could only secure nuclear reductions from a position of strength.

Reagan met with the joint chiefs in late December 1982 and posed a short question that would lead to the establishment of one of the most controversial programs of his presidency. He asked the service chiefs, "What if we began to move away from our total reliance on offense to deter a nuclear attack and moved toward a relatively greater reliance on defense?"[116] The chiefs were not quite sure what to make of Reagan's question, but they began examining strategic defense options. Reagan had, however, likely already made up his mind that strategic defense was indeed possible. He was enthusiastic about missile defense going back to his days on the campaign trail when Graham presented his concept for space-based missile defense. In September 1982, the president had met with Teller, who told him that due to technological advances, strategic defense was indeed feasible.[117] The meeting with Teller has been identified as a key moment for the birth of SDI, but this was not the case in reality. As noted previously, members of Reagan's Kitchen Cabinet had already advised the president to move forward with strategic defense. Teller later acknowledged that he was not "particularly influential" in Reagan's decision to pursue SDI.[118] Teller's missile defense framework, involving a nuclear-powered laser, would not find a welcoming audience in the White House or the Pentagon. The scientist's comments did, nevertheless, likely reinforce Reagan's resolve to establish a strategic defense program.[119]

Robert McFarlane, then deputy national security advisor to Reagan, consulted several scientists who informed him that substantial investment in strategic defense research "might really turn something [sic] that within our lifetimes could make a difference."[120] In January 1983, Admiral James Watkins, the chief of naval operations, met with Teller. The admiral wasn't impressed by Teller's X-ray laser. He said, "Forget it, Edward it isn't going to sell politically," but came away "impressed with the overall prospects for missile defense, especially for space- and sea-based systems, in the long term." Most of the joint chiefs' attention was on the feasibility of directed-energy weapons for missile defense. The infrastructural requirements for a functioning system were not a significant consideration at this stage. The demand that strategic defense would place on the nation's space infrastructure was an afterthought at best.

After quick consultations with a limited group of scientists, not including the president's science advisor, on February 11, 1983, the joint chiefs informed Reagan that they supported missile defense research and development. That evening, the president wrote in his diary, "What if we were to tell the world that we want to protect our people not avenge them; that we are going to embark on a program of research to come up with a defensive weapon that could make nuclear weapons obsolete?"[121] It must be emphasized that at this point, there was an ideological framework for SDI premised on lessening dependence on nuclear deterrence, but the technological details were ambiguous. There were no concrete ideas about what a technological system for strategic defense might look like beyond the fact that it would likely need to be space-based to destroy ballistic missiles in their initial phase of flight. Paul Lettow observes that "Reagan, unlike anyone else involved in the inception of SDI, intended that missile defense could and would help bring about the total elimination of nuclear weapons."[122] At this point, Reagan's advisers were concerned (and would remain so) about his linkage of SDI and nuclear abolitionism.[123]

The president informed McFarlane and William Clark, the national security advisor, that he wanted to announce a missile defense initiative during his scheduled address to the nation on March 23. Clark supported the idea, but McFarlane thought it was inadvisable. In a rare moment of directness with his staff, Reagan made it clear that he was giving an order and instructed McFarlane to keep the speech under wraps.[124] Reagan made McFarlane the lead for drafting the speech. He was assisted by Admiral John Poindexter, a PhD physicist who served as Clark's military deputy, and Colonel Gil Rye, the head of space policy on the NSC. McFarlane and his small team showed Jay Keyworth, the president's science advisor, a draft of the speech four days before the televised address. Even though he was skeptical of the feasibility of missile defense, he became one of SDI's most vocal supporters. Keyworth later said that he put aside his concerns about missile defense because "he felt that the president's political instincts were superior to his own scientific uncertainties, and his job was to ensure that the science was up to the task."[125] (One could argue that it was his job as science advisor to ensure that the president was made aware of the technological uncertainties.)

Weinberger and Shultz were kept out of the speechwriting process.[126] The president's decision to launch a strategic defense effort that would require a radical change in the security foundation of the Western alliance took place outside of the established policymaking process. When Shultz learned of the speech within forty-eight hours of Reagan's televised address, he was horrified. The secretary of state told Clark that "it could hit the allies right between the eyes. This is the year when we especially need a cohesive alliance in our negotiations with the Soviets." Shultz questioned the ability of the joint chiefs to weigh in on such a technical matter as missile defense, asking Clark, "Why place so much confidence in the Joint Chiefs of Staff? They are in no position to make what amounts to a scientific judgment."[127] Shultz was especially worried about the implications of the president's speech for the ABM Treaty.[128] Strikingly, members of the administration did not at all consider the implications of SDI for space policy. Extended interagency deliberations occurred before Ford and Carter issued their proclamations on military space policy and strategy; no such discussions took place prior to SDI.

In the afternoon of the president's address to the nation, Shultz (with White House permission) showed a draft of the speech to Anatoly Dobrynin. The Soviet ambassador predicted that Reagan's missile defense initiative would unleash "a new phase in the arms race."[129] Later that day, the president shocked the world when he called for American scientists to develop the capability to render strategic nuclear weapons "impotent and obsolete."[130] A few days later, Reagan told reporters that he would be willing to share a strategic defense capability with the Soviet Union "to prove that there was no longer any need for keeping these missiles."[131] The president would repeat his desire to share SDI with the Soviet Union; both his advisors and Soviet leaders doubted that this would ever happen. Reagan nevertheless believed that a strategic defense project could be a mechanism for promoting peace and eliminating nuclear weapons; Soviet leader Yuri Andropov, however, viewed it as an unprecedented provocation.[132]

Two days after the speech, Reagan signed NSDD-85 "Eliminating the Threat from Ballistic Missiles" and directed "the development of an intensive effort to define a long-term research and development program aimed at the ultimate goal of eliminating the threat posed by nuclear ballistic missiles."[133] Notably, the NSDD did not even allude to the technological basis for such a capability. Policy documents concerning SDI reveal a disconnect with presidential rhetoric. Although Reagan talked of strategic defense leading to a nuclear-free world, the program focused specifically on one type of nuclear delivery vehicle—ballistic missiles—although defending against cruise missiles would later receive attention too.

The administration commissioned multiple panels to evaluate the technological and strategic aspects of the missile defense program.[134] Former NASA chief James Fletcher oversaw the Defense Technologies Study Team, known as the Fletcher panel. The Fletcher panel concluded that "powerful new technologies are becoming available that justify a major technology development effort offering future technical options to implement a defensive strategy."[135] The Fletcher study envisaged a multilayered missile defense system with land and space components but did not foresee deployment until after the year 2000.[136] The survivability of the space-based components was a critical issue. Richard DeLauer explained to lawmakers that the space components of any strategic defense system would be "fragile."[137] The survivability of a missile defense system required "a combination of technologies and tactics that remain to be worked out."[138] Some SDI advocates used the Apollo program as evidence that the US could overcome the technical hurdles associated with a complex space-based missile defense. DeLauer pushed back on the Apollo analogy, saying that any one single area of a space-based missile defense system had "greater complexity than the programs [like Apollo] that have been loosely talked about."[139] The very fact that SDI entailed a research and development effort lasting ten to twenty years was quite beyond the norm for Department of Defense acquisitions cycles and therefore a source of uncertainty as well.[140]

The Hoffman panel, named after the study's lead Fred S. Hoffman, the head of a California-based think tank, examined the implications of defensive systems for future security strategy. Similar to the Fletcher study, the Hoffman panel claimed that "new technologies offer the possibility of a multilayered defense system." Since there were significant unknowns about the technical feasibility of integrating all of the required components for a comprehensive strategic defense, Hoffman's group advised that "partial systems" or "systems with more modest technical goals" might be worthwhile intermediate steps. In considering system effectiveness, Hoffman's study argued that the deterrent potential of a strategic defense system depended on Soviet "objectives and style in planning for and using military force" and that Soviet assessments of weapons effectiveness "may differ sharply from our own." In other words, strategic defense might deter the Kremlin from launching a nuclear first strike. This deterrent argument likely came from Andrew Marshall, the head of the Pentagon's Office of Net Assessment, who also contributed to the Hoffman report. He maintained that the critical factor in determining deterrent potential was the Soviet view of a weapon system's credibility.[141] Effectiveness criteria would become a source of contention within the interagency and in public debates on SDI. The vulnerability of space-based components surfaced yet again, and Hoffman's team emphasized the imperative to build a space-based system resilient to ASAT attacks.[142]

Initial studies on space-based missile defense raised a number of technical and political considerations. Although the role of strategic defense in a deterrence strategy could be debated, the challenges associated with deploying, maintaining, and defending the vast infrastructure for space-based missile defense were indisputable practical matters that had to be addressed if the program was ever going to get off the ground. A status report on SDI would later point out that "the production, transportation, support, logistics, and administrative requirements of a strategic defense system are as tremendous as the military and technical requirements."[143] An operational strategic defense system would require a gigantic expansion of US military space infrastructure that would have to be prioritized ahead of other national security and civil space projects.

In January 1984, Reagan formally established SDI as an official US government effort.[144] At this point, SDI was only a research program to develop technologies for missile defense; there was no specific concept for a missile defense *system*. To manage the program, the White House created the SDIO, reporting directly to Secretary of Defense Caspar Weinberger to ensure that interservice rivalry, a fact of life in the Pentagon, did not stymie SDI. An independent SDIO answering to Weinberger "not only highlighted the importance of SDIO; it also accorded its director unhampered access to the secretary of defense, a privilege (and power) that few others in the Pentagon had."[145] Growing SDI opposition among the military services, especially in the Air Force, made independence essential. Talking points for a meeting between Keyworth and Weinberger detail how the Air Force had "told its executing agencies (e.g., Space Division) to slow-roll 'anything and everything concerning Starwars [sic].'" Air Staff—that is, headquarters Air Force at the Pentagon—informed Major General Robert Rankine, the deputy lead for space acquisition in the Air Force, that it "was not interested [in SDI]." According to Capitol Hill staffers, the Air Force was "actively cutting knees from under [the Strategic] Defense Initiative."[146]

Weinberger appointed US Air Force Lieutenant General James "Abe" Abrahamson to lead SDIO. He had served as the associate director of NASA, was a former astronaut in the Manned Orbiting Laboratory program, and had directed the F-16 jet program. Abrahamson possessed the requisite technical management skills for taking on such a technologically complex program. The general's personality was perhaps even more important than his professional qualifications. He was tall, slim, handsome, articulate, and very charismatic. He turned out to be the best SDI promoter that Reagan could have ever hoped for. Abrahamson's effective salesmanship would convince some skeptical members of Congress to support SDI, and he was largely responsible for securing the support of UK Prime Minister Margaret Thatcher and allaying some of the concerns of other European allies about SDI.[147] Without his fully realizing it, Weinberger's selection of Abrahamson was one of the most important decisions for SDI's future.

The White House had the authority to establish SDI, but it would have to convince Congress to pay for it. SDIO's budget submission for fiscal year 1985 reveals that the majority of its funding request was associated with space infrastructure; this primarily involved the surveillance, tracking, and kill assessment (SATKA) mechanisms required for developing a missile defense system. Directed-energy weapons accounted for less than half of the requested funds for SATKA. For fiscal year 1985, SDIO received approximately

$1.4 billion out of the $1.77 billion it requested.[148] Notably, even with SDI detractors in the US Congress, such as Senator Ted Kennedy, who famously referred to SDI as a "Star Wars" scheme, SDI's budget tripled after three years and would reach approximately $30 billion by the end of the Cold War.[149]

Only a few weeks after Reagan formally established SDI in early January 1984, he made a significant space announcement during his annual State of the Union speech. The president invoked America's pioneering spirit, saying that the US would push forward into space, "the next frontier." He directed NASA to develop a permanently manned space station within a decade. The space station would initiate "quantum leaps" in medicine and scientific knowledge directly benefiting life on Earth. It was to be an international project, thereby "expand[ing] freedom for all who share our goals." Reagan pledged that just as the "oceans opened up a new world for clipper ships," the commercialization of space would lead to greater economic prosperity.[150] Both SDI and the space station represented the aspiration to use space technologies to further US defense, diplomatic, and economic goals, which reflected the ideology underlying the conservative space agenda.[151] Notably, Reagan decided to pursue both SDI and the space station against the advice of key advisors, demonstrating his personal conviction that exploiting space was inextricably tied to his vision of American greatness.[152]

SPACE BECOMES A PROBLEM

In early 1984, there were still more questions than answers about the underlying strategy and technologies associated with the president's SDI project. The strategic implication of the program quickly emerged as one of the most contentious issues. Within days of Reagan's March 23 address to the nation, SDI became a source of tension in superpower relations. Soviet leader Yuri Andropov accused the US of using the new missile defense effort to "gain a first strike capability against Soviet strategic forces," and he warned that "the USSR will not allow the US to gain military superiority."[153] He further alleged that the US was using SDI to "militarize outer space," overlooking the fact that space had long been militarized.[154] A diplomat at the US embassy in Moscow sent a cable to Washington observing that "the tone of Andropov's remarks was the sharpest we have seen from the top Soviet leader for some time."[155] KGB defector Vasili Mitrokhin explained that "the Center [KGB headquarters in Moscow] interpreted the announcement of the SDI ('Star

Wars') program in March 1983 as part of the psychological preparation of the American people for nuclear war."[156]

High-level Soviet defense and intelligence officials believed that Reagan wanted to accelerate the arms race by seeking to deploy a vast number of weapons in space. A Soviet memorandum identified space militarization, deployment of space weapons in particular, as *odnim iz glavnik voprosov* ("one of the main questions") affecting world peace and international stability.[157] Some senior Soviet officials were convinced that American space weapons could be used to destroy targets on land, at sea, and in the air, in addition to their ASAT and missile defense applications. Consequently, the Soviet Union referred to SDI technologies as *kosmicheskoye udarnoye oruzhiye* ("space-strike weapons").[158] The Kremlin's fear of SDI being used to secure strategic superiority placed space weapons at the center of the American–Soviet arms control dialogue.

While Moscow accused the US of trying to seize the high ground of space, the Kremlin sought the diplomatic high ground on space militarization. Andropov pledged to a delegation of US senators that the Soviet Union would implement a unilateral moratorium on testing ASATs "for as long as others, including the US, refrained from launching ASAT weapons of any kind."[159] He said that "the planet is saturated with nuclear weapons; now there is an effort to stuff outer space with it [sic]." Andropov urged the US to support "the full prevention of testing and deployment of any space-based weapons designed to strike targets on the ground, in the air, or in outer space."[160] If the US agreed to this, the Soviet leader promised to dismantle all existing ASAT systems and not to develop new space weapons. Andropov's proposal was in stark contrast to the Soviet Union's unwillingness only four years prior to limit ASATs; now, the Soviet leader wanted to get rid of them. The CIA believed that the Kremlin sought "to preclude the development and deployment of the US direct-ascent ASAT interceptor, while their longer term aim is to prevent the US from translating its technological capabilities into systems such as space-based lasers that could be used both for ASAT weapons and for ballistic missile defense."[161] In August 1983, Andropov removed the Soviet ASAT from active service.[162]

Many members of the US Congress wanted the Reagan administration to agree to limits on the development of space arms, beginning with ASATs. In the summer of 1983, 106 members of Congress signed a letter to the president urging him to work with the Soviet Union in preventing an arms race in space because "the US is highly dependent on its space-based military assets for vital communications, navigation, intelligence, and treaty verification." The letter called on the White House to agree to an immediate moratorium on ASAT testing.[163] The Reagan administration found a testing moratorium unacceptable, claiming that the US needed to move forward with its air-launched ASAT because a "moratorium would put the US in the position of accepting . . . a proven Soviet ASAT capability which we could not deter with a system of our own," and also citing concerns about verification and potential Soviet cheating.[164]

In a March 1984 report to Congress, the White House explained that an interagency group had completed an examination of ASAT arms control approaches and found them all to be unsatisfactory.[165] Verification was identified as a primary obstacle. Verification issues aside, the Pentagon was opposed to any limits on ASATs because it wanted "to pursue certain highly sensitive programs which would give us the ability to neutralize certain Soviet satellites in time of crisis."[166] The even more important problem was that "ASAT and SDI technology overlap is pervasive and any effective ASAT limitation would restrict SDI aspects."[167] Donald Kerr, the head of Los Alamos National Laboratory, observed in 1983 that "many of the more advanced technologies that now are being considered for anti-satellite use are virtually indistinguishable from ABM technologies."[168] Fundamentally, the White House wanted the freedom of action in space required to deploy both offensive and defensive capabilities.

The "ASAT–SDI entanglement" created a public relations problem for the president's pledge that SDI was a non-weapon defensive measure that would be shared with the world.[169] Reagan later characterized SDI as a "gigantic gas mask" that would protect the world from nuclear annihilation just as gas masks were used to protect people from poison gas attacks.[170] In reality, SDI required the placement of weapons in space that could be used to attack Soviet satellites in addition to ballistic missiles. Reagan's senior advisors would consistently point out the fact that SDI would depend on many of the same technologies as ASATs. Ed Meese stated bluntly that "the technology is the same; a treaty on ASAT testing could kill both ASAT and SDI."[171]

The fact that SDI would give US space policy a much more militaristic image seems not to have been a concern at all for Reagan when he established the program. The further blurring of the boundaries between military and civil space activities through the use of the shuttle to launch SDI components into space was apparently not even raised until well after the March 1983 speech. In early 1984, the Undersecretary of Defense for Policy Fred Iklé informed the NSC that the US space strategy, still under development, "failed[ed] to consider the one factor which will dominate virtually all future space efforts . . . [the] Strategic Defense Initiative."[172] Iklé maintained that "the major components of an effective strategic defense system will be space-based" and that it was necessary to evaluate the "potential impact of the SDI on the three [civil, military, and intelligence] US space sectors."[173] Kerr warned that "decisions made over the next year or two—particularly decisions regarding ASAT and ABM—will strongly influence the character of future space activity." He further cautioned that "many of the more detailed consequences of the militarization or demilitarization of space are not, however, well understood."[174]

A successful American ground-based missile defense test only contributed further to tension over SDI. Since the mid-1970s, the US Army had been developing a program to validate emerging technologies that might enable nonnuclear, hit-to-kill intercepts of Soviet ballistic missile warheads in outer space.[175] This effort led to the Homing Overlay Experiment (HOE) that involved destroying a target missile launched from Vandenberg Air Force Base in California with an HOE interceptor launched from Kwajalein Missile Range in the Pacific Ocean. After three failed attempts, the Pentagon announced that the fourth one in June 1984 was indeed successful. Abrahamson described HOE as "hit[ting] a bullet with a bullet" and an important step forward for the strategic defense agenda.[176] HOE technology would also have ASAT applications, although this was not publicly acknowledged by the Pentagon.[177] Unsurprisingly, Soviet officials described the June test as a "dangerous step" that could provoke the Soviet Union into taking countermeasures.[178]

Less than three weeks after the HOE test, Soviet General Secretary Konstantin Chernenko, who succeeded Andropov in February 1984, called upon the US to enter into formal talks aimed at banning space weapons.[179] A *Washington Post* article described an ASAT competition as the "very definition of instability," since attacking satellites could lead to a "terrible" outcome. The linkage between ASAT and missile defense technologies was suspected to be the real cause of the Reagan administration's position that limits on ASATs were inadvisable.[180] Advisors to the president took notice of the growing momentum in favor of limits on ASATs. As a way to meet the Soviets "half-way," some US officials favored entertaining limits on ASATs for a period of up to five years, including an "incidents in space" agreement that would specifically ban attacks on satellites, prohibit testing and deployment of high-altitude ASATs, and limit each side to one low-altitude ASAT system.[181] This policy, if adopted, would allow the US to demonstrate that it at least appeared serious about space arms control, potentially alleviating concerns in Congress, in the general public, and among US allies. Most importantly, it would not actually prevent the US from pursuing its MHV program or space-based defense, unless Congress placed limits on these efforts. This latter possibility was alarming to SDI advocates.

Henry Cooper, the assistant director of ACDA who would eventually take over the SDI program, argued against any constraints on ASATs due to doubts that the US could verify their elimination and because of the implications for SDI. He said an "incidents in space" agreement had little risk, but a limit or ban on interceptors would be problematic and "directly intersects SDI (*a la* HOE) . . . [placing] us at a net disadvantage when SDI is considered" (emphasis in original).[182] As already noted, it is unlikely that a five-year ban would have impacted SDI, since the Pentagon did not anticipate producing even an interim space-based defense system ready for deployment before the early to mid-1990s.[183] But space arms control critics viewed ASAT limits as a slippery slope that could lead to more significant restrictions on military space programs.[184]

Although tensions over space militarization ran high, there was cause for optimism that progress could be made toward securing new arms control agreements and improving relations between Moscow and Washington.[185] Simon Miles points out that in the spring and summer of 1984, "Moscow had sent a range of positive signals" indicating a serious interest in rapprochement. Soviet Marshall Nikolai Ogarkov criticized the arms buildups as "senseless" and labeled "limited nuclear war" as nothing more than "pure fantasy."[186] Outer space was, however, a key problem. Reagan responded to Chernenko's call for discussions concerning space weapons with a package proposal that would include space, INF, and START.[187] Chernenko objected to this framework, citing concerns that SDI "might be lost in the omnibus talks."[188] Although at this stage linkage between nuclear and space issues was an obstacle to getting the arms control negotiations moving, space and nuclear issues would indeed be discussed in tandem in the Nuclear and Space Talks beginning in 1985.[189]

In the summer of 1984, Reagan approved a space strategy that laid out his administration's plan for using space to further US national interests. The national security section identified the importance of curbing the flow of space technologies to the Soviet Union and enhancing the survivability of space systems. The strategy called upon the Department of Defense to use its "space and space-related programs [to] support the Strategic Defense Initiative." In an effort to calm Congressional opposition to ASATs, the White House included the stipulation that it would "continue to study space arms control options."[190] Internal discussions regarding the ASAT issue remained tense, and opponents of space arms control, such as Weinberger and Perle, worked hard to ensure that limits on ASATs were not on the table.[191]

With ASAT arms control faltering, the administration struggled to reassure European allies and the American people that its military space agenda would promote stability. France and Britain worried that a Soviet missile defense buildup in response to SDI would erode the credibility of their nuclear deterrents.[192] Shultz's early predictions about SDI had turned out to be correct: establishing it without consulting the allies undermined transatlantic cohesion at an especially delicate time. Intelligence analysts warned that "many allies, especially the INF [Pershing IIs] basing countries, fear the negative publicity about space weapons could damage recent public relations gains that NATO has made on arms control and further erode domestic support for INF and other Alliance programs."[193]

In the lead up to the 1984 presidential election, space militarization became a campaign issue. Democratic candidate Walter Mondale sought to put Reagan on the defensive about SDI. Mondale ran a television ad showing the Earth against the blackness of space with a narrator saying, "It's from up here that President Reagan, if reelected, is determined to orbit killer weapons. He'll spend a trillion dollars. The Russians will have to match us. And the arms race will rage out of control—layer on layer—orbiting, aiming, waiting. Walter Mondale will draw the line at the heavens. No weapons in space—from either side. On November 6, draw that line with him."[194] Mondale's advisors wanted to use the prospect of an arms race in space to scare voters about Reagan's strategic defense project. A campaign strategist candidly said, "We've got to get people to burst out of this bubble that the Reagan campaign has created . . . so we took them into outer space."[195]

At a presidential debate approximately one month before the election, Mondale and Reagan exchanged verbal spears about SDI. Mondale predicted that SDI would precipitate a dangerous arms race in space that would make US satellites even more vulnerable. Reagan retorted that "[he] never suggested where the weapons should be or what kind [sic]."[196] He described SDI as a defensive program and expressed his wish that the Soviet Union would "join us" and share in the benefits of a strategic defense system. Mondale described the prospect of sharing SDI technologies with the Soviet Union as dangerous.[197] Reagan was speaking truthfully when he said that no decision had been made about "where the weapons should be": that is, whether a strategic defense system would be ground based or space based. But the administration's approach to space arms control was based on the premise that accepting any limits on space militarization could negatively impact the future deployment of space-based defense, in recognition of the fact that space weapons were critical for boost-phase interception of ballistic missiles.[198] Reagan's landslide victory suggests, however, that Mondale's attempt to use the prospect of a space arms race to sway voters had very little, if any, impact.

Even though the characterization of SDI as a dangerous space weapons program did not affect the election, it foreshadowed the Reagan administration's difficulties with getting the public to support having weapons in space. An April 1984 CBS/NYT poll had found that 67 percent of respondents approved of developing a missile defense system.[199] Yet, at the same time, a Harris poll found that 82 percent of respondents were in favor of superpower negotiations to "outlaw the use of weapons in outer space."[200] The idea of a defense against ballistic missiles appealed to many people, but the prospect of orbital weapons did not. With these factors in mind, McFarlane predicted that "the most difficult issue [for SDI] would be space," and Meese advised "[distancing] the space issue from the SDI issue, that they are not the same, and that the President's idea is not simply a space question."[201] Gil Rye pointed toward the space station project as a necessary counterbalance to the perception that the administration was "militarizing space."[202]

Clearly, the administration needed to take control of the SDI narrative with a concerted public diplomacy strategy. Reagan implored his advisors to make clear that SDI was a "system which does not kill people . . . it would free the world from the threat of nuclear weapons." It would moreover be internationalized: that is, shared with the world. Strategic defense, according to the president, would obviate concerns about verification of future arms control agreements because it would serve as the ultimate insurance policy for Soviet cheating.[203]

The fact that SDI was only an umbrella for multiple research efforts compounded the public relations challenge; neither the Pentagon nor the White House had a good answer to the question of what exactly SDI was. Reagan focused on convincing people that comprehensive defense was indeed possible and desirable when he had no idea what kind of a strategic defense *system* might one day be produced. The minutes of a December 1984 National Security Planning Group meeting sheds light on the lack of coherence in the administration regarding the technological foundation of SDI. Ambassador Ed Rowney, the chief START negotiator, said about SDI that "we are not talking about putting nuclear weapons in space, only nuclear reactors [to power SDI lasers]." McFarlane corrected him by saying that the Excalibur X-ray laser "involves a nuclear explosion in space." Weinberger objected to McFarlane's characterization of SDI and explained that "we are seeking a non-nuclear system, i.e., non-nuclear kill [mechanism] for destroying the ballistic missiles." The secretary of defense then brushed aside these disagreements over the nuclear and nonnuclear aspects of SDI and said that regardless of the technologies chosen, the program's goal was to reduce "offensive systems as we evolve towards defensive systems."[204]

What specific technologies were chosen would have significant political implications. A system built around directed-energy weapons would probably not materialize before the dawn of the twenty-first century at the earliest. Such a situation was unlikely to affect arms control in the near term. Conversely, the White House moving forward with an interim system using hit-to-kill interceptors, like the ones demonstrated in the HOE, would have immediate implications for the ABM Treaty and the strategic arms dialogue with Moscow. The unknowns about the technologies that the US would pursue required the Soviets and European allies to use their imaginations and contemplate what they considered to be worst-case scenarios. US allies were frustrated with the lack of consultation concerning the technical details of SDI because they did not realize just how little consensus there was in the Reagan administration about what the course of strategic defense research should be.[205] The technological details surrounding a strategic defense system would become the subject of even greater domestic political controversy in the US and among transatlantic allies and a significant source of tension in the American–Soviet arms control negotiations.

CONCLUSIONS

Reagan's first term in office witnessed the most radical American space announcements since the presidency of John F. Kennedy. Pushing the commercialization of space, moving forward with an ASAT program, establishing SDI, and directing NASA to develop a space station were all closely intertwined. A conviction that space technologies were inextricably linked with American security and prosperity tied all of these endeavors together. Space programs were not an adjunct to broader US policy objectives but rather served as a vehicle for promoting the full spectrum of American interests. A central discontinuity between Reagan and his predecessors was that space militarization *overtly* became a key element of American statecraft.

Understanding why Reagan established SDI is a complex matter. The program emerged out of a mixture of issues. The idea that advanced technologies could provide a path away from dependence on nuclear weapons certainly appealed to the president, which was the product of Reagan's underlying techno-optimism and skepticism of nuclear deterrence. His view of Soviet military power was also a factor. Scholars of this period have tended to view US public diplomacy that framed SDI as a response to Soviet defense programs as only a convenient justification for the program.[206] Yet, already in 1982 and 1983, we do find sincere concern among Reagan and his advisors about intelligence reports suggesting that the Soviet Union might gain the upper hand in advanced technologies that could be used for strategic defense and securing military advantages in space.[207] It is unlikely that this intelligence was the decisive factor in Reagan's SDI decision, but the archival record reveals that the president increasingly viewed space as a contested "high ground" in the early years of his first term in the White House.[208]

In contrast to prior US policy decisions regarding space militarization, SDI was not the product of intensive interagency deliberations. Reagan intentionally kept discussions leading to its creation out of the formal bureaucratic process, fearing that doing otherwise would kill the program. Since SDI was the result of a relatively quick top-down decision, its consequences for US foreign policy were not thoroughly considered. Setting aside the more complex question of whether the US could produce a comprehensive space-based defense system, the political and financial implications of an unprecedented expansion of space infrastructure that space-based missile defense would require did not significantly factor into Reagan's calculus.

   Although limits on ASATs had been on the table since the late 1970s, this subject had been a largely peripheral issue in the American–Soviet strategic arms dialogue. SDI, however, quickly moved space arms control from the edge to the center of the superpower arms control agenda. While the White House did not yet have a clear idea of what a space-based missile defense might entail, key advisors to Reagan opposed even entertaining limits on military space programs due to the implications for SDI. The prospect of American space-based missile defense and ASAT deployment made space militarization a contentious issue in the transatlantic alliance as well, forcing US allies to pay much closer attention to an area that had historically been largely absent in high-level discussions within NATO. The linkages between space-based defense and ASAT technologies undermined Reagan's pledge that the program was fundamentally peaceful and defensive in nature. Managing opposition to intensified space militarization would become an especially difficult task for the Reagan administration in its dealings with the public, allies, and the Soviet Union.

3 OUT OF THE BLACK

In the summer of 1986, Pope John Paul II called upon world governments to create "joint agreements and commitments" that would ensure "the peaceful uses of space resources."[1] The Pope's statement came in the midst of intense argument about SDI and the military uses of outer space. Both the US and the Soviet Union had attempted to persuade the Holy See to adopt their respective positions on the controversial space-based missile defense program.[2] Because the Reagan administration presented SDI as a moral alternative to the threat of nuclear annihilation, elements of the US government sought the endorsement of the leader of the Roman Catholic Church, who had established a reputation as a staunch anti-communist.[3] In an effort not to be seen as partial to either the US or the Soviet Union, John Paul II refused to publish a report produced by the Pontifical Academy of Sciences that was critical of SDI.[4] The involvement of the Pope in the SDI debate was yet another sign of how the military uses of space had become a central issue in the international arena of high-stakes diplomacy. Controversy over SDI mobilized a wide range of actors to take part in debates over the military use of space in the 1980s, and the Pope's participation is perhaps one of the most prominent symbols of the diversity of actors who shaped the dialogue on strategic defense.

Almost immediately after Reagan made his March 23 SDI speech, scientists, national security analysts, journalists, and politicians began to make arguments for and against space-based missile defense. The Reagan administration was not prepared for SDI becoming a topic of discussion among defense intellectuals and the general public. Even though the Department of Defense did not yet have a firm idea in 1983 of what kind of strategic defense system would ultimately be pursued, several scientists immediately began publishing articles detailing how a comprehensive space-based missile defense would be prodigiously expensive, an unparalleled technological challenge, and a source of even greater insecurity in superpower relations.

There is a growing body of scholarship that explores the role of scientific expertise, along with the questions surrounding technological feasibility, in the public debates over SDI.[5] But the discourse on SDI was about much more than technical and strategic disagreements; it was also defined by divergent views on humanity's future in the cosmos. Arguments about infrastructure for strategic defense, SDI's impact on arms control, and the feasibility of space-based missile defense simultaneously advanced different views about the utility and desirability of military activities in space. The critics of SDI argued that the "sanctity" of space could and should be preserved through arms control, while many opponents believed that the "weaponization of space" was unstoppable and in the national interest.

Since the dawn of the Space Age, the US government consistently stressed the idea that the American space program was devoted to the peaceful uses of outer space. Even though space had been militarized from the beginning of human activities in the cosmos, in the 1960s and 1970s, neither the US nor the Soviet Union drew attention to their military space programs. However, building support for SDI required bringing aspects of US military space activities, which lived in the "black" world of special-access programs protected by code words, into the public sphere. But releasing more information about US military space activities would not be sufficient for allaying the intensifying international concerns about a space arms race.

In an attempt to pacify the anti-SDI lobby, Reagan tried to separate space-based missile defense and offensive space weapons; as has already been noted, such a distinction was rhetorically powerful but artificial in reality. Concurrently, the US government embarked on an organized public diplomacy campaign that presented SDI as a necessary response to ominous Soviet space weapons. To this end, the US government released sanitized intelligence reports to highlight a menacing Soviet military space effort that was a direct threat to US national security. In promoting strategic defense, the Reagan administration fundamentally had to reorient the views of the American electorate and key allies on the uses of space for national security endeavors. In doing so, the White House had to walk a fine line in its simultaneous characterizations of SDI as a pacific endeavor to promote stability and an answer to an accelerating military space race.

MUTUALLY ASSURED SURVEILLANCE

In May 1961, John F. Kennedy implored Congress to appropriate the funds to place a man on the moon and bring him safely home before the end of the decade. Even more important than a moon mission being "impressive to mankind" would be its contribution to the "long-range exploration of space."[6] When Kennedy made his moon speech at Rice Stadium in September 1962, he emphasized the pacific nature of the US space program. The president declared: "We now look into space, to the moon and to the planets beyond, and we have vowed that we shall not see it governed by a hostile flag of conquest, but by a banner of freedom and peace. We have vowed that we shall not see space filled with weapons of mass destruction, but with instruments of knowledge and understanding."[7]

Kennedy explained that a robust US space program was essential for winning the "battle that is now going on around the world between freedom and tyranny." And he stressed that space competition was a peaceful alternative to military confrontation. In reality, the president was not at all interested in space exploration. Kennedy told NASA Administrator James Webb that Apollo was important for "international political reasons" and that he was "not that interested in space"; exploration was not a compelling justification.[8] Roger Launius has pointed out that the "symbolism of Kennedy's Apollo commitment has held special appeal for advocates of space exploration."[9] Many Americans had come to believe that the US sought to use space exploration as a *peaceful* mechanism for outcompeting the Soviet Union. The rhetoric surrounding Apollo only reinforced this idea and would have significant implications for the way in which people perceived SDI in the 1980s.

At the same time that Kennedy was presenting his moon plan, his administration secretly moved all military space programs into the "black" with the issuance of Department of Defense Directive S-5200.13 in March 1962. The top-secret document maintained that it was "impractical to selectively protect certain military space programs while continuing an open launch policy for others since to do so would emphasize sensitive projects."[10] In 1961, the director of central intelligence established the BYEMAN Control

System, which protected specific details about the mission and capabilities of reconnaissance satellites.[11]

The popular outlet *Aviation Week and Space Technology* regularly published articles on secret military and intelligence satellites, which led commentators to refer to it as "Aviation Leak and Space Mythology," but the US government remained tight-lipped about national security space activities.[12] When the Johnson administration approved the deployment of Program 437, a nuclear-tipped ASAT designed to defend against Soviet orbital nuclear weapons, US national security officials emphasized that "no public attention should be directed toward development of [US] anti-satellite capabilities" with the objective of not undermining US space reconnaissance and freedom of space.[13] The US government sought to signal that it intended space to be a domain of peace rather than a potential avenue for extending the arms race.

Beginning in the 1970s, the US government became selectively more open about certain military space programs. After launching the first Defense Support Program (DSP) early warning satellite in 1970, the Department of Defense released some information about DSP's primary mission as an early warning platform. The Pentagon hoped that disclosing the existence of a space-based early warning system would strengthen the US's ability to deter the Soviet Union from launching a nuclear first strike.[14] Showcasing advanced early warning satellites as tools of deterrence fit squarely into the narrative that the US space program promoted peace and stability between the superpowers. American officials nevertheless remained quiet about NRO systems.[15]

The superpower arms control negotiations in the late 1960s and 1970s precipitated a debate among US officials about the secrecy surrounding military and intelligence satellites. Policymakers wondered how they could convince the electorate, Congress, and US allies to support arms control treaties whose verification would depend on satellites that were highly classified. Many officials wanted to share some public information about NRO satellites, but intelligence leaders opposed any classification changes on the grounds that they would compromise sensitive sources and methods.[16] Beginning with the Interim Agreement and the ABM Treaty, Washington and Moscow agreed to use "national technical means" of verification as a euphemism for satellites, obviating the need to declassify the existence of reconnaissance satellites.[17]

Despite the intense secrecy regime surrounding the NRO, the use of satellites for arms control verification contributed to a general perception that space technologies were enabling superpower rapprochement. Unsurprisingly, information about the use of satellites for verification made its way into the media. A 1972 *Houston Post* article informed readers that "spies in the sky keep two big powers in balance." Satellites provided a means of trust building that would contribute to a an "effort to scale down . . . [nuclear] arsenals."[18] A *US News and World Report* article (also published in 1972) explained that due to satellites, "the world is now virtually an open book" and that space systems had become "vital tools of American policy."[19] These ideas dovetailed nicely with the "we came in peace for all mankind" rhetoric of the moon landing.[20] In this context, a US policy shift that included development of space weapons would appear as a dangerous challenge to the use of space technologies to make the world a safer place.

Although Ford did indeed adopt a militarily competitive space policy based on the premise that the US needed weapons to destroy Soviet satellites in wartime, the justification for this shift was not publicly disclosed. Media reports about killer satellites created the impression that the US ASAT was only a response to the Soviet ASAT program. There was no public mention of the *real* reason behind the Ford ASAT decision: that the emerging military support role of satellites made them valuable targets in a war with the Soviet Union.[21] Brent Scowcroft, Ford's national security advisor, stressed the need for a clear statement concerning the rationale for an ASAT program. Otherwise, "budgetary pressures, arms control considerations, and other international policy factors could impede progress in this area."[22] But Ford left the White House before there could be any public discussion about ASATs. Consequently, even though Ford's space policy differed greatly from that of his predecessors, the *public* presentation of US space policy did not substantially change. It was not until Reagan came into office that the idea to use ASATs to attack Soviet space systems and to deter Moscow's use of ASATs had entered into the mainstream of the government's public diplomacy on space matters.

Carter's pursuit of ASAT arms control reaffirmed the US commitment to the peaceful use of outer space. When the president decided to proceed with ASAT development, Secretary of Defense Harold Brown told the public that the US was being dragged into a space arms race because the Soviet Union would not agree to ASAT limits.[23] The fact that there were senior defense officials who wanted an ASAT program to be able to attack Soviet satellites, regardless of Moscow's ASAT program, was not public knowledge. The president's 1978 speech at Kennedy Space Center presented photoreconnaissance satellites as having benefited all humankind through their role in monitoring arms control agreements, but there was no mention of satellites supporting military functions such as precision targeting. The president constructed a progressive narrative, describing how satellites "brought us a great deal of human knowledge and . . . may also have brought us a measure of wisdom."[24]

With the superpowers both developing ASATs in the late 1970s, the 1967 Outer Space Treaty no longer appeared to be sufficient for keeping weapons out of the cosmos. The treaty signatories had agreed not to place weapons of mass destruction in space, but there was no prohibition on nonnuclear space weapons. Members of the UN Conference on Disarmament (CD) acknowledged that the Outer Space Treaty "by itself did not guarantee the prevention of an arms race in outer space."[25] The growing anxiety that "advances in science and technology had made the extension of the arms race into outer space a real possibility" prompted the CD in 1981 to commence discussions on an international agreement to Prevent an Arms Race in Outer Space (PAROS for short).[26] However, there was disagreement about how to even define a space weapon.

At a UN conference on space in 1982, Secretary General Javier Perez de Cuellar warned that an arms race in space would increase the potential for confrontation between nations. "We must oppose vigorously the increased militarization of outer space. We have time, but very little."[27] Even before SDI, the prospect of a space arms race was a source of anxiety in the public sphere. As détente was crumbling, the heavens increasingly looked like a future battleground. The presentation of reconnaissance satellites as stabilizing silent sentinels in the sky had drawn a hard line between weaponized and non-weaponized military space activities. Reagan would have to convince the American electorate, allies, and the Soviet Union that a weaponized phase in the militarization of space could somehow lead to greater security. But his administration underestimated what a prodigious task this was from the standpoint of public diplomacy.

ENTER THE SPACE INTELLECTUALS

Herman Kahn, a distinguished strategist, said in 1982 that he could "foresee the day when 'clean wars' could be fought in space." Dr. Robert Cooper of the Defense Advanced Research Projects Agency expressed the view that "as the superpowers deploy more satellites that provide crucial information for weapons on the ground, the incentive to knock them out of operation increases."[28] This latter perspective reflected the same ideas that led Ford to approve a new US ASAT program in January 1977. There were fundamentally two opposing schools of thought on space security in the early 1980s. The first argued that the US needed to pursue arms control to restrain an arms race in space. The second maintained that space weapons were vital for denying the Soviet Union access to space in wartime, destroying ballistic missiles, and defending satellites.

Although war in space had garnered media attention in the late 1970s, Reagan's presentation of his strategic defense vision to the world in 1983 added an entirely new dimension to the discourse on superpower space activities. As a result of SDI, defense analysts, arms control experts, and scientists had begun to consider space security issues more closely. The British scholar of military strategy Colin Gray published a book in 1982 called *American Military Space Policy: Information Systems, Weapon Systems and Arms Control*. This was the first comprehensive (unclassified) academic study of US thinking on the use of space to further national security interests that took into consideration the technological changes in the 1970s, especially the rise of nonnuclear ASATs and near real-time photographic reconnaissance capabilities.[29] He wanted the book "to encourage informed debate of US space policy," which was lacking in the strategic studies community. At this time, Gray was providing his expertise to the Reagan administration's General Advisory Committee on Arms Control and Disarmament.[30] In 1983, he published an updated version of his book that examined some of the consequences of SDI for military space policy. He not only addressed the ways in which the US and the Soviet Union were using space systems for national security purposes, but also laid out some of the core tenets of a theory of spacepower. In contrast to the arms control advocates, Gray viewed any treaty limiting military space capabilities as an undesirable constraint on American technological competitiveness.

Within days of Reagan's "Star Wars" speech, American physicist Richard Garwin published a piece in the *New York Times* entitled "Reagan's Riskiness," which described SDI as a dangerous undertaking. Garwin, who had substantial experience with nuclear weapons, missile defense, and military space programs, explained that space-based missile defense would require a system of "unprecedented effectiveness." Rather than providing more stability, he maintained that space-based missile defense would "lead to a war in space as a prelude to war on earth."[31] Garwin set the stage for the granular system-level analysis that would be carried out by other defense scientists, such as Ashton Carter, and scientific organizations, including the Union of Concerned Scientists and the American Physical Society.

Gray wanted to convey to the readers of his book that while the technical considerations raised by scientists such as Garwin were important, scientific arguments on missile defense oftentimes masked strategic biases. He observed that "lurking behind the frequent technical, and somewhat infrequent tactical and strategic arguments, about weapons in space are a wide range of vested interests and deeply ingrained doctrinal preferences."[32] The fact that some of "the best technical minds in the US" had doubted the practicality of ICBMs in the early 1950s served as a reminder, according to Gray, that "experts are not always strong on foresight." He further observed that applying traditional military concepts such as maneuver warfare and active defense to the space domain required a paradigm shift for military officers and defense planners who did not think about space systems as anything more than support infrastructure. Gray nevertheless lamented Reagan's establishment of a space command without first creating doctrine to guide its activities.[33] In short, there was much talk about needing to deter the Soviets in space but little coherent thinking about what a comprehensive military strategy involving space capabilities might look like.

Whereas military officers had been prominent voices in the push for a more militarized space policy in the late 1950s and in the early 1960s, this was less the case in the 1980s.[34] Even many senior military officers did not have the security clearances to access specific details about US military and intelligence satellites. Since the US government had intentionally obfuscated the details surrounding military space systems, having an informed debate, even among defense intellectuals, was difficult. There was also a general lack of interest among the military services for a greater push toward a more militarized space policy because of the potential effects on traditional defense programs. Gray pointed out that there was bureaucratic resistance to "promoting the idea of a 'space force,' in part because such a force would compete for shares of the [defense] budget." In sum, parochial service politics would make military leaders, especially in the Army and the Navy, reluctant to embrace a new way of thinking about space that would primarily benefit the US Air Force.

In 1983, the Soviet Union announced that it would unilaterally establish a moratorium on ASAT testing. Andropov's proposal to ban testing of weapons in space, leading to their full elimination, seemed to present an opportunity after the failed ASAT talks in 1978 and 1979 to make serious progress on space arms control. The subject of arms control reaching into space became one of the most contentious issues among national security commentators. Paul Stares, a political scientist, had begun to write about space security issues in the early 1980s and was the ideological opposite of Gray. Stares wrote that "1983 may well be considered as the year in which the United States irrevocably committed itself to a new course in space" that signaled a "new trend in the militarization of space." He presented the early 1980s as a rebirth for ideas about space weaponization that were challenged by the Eisenhower administration's commitment to what he calls the "passive" uses of outer space.[35] The only way forward, he believed, was to engage with the Soviet Union on limiting space weapons. However, he and Gray agreed on a key point: that the US ASAT program would not serve as an effective deterrent. But they diverged on how to respond to the Soviet ASAT threat. For Stares, arms control was the optimal way to protect vulnerable US satellites, whereas Gray maintained that ASAT arms control was not verifiable and would prevent the development of space weapons that could prove to be vital in a conflict with the Soviet Union.[36]

Skeptics of ASAT arms control oftentimes pointed to verification as the main problem. How would the US know if the Soviet Union deployed weapons in space? John Pike of the Federation of American Scientists pointed out that Soviet ASATs were launched into space using large SS-9 rockets that were "readily observable by US national technical means": that is, reconnaissance satellites.[37] In contrast, Gray argued that smaller space weapons, such as space mines, could not be detected by reconnaissance satellites, rendering an ASAT arms treaty unverifiable.[38] Underlying Gray's technical analysis was his view that the Soviet Union had cheated on arms control before and would likely do so again.[39] Another factor at play was Gray's firm belief that the US should pursue any capability that could give it a lead in the superpower military competition. Donald Hafner, a US official who had been directly involved in the American–Soviet ASAT talks, observed in a 1982 article that verification was not the real reason why key officials resisted ASAT negotiations; in reality, they sought "more exotic and futuristic weapons . . . deployed in orbit."[40] These observations in no way suggest that verification concerns were not real or valid, but rather that opposition to ASAT arms control was frequently accompanied by the viewpoint that space weapons should be pursued to secure a strategic advantage over the Soviet Union.

ASTROCULTURAL WARS

The technical nuances of the space debates among defense intellectuals were easily lost on the general American electorate. Nuclear deterrence was complex enough, but missiles at least were recognizable and familiar. Soviet ICBM launch sites and bomber bases could be placed on a map and shown on television, in movies, or in newspapers. Movies such as *Dr. Strangelove* and *WarGames* provided highly fictionalized accounts of US nuclear forces, but they at least acquainted audiences with the infrastructure of nuclear war. The "geography" of space, on the other hand, was arcane. Satellites were driven by the principles of orbital mechanics, which could be difficult for the nonspecialist to grasp. Security experts discussed and debated the dangers and merits of having ASATs that could reach satellites in low Earth orbit and geosynchronous orbit, where especially important nuclear warning satellites were deployed.[41]

Space terminology was unfamiliar, highly technical, and intimidating. Films such as *Moonraker* and *Star Wars* showed completely unrealistic portrayals of space operations that violated basic Keplerian principles. They nevertheless shaped the popular imagination of what a space conflict might look like. The fact that Senator Ted Kennedy (D-MA) referred to Reagan's grand vision as "Star Wars" and the name stuck is perhaps the clearest example of how popular conceptions of space influenced the way that people thought about SDI.[42] Even SDI supporters had conflicting views on the name "Star Wars"; Reagan resented the association, saying that it "denigrate[d] the whole idea."[43] By contrast, Phyllis Schlafly of the Eagle Forum, a conservative interest group, embraced the "Star Wars" moniker because it highlighted SDI's role in a "drama of the battle between good and evil."[44]

Debates over SDI were more than technical and strategic exercises about deterrence and strategic stability. At a fundamental level, they were about what kind of place space would be in the near future. Would space weapons transform the heavens into a battlefield? Could laser weapons usher in a new age of peace and security? Did increased militarization of the heavens portend a dark future for humanity's exploration of the cosmos? Depending on one's views of space, SDI could be either utopian or dystopian.

The founders of High Frontier, the most prominent citizen support group for SDI, clearly wanted to appeal to romantic notions of conquering and colonizing outer space. In the court of public opinion, SDI was as much an astrocultural battle as a political one. Alexander Geppert defines "astroculture" as the ways in which "human beings have used their creative powers to render the infinite vastness of outer space conceivable."[45] As politicians and commentators lined up in favor and against SDI, they oftentimes were making implicit, and at times explicit, arguments about humanity's future in the cosmos. Congressman Joe Moakley (D-MA) wrote in a December 1983 issue of *Arms Control Today* that "this generation is the first to look to the heavens and know the stars are within reach. Shall our children pursue this new destiny peacefully, in the spirit of exploration? Or shall they view outer space as yet another arena for the futile attempts of one nation to gain temporary military advantage over another? The choice is before us now. Let us work to keep space free from weapons."[46]

Moakley advocated restraint through arms control to maintain the sanctity of space as a special domain that would be untouched by the horrors of war. But not everyone believed that this was a realistic proposition. ASATs were already in existence, and advancements in electronics were going to make them more accurate and lethal. Both the US and the Soviet Union recognized that satellites provided military advantages and were therefore critical targets for enemy strikes in wartime. Senator Barry Goldwater observed in 1984 that "space is just another place where wars will be fought."[47] Gray agreed with Goldwater that "military conflict in space is not a matter for US policy choice today—the choice has already been made."[48] Satellites, according to Gray, had become too critical for modern warfighting, and the Soviet Union could not therefore be dissuaded through arms control from seeking to destroy American space systems in a conflict.

To highlight what they saw as the immediate dangers of SDI, many of the program's opponents argued that any space-based weapons would threaten the satellite surveillance regime employed by the superpowers. Famed astronomer Carl Sagan said in 1985 that he was "not against the militarization of space . . . we have been militarizing space with reconnaissance satellites since the 1960s, and they're worth their weight in gold. It's the introduction of weapons into space that worries me very much."[49] Moakley similarly invoked the stabilizing power of reconnaissance satellites, saying that "placing weapons in space that might threaten these [reconnaissance] satellites will raise rather than lower the chances of a devastating nuclear war on Earth."[50] Congressman George Brown (D-CA) distinguished between two kinds of military activities in space: information gathering, which was beneficial, and "so-called force-extending activities," which he called "the weaponization of space." Brown argued that this "emerging competition in space weapons threatens this delicate intelligence apparatus and is, therefore, destined to increase the likelihood of a nuclear exchange." ASATs would destroy the satellites that "serve as our eyes and ears in the sky."[51] Congressman John F. Seiberling (D-OH) rhetorically asked, "Should we add to that threats to satellites that monitor deployment of weapons by both sides and follow that up by spending billions in a futile and destabilizing space weapons competition?"[52]

Notably, Moakley, Brown, and Seiberling did not sit on any congressional committees that granted them access to classified information concerning US military and intelligence space programs. Their and Sagan's position on space weapons was inextricably linked to the widespread perception that satellite reconnaissance was stabilizing and that space should therefore remain a de-weaponized zone. It is impossible to know if access to US intelligence on Soviet space weapons would have changed their viewpoint. Senator John Kerry (D-MA), who did have access to top-secret information about US reconnaissance satellites and Soviet ASAT capabilities, vehemently opposed SDI in part because he believed that it would make American intelligence satellites more vulnerable.[53]

For many SDI advocates, space weapons offered a way out of the nuclear stalemate. Rather than being destabilizing, space defense could save humankind from nuclear Armageddon. Zbigniew Brzezinski co-authored a 1985 opinion piece with the astronomer Robert Jastrow and the diplomat Max Kampelman in defense of SDI. Even though Brzezinski had been an ardent supporter of ASAT arms control in the late 1970s, he maintained that serious consideration should be given to space-based missile defense because it could ultimately lower the probability of nuclear war. Instead of presenting SDI as a futuristic endeavor, Brzezinski, Kampelman, and Jastrow argued that "some of the [missile defense] technologies are mature and unexotic" and that a strategic defense system could therefore be deployed within the next decade. The authors specifically identified the critical role of kinetic space interceptors: the "so-called 'space weapons' of strategic defense are indispensable for the crucial boost-phase defense. To eliminate them would destroy the usefulness of the defense."[54]

Hugh Thomas, a British historian with close ties to Thatcher, rejected the notion that space should be preserved as a sanctuary. In defending SDI during a speech to the Conference of Science and Technology in Lisbon in 1986, Thomas said:

> The [SDI] project seems to belong to the world of Jules Verne: a conflict 300 miles above the earth is not within the bounds of experience . . . there is an understandable reluctance to contemplate the idea of conflict in space . . . because [of] attachment to the idea of primitive innocence among the stars . . . though it is attractive to entertain the idea of a virgin universe, it is no more realistic than that of the forest so loved by, if so unknown to, Jean Jacques Rousseau when he talked of the noble savage. I would myself prefer to have war in space than war in Westminster or the 8th arrondissement.[55]

Thomas's speech highlighted the contested nature of space war. For some, conflict in space would inevitably lead to nuclear conflagration on Earth. For others, such as Thomas, a war in space could ensure the safety of people below. This latter position treated space as a discrete realm that was somehow disconnected from Earth. In reality, no such distinction existed, but it nevertheless had a powerful effect on the human imagination and appealed to a deep-rooted desire to find a way back to a pre-Nuclear Age sense of invulnerability.

With humanity's future in space at a crossroads, the Pontifical Academy of Sciences hosted a conference in early October 1984 devoted to examining the impact of space exploration on humankind.[56] In reality, the scope of the conference went beyond exploration and looked at the many ways in which space activities affected human beings. Carlos Chagas, the president of the Pontifical Academy of Sciences at that time, wrote in the introduction to the conference's final report that "space exploration is a formidable lever for the improvement of the human condition all over the world . . . but in unscrupulous hands it may increase bondage to wealth and poverty, and even generate war."[57] The Pontifical Academy of Sciences had been an active participant in debates on nuclear weapons. It was therefore only natural that it would seek involvement in the international discourse on outer space as this topic, especially its moral dimensions, became a prominent issue in international security affairs.

Before the conference ended, John Paul II granted an audience to the participants in the Apostolic Palace. Before the Pope's address, Chagas emphasized the myriad roles that satellites were playing in human affairs. He described how communications satellites had "shrunk the dimensions of the world" and said that satellites allowed countries to "observe what is happening, for better or for worse, in any other part of the world as if our antipodes were in our fingertips." In particular, Chagas identified reconnaissance satellites as "an important weapon to fight the misuse of power and military strength. Most unhappily, however, they may—alas!—serve also war purposes." Chagas believed that information from satellites could alleviate poverty, improve education, and enable more effective uses of natural resources. To achieve these goals, it was necessary to prevent space from being transformed into a domain of military competition that would threaten the peaceful uses of space for the betterment of humankind.

When the Pope addressed those gathered in front of him, he echoed many of the themes raised by Chagas. He regarded the "presence in space of man [and] his machines with the same admiration as that of [Pope] Paul VI." Satellites could be used to spread culture and eradicate illiteracy. Space stations and planetary science probes would contribute to unlocking secrets of the universe. The Pope provocatively asked: "To whom does space belong?" He answered that outer space belonged to all of humanity. It was not for one country, or the superpowers, to dominate, control, and exploit.[58] From Reagan's perspective, the Pope's observations were in line with his own objectives for SDI. The president sincerely believed that space-based missile defense would save humanity from the threat of nuclear annihilation.

THE SYSTEM BECOMES A PROBLEM

In SDI's infancy, the Department of Defense did not provide any specific details about what form a strategic defense system might take. But the Pentagon's silence on specific concepts for a strategic defense capability did not prevent intense speculation on what kinds of technologies might ultimately be pursued and how they could affect strategic stability. The most controversial aspect of SDI was its space echelon. In the lead up to the 1984 reelection campaign, Democratic candidate Walter Mondale told voters that "we have more at stake in space satellites than [the Soviet Union]." Perhaps inspired by the 1983 film *WarGames*, he equated SDI with delegating "to computers the decisions as to whether to start a war." Reagan pushed back on Mondale's claims, saying that no decisions had been made about the nature of a strategic defense system.[59] Even in its infancy, the technological system that SDI might produce became a controversial subject and one that the administration was not effectively managing.

Differing opinions on the technological and strategic viability of missile defense, coupled with conflicting views on the role of space in defense strategy, created a public relations nightmare for the White House. To cultivate greater domestic political, congressional, and allied support for SDI, the administration needed a coherent messaging strategy: that is, to "get all the various players singing from the same policy music."[60] To this end, in the fall of 1984, a NSC working group crafted a road map for presenting SDI to the public, which Reagan approved in May 1985. The final document stressed that "we do not have any preconceived notions about the defensive options the [SDI] research may generate."[61] This language suggested that there was no commitment to any particular missile defense technologies, such as interceptors in space. Trying to appear agnostic about what form a strategic defense system might take was clearly an attempt to "distance the space issue from the SDI issue."[62] In other words, the White House did not want the public to think that SDI was only a space-based missile defense program. But a strategic defense system with the ability to destroy ballistic missiles in their boost phase would have to have interceptors in space. And Abrahamson had already gone out of his way to establish the linkages between strategic defense and space technologies.[63]

In January 1985, Abrahamson presented a paper entitled "SDI and the New Space Renaissance" to a Symposium on Space Weapons sponsored by the Planetary Society and the American Academy of Arts and Sciences. The general devoted special attention to what he called "space logistics," which included the "launch vehicles for emplacing space-based platforms and for servicing space components." He predicted that SDI would greatly benefit civil and commercial space activities due to SDIO's plan to reduce the "cost of taking large payloads into orbit by a factor of ten or more." Abrahamson presented SDI as a catalyst for a broad range of space activities that would directly contribute to science, exploration, and economic prosperity, in addition to national defense. SDI was part of a narrative of technological progress. Science and technology provided man the means for completing the "emancipation from the Middle Ages," and this space renaissance would expand "our understanding of human life."[64]

Setting aside his ahistorical representation of the relationship between the Middle Ages and the Renaissance, Abrahamson presented a vision for humanity's future that was in line with the space boosters of the 1970s who believed that military space programs would benefit human space exploration (among other space endeavors).[65] An expanded militarization of space would lead to the rebirth of the American space effort and inaugurate a new era in the exploration of the universe, providing tangible benefits for all humankind. This appeared to be a return to the technocratic thinking behind Apollo and the belief that technological progress could fix, or at least substantially alleviate, social issues such as racism, gender inequality, and poverty.[66] Security studies expert Columba Peoples writes that "Abrahamson also joined Reagan and Jastrow in invoking America's technological legacy as a response to skepticism from some parts of the scientific community."[67] But Abrahamson was doing much more than pushing back on scientific doubt. He sought to demonstrate that space-based missile defense was a force for good rather than evil and therefore aimed to reframe the debate about space weapons and transforming the heavens into a battlefield.

In the first few years of SDI's existence, it was solely a research program examining different technologies, but people could not help but speculate about a future strategic defense *system*. When Garwin published his op-ed in the *New York Times* only seven days after Reagan's SDI speech, he specifically addressed what he saw as the problems inherent in a strategic defense system. Garwin believed that placing weapons in space would not only create more insecurity in superpower nuclear relations, but also that the vulnerability of a space-based strategic defense system would be a source of instability. He maintained that "such a system could not be made sufficiently effective, reliable, and secure against jamming and other countermeasures." Because the infrastructure of space-based strategic defense would not be robust, Garwin predicted that the Soviet Union would invest in more ASATs and space mines that could increase the likelihood of conflict in space. In a word, the very vulnerability of a strategic defense system increased the probability of war.

The Office of Technology Assessment (OTA) contracted Dr. Ashton Carter, a physicist and defense policy expert at MIT, to produce an unclassified background paper on directed-energy weapons in space. Notably, Carter had full access to classified information and studies performed for the Executive Branch.[68] In April 1984, Carter published his findings. He devoted significant attention to the vulnerability of space-based strategic defense systems. He said that this was necessary because "all boost-phase intercept BMD concepts have crucial components based in space . . . vulnerability of these satellites is a cardinal concern because their orbits are predictable (they are in effect fixed targets)."[69] Carter described a full range of options available to the Soviet Union for degrading and destroying the space-based elements of a strategic defense system. Carter also highlighted the infrastructural challenges, especially the cost for deploying so many components into space. He concluded that more than a thousand shuttle missions would be required to place the necessary components in space; he did not even attempt to estimate the quantity and cost of space missions for maintaining space-based infrastructure. His message was nevertheless clear: space-based defense required prodigious investment in an exorbitant infrastructure that would be vulnerable to many different kinds of countermeasures. According to Carter, even if space-based missile defense turned out to be feasible, it was not defensible.

The same month that Carter released his background paper, the Union of Concerned Scientists published its report entitled "Space-Based Missile Defense." Richard Garwin, Hans Bethe, Carl Sagan, Victor Weisskopf, and several other distinguished scientists and strategic studies experts were members of the study panel. Like Carter, the authors of this study pointed out that any defense designed to attack Soviet missiles in their boost phase of flight would have to be space based.[70] They also highlighted the great cost and difficulty associated with placing such an immense infrastructure in space. Their comprehensive assessment included the various kinds of interceptors that the US could develop and an examination of the challenges associated with integrating such a large system. Addressing vulnerability, the study members concluded that "space-based battle stations are intrinsically fragile" and vulnerable to a wide variety of countermeasures.[71] Even though Reagan had said that SDI was only defensive, Garwin and his colleagues pointed out that SDI technologies could be used as ASATs and would therefore be "highly detrimental to US interests given the critical national security functions of satellites."[72] In sum, even a marginally effective strategic defense system would be destabilizing because of the offensive applications of missile defense interceptors.

Remarkably, the discussion about the vulnerability of space systems had moved into a public forum. If the space-based components of a strategic defense were in predictable orbits and therefore vulnerable, so too were reconnaissance satellites that had very limited maneuverability and were trackable by Soviet space surveillance. Any details about the limitations or weaknesses of military hardware are closely held secrets, and in light of this security concern, the CIA warned in a top-secret memorandum that "as a result of the public debate on SDI," key NRO technologies "will be compromised in the near term."[73] CIA security personnel recommended strengthening safeguards within SDIO to ensure that sensitive information about US space capabilities did not leak into the public arena as a result of its efforts to defend space-based missile defense. Despite this need to protect sensitive space technologies, the lid on military secrecy in space was lifting to a degree.

DUELING PROPAGANDA

Shortly after Reagan announced SDI, the Soviet Union commenced a concerted information campaign accusing the US of "militarizing outer space." Stanislav Levchenko, a KGB defector, said at a Heritage Foundation forum in 1985 that "SDI is the no. 1 point [sic] of Soviet propaganda now" and that "it will be for a very long period of time."[74] The Kremlin published pamphlets that painted SDI as a dangerous effort to weaponize space and one that would only increase the likelihood of war. Prominent Soviet scientists such as Evgeny Velikhov and Roald Sagdeev, both of whom had been involved in Soviet space and strategic defense research, attempted to mobilize scientists around the world to oppose SDI. In response, the US government executed its own influence campaign that highlighted an expanding Soviet military space threat.

Intelligence reports on Soviet military space programs were declassified and published alongside evocative depictions of Moscow's strategic weapons in an effort to demonstrate to the American people and allies that SDI was a necessary response to the Soviet Union's own military space efforts. According to Erik Pratt, the Reagan administration used "mounting fear that the USSR would . . . subjugate the world to a 'Pax Sovietica' [in space]" as way to build support for space-based missile defense.[75] Ellis argues that "to deflect criticism of the SDI as an unwarranted and destabilizing provocation, the Reagan administration began framing the policy as a prudent response to the Soviets' menacing 'Red Shield' BMD program."[76] In 1984, a joint UK MoD–FCO report for Thatcher noted an "increasing [American] emphasis on Soviet efforts in this [strategic defense] field as the justification for [SDI]."[77] British officials explained that the use of the Soviet strategic defense program as an argument in favor of SDI emerged in the fall of 1983 as the administration was coming under more criticism over SDI.[78] UK analysts noted that Soviet propaganda on military competition in the cosmos represented a "real [Soviet] anxiety about an arms race in outer space."[79]

There is a sense of déjà vu in the rhetoric of pro-missile defense advocates in the 1980s. In early 1963, Barry Goldwater warned of Moscow's objective to secure superiority in space.[80] Ellis writes that even though Goldwater's space alarmism did not resonate with a large portion of American voters, he nonetheless "tapped into something powerful" with his argument that the Cold War could be "won through space control."[81] Members of the High Frontier similarly painted space as the decisive battlefield of the future. The Citizens Advisory Council's Committee on Space War authored a report that predicted Moscow would use all of its resources to prevent the US from fully realizing its potential to develop and dominate the space frontier.[82]

In 1983, the Soviet Union published a pamphlet entitled *Keep Space Weapons Free* that described how the alleged US aim to transform space into a "theater of war" was based on the mistaken belief that Earth could be made safe by moving wars into outer space.[83] Senior US military officers provided significant fodder for Soviet propaganda. For example, Major General Thomas Brandt of the Joint Staff said in 1984 that "the US cannot and will not ignore the value of the military use of space and allow the Soviet Union to dominate the 'ultimate high ground.'"[84] Strategic arguments aside, Soviet propagandists argued that "space death merchants" were a powerful force behind American designs for an arms race in space because they sought lucrative defense contracts. For them, "an arms race in space means big money and, hence, big profits."[85]

US intelligence assessments maintained that the Soviet Union was using its anti-SDI propaganda campaign to communicate its own capabilities for countering a deployed space-based missile defense system.[86] Analysts noted that "it is not surprising that the Soviets are taking steps to refocus technical efforts to begin to counter SDI, if it were to be developed and deployed. It is surprising, however, that they provided such explicit information on what they are doing. The Soviets clearly wish to convey serious intent to match any new US military capabilities in space, and to imply that the US will not gain any net advantage from its efforts" (emphasis in original).[87] The Soviet information campaign targeting SDI had three primary messages: (1) SDI was technically unachievable, (2) the Soviet Union would develop countermeasures to make SDI fail, and (3) the US was upsetting the strategic balance and planning for a "nuclear war-winning capability."[88] At first glance, these ideas appear to be contradictory. If SDI was not technologically feasible, how would it upset the strategic balance, and why develop countermeasures? The Kremlin's propaganda reflected Soviet ambivalence about SDI. KGB reports reveal that the Soviet intelligence services did not have a clear idea of the overarching objective of SDI or what kind of missile defense system might ultimately emerge.[89] But even a partially functioning strategic defense could have undermined Moscow's nuclear forces, at least to a degree, and space-based interceptors of any kind would have been able to shoot down Soviet satellites.

US officials worried that the Kremlin's anti-SDI propaganda might be gaining traction when the CIA discovered that ideas from a skewed Soviet study of SDI made their way into an article published in a January 1985 issue of the *Washington Post*. Analysts explained how this Kremlin-backed investigation asserted that "space-based [chemical laser] SDI systems are too technically complex" but emphasized that the Soviets only examined "one possible SDI variant" and excluded simpler concepts such as space-based kinetic interceptors. There was also a glaring error: the equation for calculating the potential kill range of a nuclear-driven X-ray laser was incorrect. Intelligence personnel maintained that the report was primarily written for propaganda value, and they noted that it was widely distributed in the West.[90] It is highly doubtful, however, that this Soviet report had a significant impact on the SDI discourse in the US. Prominent organizations such as the Union of Concerned Scientists and the American Physical Society had already produced highly detailed reports on SDI feasibility—examining a broad range of potential system concepts—that were overwhelmingly negative.[91]

Soviet propaganda about SDI contributing to the "militarization of space" played to the fears of many people who harbored dystopian notions of death rays in the cosmos wreaking havoc on Earth.[92] Even though space had been militarized from the beginning of the Space Age, SDI vividly contradicted the image of American space activities that was familiar to US citizens and people around the world. Many observers could not really understand the nuances of the technical discussions over SDI feasibility and the strategic debates over the impact of space-based defense on deterrence theory. But nonspecialists could indeed grasp that SDI represented a new US vision for space that involved weapons in orbit. Advocates of space weapons had been around for a very long time, but they now had a patron in the Oval Office, and the prospect of war in space seemed real.

The intensity of Soviet research into space-based weapons technologies was, however, lost on the general public in the US and Western Europe. US national security officials who were aware of intelligence reports on Soviet strategic weapons programs were angered by what they saw as Moscow's hypocritical messaging that it was dedicated to peaceful uses of space while the US was seeking to develop and deploy space weapons. One way that the US government pushed back on the Soviet Union's anti-SDI propaganda campaign was through publishing classified intelligence on Soviet strategic defense and military space programs.

Beginning in October 1981, the Department of Defense had begun publishing a series called *Soviet Military Power* that was based on sanitized intelligence reports about Soviet military capabilities and included evocative images of strategic weapons. The series was motivated, at least in part, by requests from NATO allies "for declassified material to document the [Soviet] threat."[93] It became an annual publication in 1983 and was printed every year until the collapse of the Soviet Union. To reach a broader audience, the US government translated it into German, French, Japanese, Italian, and Spanish. *Soviet Military Power* became a primary mechanism for the Reagan administration to highlight Moscow's military programs and thereby justify American investment in strategic defense. Weinberger openly said that *Soviet Military Power* was intended to garner greater support for SDI among American lawmakers and European allies.[94]

Larry Gershwin, the national intelligence officer for strategic programs, had growing concerns in late 1983 that the US had not done an effective job of "refuting Soviet claims" about the US "militarizing space."[95] Senior members of the administration feared that US lawmakers and European allies would be swayed by Soviet propaganda that portrayed the US as "a producer of space war machines."[96] Robert Gates, the deputy director of CIA, believed that the administration would "win or lose SDI in the media" and that the US needed a concerted strategy for getting the press on their side, especially in the lead up to renewed arms negotiations in Geneva.[97]

Soviet Military Power served as a vehicle for attempting to change the public narrative on SDI. However, the potential benefits of releasing intelligence about Soviet military programs had to be carefully weighed against the risks. Protecting sensitive intelligence sources and methods was a vital responsibility of CIA Director Bill Casey, and he was not thoroughly convinced that *Soviet Military Power* was indeed valuable. This situation pitted the CIA against the Department of Defense. Larry Gershwin openly worried that the Pentagon's "effort to institutionalize the periodic publishing of . . . a paper—much like *Soviet Military Power*" could compromise sensitive "sources and methods."[98] Even earlier, Casey expressed concern that "cumulatively the various editions [of *Soviet Military Power*] had exposed US intelligence sources and methods."[99] Despite these anxieties, the White House clearly believed that the potential rewards of *Soviet Military Power* outweighed the risks of continued publication.

Whether *Soviet Military Power*, and similar US government publications, really swayed public and foreign government opinions is debatable. There is no solid data on how *Soviet Military Power* affected public and elite view-points in the US and Europe. In early 1986, Teller implored Casey to release more intelligence concerning Soviet strategic forces. Casey replied that "the problem is not that we are providing insufficient information, but rather that the information is not being read and is given insufficient attention by the media . . . Providing more detailed information . . . is hardly likely to solve that problem."[100] Opposition to SDI largely held constant in the US and in Europe, even after *Soviet Military Power* began to highlight the Soviet Union's strategic defense and space weapons efforts more directly.

The Reagan administration's contradictory signals regarding SDI's potential impact on stability certainly did not alleviate opposition to space-based missile defense. Weinberger claimed that the world would be subjected to "Soviet political blackmail" if Moscow succeeded in "beating the United States to space with a deployed strategic defense system." In the same statement, he claimed that SDI was not "even a weapon," it was a "harmless means of destroying weapons."[101] In other words, strategic defense was peaceful only if the US led in the technologies on which it was based. Soviet media rhetorically asked why the US was so worried about Moscow's alleged strategic defense research if SDI was indeed a pacific effort.[102]

SDI THEOLOGY

Senior advisors to Reagan recognized that members of the administration were putting out conflicting signals regarding SDI and its ultimate aims. To remedy this situation, Colonel Robert Linhard (later promoted to major general), the director for defense programs and arms control in the White

House, advised McFarlane to establish an informal policy coordination working group for SDI. He described it as "a 'mafia' like those we have used effectively to work the 'dual-key' and nuclear security issues in the past."[103] The objective of the "SDI mafia" would be to create an "SDI Bible" that would "get all the various [administration] players singing from the same piece of policy music." This "SDI Bible" would serve, in effect, as the foundation for the administration's public diplomacy on SDI.[104]

The "SDI Bible" explained that "SDI is not a system development or deployment program" and that the project aimed to enhance, not replace, deterrence. The "SDI mafia" kept the details of system effectiveness vague, saying only that a strategic defense capability would substantially reduce or eliminate the ability of ballistic missiles to destroy a "militarily meaningful portion of the US and allied military target base." Comprehensive defense was also a long-term objective; Linhard and his colleagues acknowledged that a strategic defense system "could not be deployed overnight" because some technologies would "become available sooner than others."[105] In other words, SDIO had adopted a building-blocks approach in which near-term options (e.g., kinetic interceptors) would be deployed first, followed by more exotic technologies in the next century. In May 1985, Reagan signed NSDD-172, "Presenting the Strategic Defense Initiative," which reflected the core tenets of the "SDI Bible."[106] Notably, when the administration finally released its SDI pamphlet, it mentioned space technologies only once and in an annex that refuted the claim that SDI was "militarizing space."[107]

As a part of the public diplomacy strategy, the "SDI mafia" identified potential spokespeople from various US government agencies who could give speeches to cultivate greater support for SDI.[108] Robert Gates, the deputy director of the CIA, was one such individual, and over the next several years, he gave many talks to a wide variety of private groups explaining how SDI served the national interest.[109] In addition to speeches, the SDI public diplomacy working group wanted strategic defense advocates to publish articles rebutting critical commentaries.[110] They were, in effect, planning an information warfare campaign.

During a February 1985 speech to the World Affairs Council in Philadelphia, Paul Nitze laid out three conditions that had to be met for strategic defense to be successful: (1) a system had to work, (2) it had to be able to survive attacks against it, and (3) it had to be "cost effective at the margin." The latter point meant that the cost to retain the system's level of effectiveness had to be less than the cost of offensive capabilities it was designed to defend against. Collectively, these became known as the "Nitze criteria."[111] Whether SDIO could indeed produce strategic defenses that were cheaper than Soviet capabilities for countering them remained a subject of bitter controversy through the end of the Cold War.

A MORE MORAL DEFENSE

The Reagan administration argued that SDI provided a more moral defense strategy than one based on the threat of nuclear annihilation.[112] Consequently, Reagan sought the SDI endorsement of the one person who was not only a visible symbol of moral authority on the world stage but also a global representative of anti-communism: John Paul II, the first Polish Supreme Pontiff of the Roman Catholic Church.[113] The US had formally established diplomatic relations with the Holy See only in January 1984. One year later, the US government sent multiple representatives to brief Vatican officials on SDI and to provide intelligence on Soviet strategic defense; much of this information was likely taken from *Soviet Military Power*. During a January 1986 trip, William Wilson, the American ambassador to the Holy See, implored US officials not to risk rekindling "Vatican uneasiness" due to "SDI 'salesmen' who have briefed the topic over the past 12 months." Wilson said that "each visit seemed to the Vatican to be an attempt to elicit public support for the [SDI] program." Representatives of the Pope told SDIO officials that they could never support a military program that "potentially takes food from the mouths of the poor."[114]

Even though the US was not able to secure John Paul II's endorsement of SDI, the Pope was indeed helpful on the public relations front for strategic defense. In early 1985, the Pope had asked the Pontifical Academy of Sciences to produce a study of SDI. This study alarmed the White House in light of the statement by Chagas, the Academy's president, that "the weaponization of space is the next step in this terrible, incredible race . . . to annihilate the whole world."[115] The Reagan administration feared that the report would be interpreted as papal condemnation of space-based defense. Moscow, on the other hand, saw this as an opportunity to use the Vatican in its own anti-SDI propaganda efforts.

For the first time in nearly six years, the Kremlin sent Foreign Minister Andrei Gromyko to Vatican City "to drum up a little opposition to President

Regan's Star Wars initiative to persuade the Pontiff to oppose the US plan."[116] John Paul II would recall years later that Gromyko was "very worried about the American Strategic Defense Initiative" and wanted "the Church's help against the United States."[117] The American embassy to the Holy See received phone calls "day and night" from Washington "to somehow convince the pope's advisors of our defensive point of view [on SDI]."[118] To the relief of the Reagan administration, the Pope decided not to have the SDI report made public. Did John Paul II do this because of Washington's advocacy or to contribute to a so-called holy alliance with the US?[119] The Pope had told Weinberger that "we are for peace of course, but we are not for pacifists—unilateral pacifists."[120] John Paul II knew that any negative statement about SDI would been seen as an endorsement of the Soviet anti-SDI position, which is something that he would have refused to do, given his political convictions. The Pope's silence on SDI was a victory for the US in its propaganda battle with the Soviet Union over military activities in space, but the intense debate over weapons in space would continue unabated until after the Cold War ended.

CONCLUSIONS

While many of the technical nuances of the expert-level arguments over military space technologies and strategy were lost on the general public, the very fact that an open debate on military space activities emerged in this period is notable. For much of the Cold War, American and Soviet military space programs were shrouded in secrecy. What little information was released in the 1970s about American national security space systems only reinforced the idea that the US was committed to the peaceful uses of outer space. Notions of exploration, scientific investigation, and transnational cooperation still dominated how people thought about space in the public sphere of the 1980s. For the US and the Soviet Union, their competing narratives about SDI and the military uses of space more broadly prompted them both to release more information about the "black world" of military space operations than in any earlier time period. The willingness of Washington and Moscow to modify information classification practices to push their respective military space agendas further highlights the weight attached to military spaceflight in superpower relations during the final decade of the Cold War.

Even though SDI most directly affected superpower relations, it brought together defense experts, space enthusiasts, intelligence officers, diplomats, scientists, clergy, and citizen activists into a transnational debate arena. Scientific expertise, military strategy, high-stakes diplomacy, and astroculture all converged to shape the discourse on weapons in space. The Reagan administration quickly learned that the popular imagination of spaceflight had a much more dramatic impact on the way in which people perceived SDI than antiseptic technical analyses about lasers, kinetic interceptors, and command and control software. And the very lack of concrete information about what form a specific strategic defense system might take only created more room for the public imagination to run wild about space-based defense.

The involvement of the Holy See, including the Pope himself, in the debate over SDI not only reveals how the prospect of weapons in space was a geostrategic issue, but also raised fundamental moral questions about the future of human activities in the cosmos. The fact that the Pentagon sent its emissaries around the world to convince heads of state that space-based missile defense was indeed necessary and desirable shows how the Reagan administration's SDI information campaign was a truly global effort. Clearly, however, highlighting Soviet military space programs and strategic defense research was insufficient for cultivating the support that the White House needed to move the program forward. Reagan needed the endorsement of America's closest friends and allies on the other side of the Atlantic. But the president would soon find that many of the same anxieties and fears in the public sphere about weapons in space were fast emerging as divisive issues in the transatlantic alliance as well.

4 "EUROPE MUST NOT LEAVE SPACE TO THE AMERICANS"

Reagan's establishment of SDI made military space policy a more prominent and contentious issue in US relations with Europe. Lack of consultation with allies on military space matters, especially SDI, only contributed to preexisting "chronic tensions" in NATO.[1] Pro–space arms control statements by Western European politicians created the appearance of widespread hostility toward SDI on the other side of the Atlantic, but their views on space militarization were not uniform.[2] SDI would intensify preexisting European anxieties about space becoming yet another domain of potential superpower military confrontation. As the only Western European nuclear states, France and Britain had unique interests in the impact of space-based missile defense on the future of nuclear deterrence. SDI moreover catalyzed European concerns about being left behind, as military space capabilities played an increasingly critical role in superpower military strategy. But since the European Space Agency (ESA) generally eschewed any direct ties to military space projects, Western Europe lacked any formal mechanism for military space cooperation. SDI would, moreover, force European officials to link civil and military space policy more directly as they devoted more attention to Europe's autonomy in space.

Approximately one year after Ronald Reagan made his March 1983 "Star Wars" speech, French President François Mitterrand took a three-day trip to California. He quickly concluded that the Golden State was at the "center of a technological revolution" that was driving American innovation and widening the technology gap between the US and Europe.[3] The French president feared that SDI would serve as a catalyst for American high technology, thereby pushing Western Europe even further behind the US in economic competitiveness. Consequently, France, along with other Western European states, came to view the prospect of a space arms race not only as a strategic military issue but also as an economic industrial matter.[4] With both military and economic considerations in mind, Mitterrand would warn his colleagues that "Europe must not leave space to the Americans."[5]

In 1985, Reagan invited allies to become involved in SDI with the hope of securing their support for his controversial program. For Western European politicians, deciding whether to participate in SDI quickly became entangled in the politics of European integration and in policy debates concerning European autonomy in space. The divergent space interests of Western European states created an impediment to Mitterrand's push for a single European policy on SDI. In particular, Prime Minister Margaret Thatcher eschewed any common European policy position on SDI that might upset Britain's privileged access to American military and intelligence satellites through its defense cooperation with the US. Despite substantial reservations about expanded space militarization, many European politicians determined that the potential benefits stemming from involvement in SDI outweighed the political risks. Rather than developing a uniform European position on SDI, Western European states ultimately competed with each other to secure favorable SDI technology transfer arrangements.

SPACE AND THE TRANSATLANTIC ALLIANCE

SDI's establishment intensified European debates concerning space militarization, but space policy had long been a source of contention both among the European powers and between Europe and the US. European states pursued cooperative and national space projects to expand economic prosperity, strengthen national security, bolster alliance bonds, and enhance prestige. Even though the US and Europe formed multiple joint space projects, US protectionism with regard to space technologies served as an uneasy reminder of Europe's status as a junior partner and created momentum for European independence in space. The formation of ESA in 1975 provided a centralized mechanism for civil space coordination in Europe, but it in no way represented a pan-European consensus on space policy.[6] Despite ESA's eschewal of military ties, national security and civil space considerations were intimately linked in European space policy. By the early 1980s, European states were pursuing greater autonomy in space capabilities and signaling a desire to have a voice in superpower space deliberations. But unity of effort among the European powers would prove to be illusive.

The road to a combined European space organization in the form of ESA was long and arduous. There was a constant tug-of-war between aspirations for European collectivism and national(ist) agendas on space matters. For the European nuclear powers, Britain and France, space technologies served strategic purposes; rockets that could lift satellites into space could also launch nuclear warheads over great distances. In addition to practical applications, Charles de Gaulle viewed space as an arena in which France could demonstrate its grandeur.[7] France did not, however, possess the financial resources to compete unilaterally with the US and the Soviet Union in space. Consequently, French officials sought to lead Europe in achieving greater autonomy in space. British officials, in contrast, did not prioritize space prestige projects and sought to leverage their defense and intelligence relationship with the US to secure privileged access to American space technologies.[8]

Since ESA eschewed any overt military ties, defense and intelligence space endeavors primarily took place within the confines of national European space programs. Yet, all NATO members were benefactors of American military space technologies, reconnaissance satellites in particular. Intelligence derived from US reconnaissance satellites served as a keystone of the alliance's military planning. By the mid-1960s, NATO sponsored a satellite communications program, built around US satellite technologies, to ensure that the alliance had a secure means for command and control of combined forces in the European theater. Separately, Britain cooperated with the US in the development of its own communications satellite program, called "Skynet," to be able to maintain sovereign command and control of British forces deployed far from home. Despite the fact that all NATO members were consumers of US space capabilities, the US made decisions on military space matters without significant allied consultation. There was no space equivalent of the NATO Nuclear Planning Group (NPG) that gave nonnuclear NATO members a say in the alliance's nuclear decision making. This lack of US engagement on military space matters created yet another stress point in the transatlantic alliance as the US pushed forward with ASATs and then space-based missile defense research.

By the early 1970s, the space technological gulf between the US and Europe appeared to be unbridgeable. The US had secured a monopoly over the satellite communications market through the establishment of Intelsat, a global satellite communication system open to all countries but with the US as the dominant partner. US protectionism of its lead in the growing international communications satellite market aggravated Western European states that possessed the technical know-how to produce their own communications satellites. The Apollo program became one of the most visible symbols of an overwhelming US lead in technical expertise, especially in systems engineering. The perception that Apollo was somehow a missed opportunity in Europe would make many European industrialists and some government officials more favorable toward participation in SDI research.[9] In 1985, the UK ambassador in Washington, Oliver Wright, candidly observed in a cable to London that Britain had "missed the [Apollo] space bus" and should not make the same mistake with SDI.[10]

President Richard Nixon had provided European partners opportunities to become involved in US space activities, as he looked to use space as a means of promoting US foreign policy objectives in the context of superpower détente. Notably, he reframed space as an arena for cooperation rather than competition. Even though European technological capabilities were no match for those of the US, Europe did indeed have a "financial, technological, and industrial contribution to make."[11] Not everyone in the Nixon administration was, however, enthusiastic about cooperating with the Europeans; one American official observed that "we are giving the Europeans too much technology for too little return."[12] There was also concern about Western European firms leaking sensitive technologies to the East. This anxiety over technology safeguards would resurface during negotiations over allied participation in SDI.

NASA offered the Europeans the opportunity to contribute to the development of a space tug that would deliver payloads from the shuttle's low Earth orbit to higher orbits and to become involved in a project called Spacelab, a laboratory to be housed in the shuttle's payload bay.[13] Through involvement with NASA projects, the Europeans found that restrictive US technology transfer policies limited the sharing of both advanced hardware and technical know-how. European officials also found themselves at the mercy of unpredictable bureaucratic decisions in Washington. After ESA had committed to developing the space tug, the US decided to cancel the project with little warning.[14] Consequently, France in particular pushed for greater European autonomy in space.[15] This aspiration for independence in space operations resulted in the production of a European launch system called "Ariane."[16]

With the development of an independent European launcher, the Reagan administration viewed Ariane as a direct threat to its economic interests in the growing satellite launch market. A letter from the head of NASA to Reagan's national security advisor lamented Ariane ending the US monopoly on launch services.[17] Newt Gingrich similarly warned that "the US may lose a lucrative market."[18] US officials were also concerned about the potential entry of Japan and the Soviet Union into commercial space launch services.[19] Since Reagan intended to maintain American technological leadership in the high-visibility space sector, the shuttle's launch services had to be competitively priced to stay ahead in a market that was only beginning to take off.

For the French government, independent European access to space served as a visible symbol of European technological prowess. French Prime Minister Pierre Mauroy stressed to his British counterparts that Ariane symbolized the "common [European] will to master advanced technologies and to take their share of the world space market."[20] As the only other nuclear power in Europe, French space aspirations were inextricably linked to national security considerations. France's withdrawal from the NATO command structure in 1966 only reinforced its need for sovereign control over critical satellite resources for reconnaissance and communications. Like the UK, France did not have an independent satellite reconnaissance capability, but unlike the British, French defense planners could not depend on special access to US satellite intelligence.[21] Beginning in the late 1970s, the French Ministry of Defense conducted preliminary studies on a mapping and reconnaissance satellite called *Satellite Militaire de Reconnaissance Optique* (SAMRO).[22] This satellite project would give France enhanced global intelligence capabilities that could support nuclear targeting and out-of-area operations in Africa, among other places.

Due to budgetary constraints and SAMRO's limited utility as a low-resolution satellite, by 1982, the MoD endorsed the French civilian Earth observation satellite program *Système Probatoire d'Observation de la Terre* (SPOT) in place of SAMRO.[23] With minimal support from French defense officials, Mitterrand tried to salvage SAMRO as a joint project with West Germany. Reflecting the French approach to civil space projects, Mitterrand looked for a cooperative European solution to achieve greater independence in military space technologies but with France as the senior partner.[24] France and West Germany had already successfully collaborated on the Symphonie communications satellites, Europe's first such capability in geostationary orbit.[25] West Germany carefully considered both the security and economic implications of SAMRO. Having the ability to conduct treaty verification independently appealed to Bonn's broader foreign policy goals, but West German officials recognized that France could use SAMRO data to support its nuclear targeting mission as well, which would create issues with the West German anti-nuclear lobby. The German aerospace industry was enthusiastic about the project, since its executives were convinced that the civilian Earth observation satellite market was set to grow and West Germany could be left at a disadvantage compared to its French and American counterparts.[26]

Broader transatlantic alliance considerations soon clouded Franco-German discussions about the joint reconnaissance satellite project. The West German defense ministry worried that Washington would not look favorably upon Bonn's collaboration with France on a military satellite.[27] West German Chancellor Helmut Kohl agreed with the defense ministry and respectfully declined to become involved in SAMRO. Mitterrand was nevertheless undeterred and, over the next couple of years, continued to encourage Kohl to collaborate with France in the space arena. Mitterrand appealed to West Germany's desire for prestige, stressing that space offered Bonn a path to achieving "major power" status.[28] The French president was echoing earlier arguments made by advisors to Lyndon Johnson who advocated using civil space cooperation as a "prestigious substitute" for a West German nuclear weapons program.[29] But a collaborative military space project with France was too politically sensitive. Prior to the public debate over SDI and a space arms race, even discussion of military space activities had been taboo in the West German press.[30]

Although SDI would force space militarization into a more central position in the transatlantic dialogue on strategic issues, French officials were already voicing opposition to space weaponry before SDI. In the late 1970s, the Soviet Union had an operational ASAT, and the US had initiated a new ASAT program that would come to maturity under Reagan. During a 1978 speech in front of the UN General Assembly, then French President Valery Giscard d'Estaing implored the superpowers to halt production of ASATs.[31] In the late 1970s and early 1980s, French officials proposed the creation of a UN satellite agency to monitor military developments around the world. Science fiction writer Arthur C. Clarke called this French initiative "Peacesat" and argued that it would "present mankind with an alternative to an arms race in space" by exploiting space systems to enhance international transparency.[32]

As France pushed for greater European autonomy in civil and military space technologies, Britain learned just how dependent it had become on US military space systems. During the Falklands War, Britain had to request satellite communications support from the US because its own Skynet satellite constellation lacked coverage over the Atlantic.[33] To remedy this situation, Thatcher directed the MoD to expand Britain's Skynet communications satellite constellation and develop a reconnaissance satellite; the latter would, however, ultimately be cancelled due to cost constraints.[34] Ilaria Parisi observes that "the Falklands War abruptly reminded Europe of its satellite gap [with the US]."[35]

Thatcher's decision to use the American shuttle, rather than Ariane, to launch new Skynet 4 series satellites became an irritant in Anglo-French relations. French officials regarded "a decision by [the UK government] to use [the] shuttle as a substantial threat to Ariane's status as a commercial alternative to America launchers."[36] Even if Ariane had been priced more effectively than the shuttle, it is doubtful that Thatcher would have chosen the French launcher.[37] Although the Falklands provided British officials compelling justification to pursue greater independence in military space capabilities, partnering with the French at the expense of the US was not an option. Thatcher eschewed any policy option that might in any way undermine UK access to American space technologies.

By the time that Reagan announced SDI in March 1983, European officials had already concluded that space capabilities were increasingly important tools of national power. But there was no coherent pan-European space policy. Although all NATO members relied on US space technologies to varying degrees, no formal alliance-wide consultative process for military space policy existed. Moreover, for Bonn and London in particular, any proposal for European cooperation on military space matters had to be weighed against possible consequences for their defense relationships with the US. Greater autonomy in satellite reconnaissance and communications would also make Europe more competitive in commercial remote sensing and space-based communications, but sharing dual-use space technologies that were increasingly lucrative would not be without serious difficulties. Finally, moving into the early 1980s there was no European consensus on space arms control. Fundamentally, political, military, and industrial considerations were inseparable in European space policy deliberations. All of these factors would play a considerable role in shaping the European response to SDI and subsequent debates about European participation in the controversial project.

RESPONDING TO REAGAN'S "STAR WARS" SPEECH

French diplomat Benoit d'Aboville characterized the European response to Reagan's March 1983 "Star Wars" speech as a mixture of "skepticism, bewilderment, and embarrassment."[38] Initially, European heads of government generally remained silent about Reagan's newfound public interest in missile defense throughout the remainder of 1983, but privately there was both anxiety and consternation about the president's call for a high-technology initiative that challenged existing nuclear strategy.[39] The cold response of European allies to the president's SDI speech should not have come as a surprise to the White House, especially since none of them were consulted ahead of time.[40] Lack of clarity about SDI technologies and their implications for Western security only further complicated this situation. US allies could not have been expected to endorse a project that had not been well defined. These realities aside, the White House believed that getting the allies onboard would bolster Congressional support for the program.[41] Allied solidarity was also critical for countering Soviet efforts to use SDI to divide the transatlantic alliance.

Reagan's stated desire to render nuclear weapons "impotent and obsolete" was especially a cause for concern in Britain, since Thatcher had only recently secured an agreement from the US to replace Polaris with Trident as the mainstay of the UK deterrent.[42] In light of the many technological uncertainties surrounding strategic defense, Michael Heseltine, the UK defense secretary, advised the prime minister to present a "cautious and non-committal" initial response to the president's call for a missile defense program.[43] Raising concerns about Reagan's speech could also have created problems in the transatlantic alliance during an especially politically sensitive period when US Pershing II and cruise missiles were being deployed in Europe."[44]

For the remainder of 1983, officials in Europe could not be certain if Reagan's anti-nuclear ambitions expressed in his March speech would be transformed into a formal defense program.[45] When, in January 1984, Reagan signed NSDD-119, directing the Pentagon to begin a long-term research effort "to develop and demonstrate key technologies associated with concepts for defense against ballistic missiles,"[46] Europeans were forced to take SDI seriously. British officials quickly concluded that SDI was "an issue of fundamental importance which would not go away in the near future."[47] A large-scale US missile defense project had direct bearing not only on superpower relations but also on the future of the transatlantic alliance.

The implications for Europe of a US transition away from reliance on nuclear deterrence were unclear but a cause for concern. Even if the technologies for strategic defense did not mature for many years, SDI had the potential to create political problems for European officials in the near term. Kohl was a staunch proponent of Pershing II deployments and worried that Reagan's position on nuclear weapons would play into the hands of the West German anti-nuclear movement.[48] Any SDI support from Thatcher's government could exacerbate domestic political tensions over nuclear modernization. In addition to the negative implications for the French nuclear deterrent, Mitterrand saw in SDI the potential to unleash technologies that would produce a "revolution in military affairs," thereby giving the US an even greater lead in advanced technologies with both commercial and defense applications.[49] There was significant controversy, moreover, surrounding the prospect of stationing missile defense weapons in space.

In the spring of 1984, Mitterrand instructed his Minister of Foreign Affairs Claude Cheysson to work with Minister of Defense Charles Herno in developing a diplomatic framework for making progress on space arms control.[50] At a Western European Union (WEU) meeting in Paris devoted to military developments in outer space, Cheysson tabled a proposal for a five-year renewable ban on the testing and deployment of directed-energy weapons that could be used in an ASAT or missile defense role and advocated "severe restraints on other ASAT potential systems, especially those threatening high-altitude satellites." Hans-Dietrich Genscher, the West German Minister of Foreign Affairs, responded very positively to Cheysson's proposal. UK Foreign Secretary Geoffrey Howe did not offer any comments, only because the British government had not yet settled on an official military space policy. He nevertheless noted in a trip report shared with the prime minister that he supported the French position.[51]

The French president's opposition to US military space plans came into the open when, at a June 1984 session of the UN CD, French officials proposed a ban on ASAT systems and a prohibition on testing kinetic energy systems in space.[52] Immediately thereafter, the White House sent a note to the Élysée that expressed "regret" about the French space arms control proposal and urged that such matters be handled in private channels moving forward.[53] Approximately one month later, Thatcher delivered a speech to the European Atlantic Group, underscoring the "urgent challenge of arms control in outer space" and warning that "space [could be] turned into a new and terrible theater of war." The only solution to this problem, she explained, was "negotiation and mutual restraint."[54] Since ASATs and missile defense shared many of the same fundamental technologies, key European officials concluded that pushing for ASAT arms control in the near term could restrain the development of strategic defenses over the long run.

The growing European momentum in favor of space arms control grabbed the attention of the Reagan administration. The CIA noted in an intelligence report that Western European states had become "vocal on outer space arms control" and that "even the UK, customarily the closest to the US on arms control issues . . . is considering support for limits on ASAT weapons." French, Dutch, and Italian officials stressed the linkage between missile defense and ASAT technologies, once again showing that they saw ASAT arms control as a means of constraining missile defense research and development. Intelligence analysts maintained that European allies were anxious about the implications of SDI for the credibility of the US nuclear umbrella and American defense commitments in Europe.[55]

Thatcher's public support for space arms control at the European Atlantic Group did not in fact represent official UK policy. In contrast to Mitterrand, she had not yet made an official decision on military space policy and requested detailed studies on this subject from the FCO and MoD.[56] Both Howe and Heseltine strongly advocated a space "arms control regime which hampered the development of BMD [ballistic missile defense] on both [US and Soviet] sides." Similarly, Percy Craddock, who would soon chair the Joint Intelligence Committee, tried to convince Thatcher that the ABM–ASAT linkage provided the *real* value of ASAT arms control. He maintained that it was in the "UK interests to promote control of ASATs both for its own sake and also since it would make it harder for the US to go too far down the SDI path."[57] Notably, Craddock, Howe, and Heseltine all warned that a rift over SDI had to be avoided because it could negatively affect UK access to American "space-derived intelligence" and harm broader Anglo-American defense cooperation.[58]

In the fall of 1984, Thatcher firmly rejected the advice of her senior advisors to encourage the US to pursue space arms control negotiations with the Soviet Union.[59] Since Thatcher had already publicly endorsed space arms control, why did she prohibit her senior government ministers from trying to sway the US to contemplate limits on military space activities more seriously? Most fundamentally, alliance politics were embedded in these highly technical analyses concerning ASATs and SDI. Euroscepticism was a key tenet of Thatcherism,[60] and it is not surprising that Thatcher was generally suspicious of an agenda that entailed rejecting US policy and embracing a common European position on space militarization. Such a move ran contrary to her political objective of staying close to the US, even at the expense of more harmonious relations with continental Europe.[61] She explicitly warned her cabinet that pushing for space arms control would "risk annoying [the Americans] needlessly."[62]

Keenly aware of the military balance, Thatcher had concluded that achieving parity with the Soviet Union in low-altitude ASATs would have military utility. An FCO–MoD study identified space as the "high frontier" of "military operations and economic competition and warned of a 'Pax Sovietica' based on the domination of space, just as the 'Pax Britannica' formerly rested on the control of the High Seas."[63] The prospect of a Pax Sovietica based on the domination of space was more hyperbole than a realistic proposition. Nevertheless, Thatcher did seriously worry about the Soviet Union securing any high-technology advantages. She was convinced that "you could not ultimately hold back research into new kinds of offensive weapons. We had to be the first to get it. Science is unstoppable: it will not be stopped for being ignored."[64] Thatcher adopted the view of Charles Powell, one of her closest advisors, that Britain's goal should be "to manage the new technology . . . to add to the West's overall security." In a memorandum from Powell to the prime minister, she double underlined and placed a checkmark next to Powell's conclusion that "the Americans have no option but to push ahead in this [SDI] area."[65]

Even though the prime minister rejected the MoD–FCO push for space arms control that would hamper both space-based missile defense and ASATs, she remained skeptical about the feasibility and desirability of strategic defenses. Importantly, Thatcher distinguished between a missile defense research program and any plans to *deploy* a strategic defense system. Missile defense research would serve as a hedge against a Soviet lead in advanced defensive technologies, but she recognized that any significant technological breakthrough was unlikely in the near term. The Joint Intelligence Committee observed that "it is unlikely that exotic ABM defense will threaten the credibility of a Trident based deterrent over the next two or three decades and it may never do so."[66] Technical considerations aside, the prime minister did not want to make any statements in support of a future strategic defense system that might give Reagan the false impression that she approved of his anti-nuclear rhetoric associated with SDI, which would have played into the hands of her political enemies who were opposed to nuclear modernization. Moreover, outward support for SDI and ASATs would have only drawn criticism due to the growing hostility in Europe toward a space arms race.

In Europe, the lack of a formal body for discussing military space policy complicated SDI deliberations. ESA had many experts in various kinds of space technologies (e.g., remote sensing and communications satellites) with defense applications, but the organization's discomfort with military projects made it an unlikely place for European military space collaboration. Fortuitously, the relaunching of the WEU in 1984 provided a forum for considering how European states could harness space technologies for military aims. In October of that same year, the Rome Declaration called upon WEU member states to use it for strengthening their security and defense ties.[67] Shortly before this, a WEU report identified the growing importance of the military use of space for Europe. The report's introduction maintained that "space [capabilities] will be a key determinant in future warfare" and that Europe should "not only take note but act upon this fact."[68] But closer integration of European industrial resources to produce intelligence and communications satellites would prove to be only aspirational due to political obstacles.

The British were reluctant to cooperate with European allies on military space projects out of fear that doing so could negatively impact their satellite communications industry and defense cooperation agreements with the US. The French specifically proposed working jointly with Britain to relieve "embarrassing dependence on the Americans for satellite reconnaissance photographs," in addition to developing a new satellite communications system.[69] UK officials immediately suspected that the French might try to poach British satellite technologies. MoD and FCO leadership worried that "increased collaboration in military space, with say France and Germany, would prejudice this flow of [satellite-derived] data and our special relationship [with the US]."[70] This British perspective once again underscores the inseparability of industrial and political considerations in the formulation of military space policy. Britain would not, moreover, place its access to US defense and intelligence satellites in jeopardy in order to pursue closer military space ties with European allies.

Reagan's November 1984 landslide electoral victory added urgency to European deliberations over SDI. Shortly before the election, Reagan described SDI as a "moral obligation," making clear his commitment to the project's moving forward.[71] To reduce transatlantic tension over strategic defenses, Thatcher set out to convince Reagan to make a statement that SDI would be a long-term research program and that there were no set plans to change NATO's deterrent posture anytime soon. The president and prime minister met at Camp David in December 1984 to discuss SDI among several other key issues. Thatcher plainly stated her opposition to the idea that nuclear deterrence might be made irrelevant, and she questioned the feasibility of strategic defenses. According to Robert McFarlane, Reagan respected Thatcher "above all others" and "was very sobered by" her arguments.[72] Yet, Thatcher's strategic defense doubts did not in any way curb Reagan's optimism that SDI would indeed produce a workable strategic defense system.

Thatcher achieved her central goal for the trip when Reagan made a statement to journalists gathered at Camp David, pledging that: (1) the aim of the West was not to achieve superiority but balance, (2) SDI deployment (but not research and testing) would be a matter for negotiation, (3) the program's aim was to enhance not to undercut deterrence; and (4) both East and West should reduce dependence on offensive nuclear systems.[73] Despite Reagan's assurances that SDI would not precipitate any near-term changes in NATO strategy, Europeans remained worried.

To the surprise of European leaders, US officials alluded to the possibility of allied involvement in SDI but offered few details. Abrahamson told European officials that "joint programs between NASA and ESA . . . provided encouraging precedents" for collaboration on SDI.[74] Surely European officials did not find prior space cooperation with the US as encouraging precedents for collaboration on SDI.[75] Most importantly, it quickly became clear that Abrahamson and his staff had "no settled views or answers on many questions [regarding SDI's technological feasibility]."[76] The general explained that the US would not pursue a "crash program" to produce a limited strategic defense capability, with a more expansive one planned for the future, because the Soviets could too easily overwhelm such a system by building more ballistic missiles.[77] (Notably, as SDIO ran into both technological and political obstacles, it would indeed pursue a more limited program to demonstrate a capability up front with the hope that a larger, more effective system would come later.)[78] European officials expressed concern that ASATs would make SDI vulnerable—a problem that Abrahamson acknowledged without offering any solutions.

European leaders grew frustrated with US security classification policies that complicated information exchanges on SDI. European defense experts had to rely on reports from the Union of Concerned Scientists, articles in *Scientific American*, and statements by prominent US scientists such as Hans Bethe, Kosta Tsipis, and Herbert Lin.[79] Even the UK with its close cooperation with the US on intelligence and nuclear matters was left in the dark on key aspects of SDI.[80] British officials reported back to London that during a visit to the Pentagon, they were "allowed to attend all the [SDI] briefings at the secret level but [were] excluded from a session involving nuclear driven devices."[81]

INVITING THE ALLIES TO PARTICIPATE

Even after Reagan agreed to Thatcher's four points at Camp David in December 1984, European attitudes toward SDI appeared largely unchanged. CIA officials reported in January 1985 that SDI "poses both near term and longer term problems for our NATO allies."[82] SDI conjured up fears about a "fortress America" mentality that presaged a decoupling of US security interests from the European continent. Both France and Britain remained anxious about the effects of SDI on their relatively small nuclear deterrents. Intelligence officials maintained that while SDI created unique problems for the European nuclear powers, the West Germans and Italians "may even look to SDI as a way to reduce the distinction between NATO's nuclear and nonnuclear members."[83] The Italian Ministry of Foreign Affairs indeed concluded that SDI offered an opportunity for *riequilibro* ("rebalancing") the political and military standing of European countries "in favor of those who do not possess nuclear weapons and missile forces of their own."[84] Nuclear politics aside, European officials could agree that SDI would have significant implications for advanced technology industries, even if the US decided not to deploy a strategic defense system. If European politicians determined that they should become involved in SDI research, the more contentious question that had to be answered was whether they would develop a collective policy on SDI participation or become competitors for SDI contracts.

Reagan's advisors believed that internationalizing SDI was necessary to build greater support for the project at home. Kenneth Adelman, the head of ACDA, optimistically predicted that participation could reduce "allied hostility to the SDI program" and provide the "allies a stake in evolving a strategy toward more reliance on defense."[85] In practice, White House officials had no interest in giving allies a voice in the management of the SDI program. In the near term, offering allies the chance to get access to more advanced technologies was at the center of the US strategy for attracting foreign partners. Adelman argued that the European anxiety that "SDI research, like our space program of a previous era, will result in technology breakthroughs . . . [making the Europeans] slip even farther behind us" would serve as a useful lever for securing allied participation.[86] He nevertheless cautioned that US officials should not oversell the idea that SDI involvement would shower allies with funds for technologies with broad commercial and military applications.[87] Noticeably absent from White House discussions was any concern about intellectual property rights. Disagreements over ownership of intellectual property had been a source of tension between the US and European partners in other cooperative projects[88] and would reemerge as a contentious issue during discussions with both West Germany and the UK regarding their participation in SDI.

On March 26, 1985, Weinberger sent letters to fifteen American allies, predominantly NATO countries but also Australia, Japan, Israel, and the Republic of Korea, inviting them to become involved in SDI research and development.[89] The offer lacked substantive details, only saying that countries could take part in "cooperative research . . . in areas of technology that could contribute to the SDI research program."[90] European allies were offended by Weinberger's stipulation that they respond within sixty days of receiving the secretary of defense's letter, which they interpreted as an ultimatum.[91] Consequently, Weinberger dropped the sixty-day requirement. For the White House, however, time was of the essence.

Inviting the Japanese to get involved in SDI was also tied to US efforts to get the transatlantic alliance behind the project. Reagan's advisors were well aware that European politicians increasingly viewed Japan as a technological, and therefore economic, threat. A set of talking points for CIA Director

William Casey noted that any Japanese interest in SDI would "stimulate and almost ensure European participation as well."[92] In many ways, getting the Japanese onboard with SDI was a far more challenging task than securing European support. Japanese politicians were uncomfortable with SDI because of its widespread perception as a destabilizing weapon.[93] Japanese Prime Minister Yasuhiro Nakasone nevertheless believed that these complexities could be circumvented, and his interest in the US offer would not go unnoticed by European leaders.[94]

Internationalizing SDI served multiple purposes. Getting allies into the program could help win the support of members of Congress who were on the fence, especially before the Pentagon submitted the SDIO budget for the next fiscal year. More importantly, foreign involvement would strengthen the US position on SDI going into the arms control negotiations with the Soviet Union. Even though political considerations were paramount, SDIO did want to benefit technologically from its allies in areas where they excelled. SDIO officials created a list of specific allied capabilities and firms that had competencies of interest.[95] American defense experts concluded that French, German, British, Italian, and Japanese firms could meaningfully contribute to research in optics, sensors, computer software, kinetic energy weapons, directed-energy weapons, satellite subsystems, and other key areas.[96] Notably, Reagan's advisors consistently underestimated the complexities of SDI-related technology transfer. This situation is ironic, since combatting the "technological hemorrhage to the East" was a top priority for Reagan. The US government went to great lengths to ensure that adequate technology safeguards were in place, since the Soviet Union viewed many European institutions as "softer" targets for technology theft than their American counterparts.[97]

The Coordinating Committee for Multilateral Export Controls (COCOM) served as the primary mechanism for developing policies to restrict the flow of sensitive technologies to the East. Established in 1949, COCOM included all members of NATO, except Iceland, and Japan joined in 1952.[98] Although officially a coordinating agency for protecting the West's technological advantages, several European participants believed that the US used COCOM to further its own interests at the expense of theirs. Horst Ehmke, a leading German Social Democrat, expressed frustration with the wide range of technologies subject to COCOM restrictions. He claimed that "50% of new products; from children's toys to toasters to satellite technology, were covered by US trade limitations."[99] In the early to mid-1980s, the CIA reported that COCOM restrictions had become a source of tension between the US and its European allies.[100] In this context, why would the Europeans believe that the US would share its most sensitive SDI-related technologies without substantial restrictions?

UNITED OR DIVIDED?

Rather than leading to greater transatlantic cohesion concerning SDI, the invitation to participate in the project made it an even more controversial issue. Weinberger's SDI invitation letter loomed large at the thirty-seventh NPG on March 26 and 27, 1985. In 1966, NATO had established the NPG, which gave nonnuclear members of the alliance a greater stake in nuclear issues.[101] European officials shared the conviction that any involvement in SDI should give them a say in decisions about the program's future. West German Defense Minister Manfred Wörner demanded that any European industrial involvement in SDI must be complemented by more in-depth politico-military dialogue.[102] Wörner's comments reflected broader European anxiety about being blindsided by the US yet again on strategic matters. This NPG meeting underscored the lack of allied unity on SDI at an especially tense period in NATO's history; SDI only added fuel to the fire as the anti-nuclear movements across Europe intensified their campaigns against the alliance's nuclear modernization.[103] NATO Secretary General Peter Carrington tried to persuade the press that the NPG "meeting had been harmonious and fruitful."[104] But SDI remained a significant point of tension in large part because officials were forced to consider the political implications of a program whose technological maturity was a distant prospect while attempting to manage the immediate backlash to the very *idea* of strategic defense.

On the sidelines of the NPG, Heseltine and Wörner discussed the various options for responding to Weinberger's invitation letter. At this stage, the German defense minister favored a "concerted European effort."[105] Carrington and Italian Minister of Defense Giovanni Spadolini similarly believed that there should be a joint European response to Weinberger. In a memorandum to the prime minister, Heseltine advocated a "shared European approach" to SDI. He explained that this would "counter pressure on individual countries to be sucked in to support for the SDI going beyond . . . [the] four points [at Camp David] because of the lure of . . . technologies of the future." Additionally, a joint response would ensure that the European powers "could share the benefits of their collaboration." Heseltine identified CERN, the European Organization for Nuclear Research, as a potential model for joint European contributions.[106]

Charles Powell quickly intervened and attacked Heseltine's pan-European approach to SDI, writing to Thatcher: "Perhaps I am too suspicious. But it seems to me that *one reason* why the Defense Secretary proposes a *joint European response* on SDI research is that he wants [to] build up a body of opinion *skeptical of SDI*." There is no good industrial or scientific reason for a *joint response*: so it *must be political*" (emphasis in original).[107] Not only was Heseltine the most vocal European integrationist in the Cabinet, Powell was also alarmed by the fact that Heseltine and Weinberger were "fervently at loggerheads" and "there [was] no love lost between them."[108] Heseltine's views on European integration and his difficult relationship with Weinberger did not bode well for Anglo-American cooperation on SDI.

Although it did not take much convincing, Thatcher firmly sided with Powell, concluding that Britain "would be more likely to lose from [a collective European approach to SDI]" and that the French would be unlikely to share scientific knowledge or technology in areas where they were ahead.[109] From her perspective, the only way forward was a strictly bilateral arrangement with the US. Underscoring the sensitivities attached to SDI and Anglo-American relations, Thatcher made it clear that she was "running SDI policy directly."[110]

Mitterrand complicated European deliberations over SDI when, in April 1985, he established the European Research Coordination Agency, "Eureka" for short. Officially, Eureka would pool European resources, with the ultimate goal of closing the science and technology gap with the US. Although the French president claimed to have been contemplating Eureka for some time, the CIA argued that the "timing and hasty packaging of the initiative were almost certainly prompted by the US invitation to its allies to participate in SDI research."[111] Robin Nicholson, Thatcher's science advisor, was convinced that the "US proposals for SDI [had] sparked French fears about the civil spin-off within the US which might give the US an unassailable lead in certain technologies," ultimately leading to Eureka.[112] Soviet officials similarly concluded that Eureka was "a technological reply to the American [SDI] challenge" and might "eventually become the European version of the American 'Star Wars' project."[113] With Eureka, Mitterrand built on the Gaullist tradition of seeking greater independence in science, technology, and military matters.[114] Yet, he did not develop an effective Eureka marketing strategy to capture the public imagination or generate significant enthusiasm among his European colleagues.[115]

Problematically, Eureka did not offer European governments an avenue to exert greater influence over SDI. Quite the contrary, eschewing involvement in SDI in favor of Eureka would have only exacerbated tensions with the US concerning Reagan's project. Even if officials in capitals across Europe did not support SDI, they feared distancing themselves to the point that they eliminated any opportunities to shape US decisions regarding the future of the program and to benefit from it technologically. There was also little guarantee that a pan-European high-technology cooperative would be free of technology transfer issues. In considering barriers to more seamless technology cooperation in Europe, Robin Nicholson identified "France [as] the worst culprit."[116]

Like Thatcher, West German Chancellor Helmut Kohl concluded that West Germany needed to be able to influence US policy on strategic defense while also benefiting from any technological "spin offs" from SDI research. In early 1985, Kohl broke almost two years of official silence on SDI and expressed interest in becoming involved. The UK embassy in Bonn reported that the "German line on SDI [was] becoming more favorable, Kohl and Wörner [were] speaking publicly of European participation in the SDI research program, on certain conditions."[117] The British believed that anxieties about "Europe missing out on technological leap forward" was a significant factor.[118] Edoardo Andreoni correctly observes that the German chancellor's public warming to SDI "did not reflect a consensus within the Federal Government nor put an end to the contradictions of West German policy towards Reagan's initiative."[119] Members of the Christian Democratic Union (CDU), in particular, worried that a shift toward deterrence based on defensive systems entailed the erosion of the American nuclear guarantee.[120] Kohl's CDU was generally more supportive of involvement, whereas members of the Social Democratic Party (SPD) were completely opposed to SDI from the very beginning.[121]

Like Heseltine, West German Vice Chancellor Hans-Dietrich Genscher preferred a common European response to SDI and wanted to ensure that collaboration would not be a "technological one-way street"; Bonn had been especially disappointed with technological cooperation with the US on projects such as Spacelab and the space tug, and was therefore skeptical about US promises to share sensitive technologies.[122] In April 1985, the *Bundesverband der Deutschen Industrie* (Federation of German Industry) organized a conference in Bad Isenberg in which representatives of SDIO briefed chief executives of major German companies on the kinds of contracts that would be available to West German firms.[123] American engineers and scientists within SDIO believed that West German companies were especially well positioned to contribute to SDI in key high-technology arenas such as optics and microwave sensors.[124] West German firms certainly welcomed SDIO's interest in their expertise.

In Bonn's deliberations over SDI, there were military factors at play as well. In April 1985, Franz-Josef Strauss and Alfred Dregger, both prominent conservative West German politicians, proposed a European Defense Initiative (EDI) as a complement to SDI.[125] EDI was intended to be a ground-based missile defense designed to safeguard Western Europe against "conventional Soviet missile capabilities" not covered by SDI.[126] The EDI concept received mixed reactions from European politicians. The Italians, for example, had already shown interest in theater missile and air defense.[127] Unsurprisingly, the SPD opposed it on the grounds that it would be "tantamount to accepting SDI."[128] Technologies developed for a pan-European missile defense system would have a broad range of military and civil aerospace applications. Although a ground-based system, EDI would still need space-based remote sensing platforms for acquiring data on incoming missiles and aircraft. Command and control infrastructure for such a sophisticated system could also be used for civil space projects, once again underlining the close linkages between civil and military space policy.[129]

French Defense Minister Charles Hernu did not want to limit Europe to a system that functioned, in effect, as an adjunct to American space-based defenses. He therefore endorsed "European countries [getting] together to see if there is the possibility at the government level to launch a European research initiative in the domain of space-based defense" and that France should be willing to do it alone if other European partners did not want to get involved.[130] Hernu established an office in the MoD to examine space policy, saying candidly that "we need to prepare our presence in space."[131] Jen-Luc Lagardere, the president of Mécanique Aviation Traction (Matra), a conglomerate of industrial enterprises, declared that "without a military program in space, neither France nor Europe can expect a seat in the front row."[132] The question that European politicians had to answer was whether they would attach themselves to SDI, a common European program, or a middle ground between the two. Although French officials tried to deflect comments about Eureka's military implications, Hernu admitted that French officials viewed it as "a common technological trunk which feeds both the civilian and military domains."[133]

Hernu's proposal to explore European space-based defenses further was not a one-off comment but rather reflected Mitterrand's conviction that France needed to keep pace with advanced technological developments. Even though the French government had been vociferously advocating space arms control, Mitterrand alluded to developing space capabilities that went beyond "passive" functions such as intelligence and communications. He said, "Let me give an example: that of the conquest of space. Europe should be capable of launching a manned space station that would enable it to observe, transmit and therefore counter all eventual threats. That way, it will have made a great step forward in its own defense."[134] Being able to "counter all eventual threats" more than implied some kind of system with weapons, but he did not clarify what specifically he was talking about. French policy was of course opposed to any changes to the 1972 ABM Treaty, and there was serious concern about what the implications of a superpower missile defense competition might be for the French nuclear deterrent.[135] Mitterrand's statement pointed toward French anxieties that if space-based missile defense technologies were going to take on greater importance, Western Europe—France in particular—could not be left behind.

Mitterrand's space station idea was also a response to Reagan's 1984 announcement during his State of the Union address that the US would offer allies the opportunity to participate in a civilian space station project. France was, however, wary of becoming involved, seeing it as yet another avenue for increasing European technological dependence on the US.[136] With France as the leader of the space industry in Europe, Mitterrand preferred that ESA focus on an independent West European space program.[137] Western European officials viewed the space station as a litmus test for American willingness to share advanced technologies across the board. CIA analysts candidly observed that "failure to reach agreement on the technology [transfer] issue for a civil program like the space station almost certainly would jeopardize Western European willingness to participate in SDI development."[138]

ESA wanted broader technology exchange than in past joint projects, such as Spacelab, seeking the status of equal partner rather than that of a

subcontractor.[139] Although restrictive technology transfer only seemed to justify Mitterrand's call for independent European space projects, the technology transfer difficulties experienced by ESA likely further convinced Bonn
and London, perhaps Rome too, that they would have a better chance of
securing favorable terms on SDI participation on a bilateral basis rather than
through a multiparty cooperative framework. Moreover, space projects with
defense applications aroused controversy in the European space community
that traditionally eschewed overt ties with military projects.[140]

At the Bonn Economic Summit in May 1985, Thatcher informed Reagan
that the UK wanted to get involved in SDI and that she hoped "we shall do
it in the same way as we have collaborated since we worked together on the
atomic bomb." Participation had to be a two-way street, she emphasized,
and the British wanted to be "partners not subcontractors." She further highlighted the need for a bilateral arrangement rather than a common European
approach.[141] In her discussions with Mitterrand at the summit, she avoided
both Eureka and SDI. Thatcher already knew that the French were not going
to pursue a government-to-government agreement with the US on participation, and Mitterrand's silence on this issue staved off a potentially difficult
conversation.

NEGOTIATING THE TERMS OF INVOLVEMENT IN SDI

The industrial and alliance considerations were significant factors in West
Germany's decision to participate in SDI research. Senior officials in Bonn,
such as Horst Teltschik, were skeptical that SDI would produce a functioning
missile defense system that would ever be able to replace nuclear deterrence
as the mainstay of NATO's strategy, but in 1985 and 1986, there was still
hope that technological cooperation could be beneficial for West German
industrial and political agendas.[142] The SPD expressed concern that West German scientists who participated in SDI research would lose their "scientific
autonomy" due to the security classification of their work.[143] Taking a politically pragmatic stance, other West German officials primarily saw involvement as a way to secure greater influence over Washington's policies on SDI.

Despite Soviet protests to Bonn about West Germany directly participating
in SDI, Kohl remained outwardly supportive of SDI research, and he wanted
to move ahead with securing industrial contracts.[144] Genscher was steadfast in
his position that becoming directly involved in SDI was bad for West German

interests and European security more broadly. He was also skeptical about the level of technology transfer that the US would indeed be willing to facilitate. Abrahamson met with representatives of the West German aerospace firm Messerschmitt-Bölkow-Blohm and introduced them to a three-tier model for Bonn's participation: tier 1, the US would finance research into core functions such as boost-phase, midcourse, and terminal defense; tier 2, research into defenses against long- and short-range theater nuclear forces would be co-financed with European allies; and tier 3, there would be development of air defenses and countermeasures for cruise missiles, which was of particular interest to Bonn.[145] As a sign of his commitment to German industrial involvement in SDI, Kohl sent one of his closest advisors, Horst Teltschik, to Washington to discuss the terms of formal participation in SDI.[146]

At this time, Thatcher was growing frustrated with the fact that West Germany was moving ahead with involvement in SDI research, while the UK MoD seemed unconcerned. In May 1985, Heseltine reported that Britain needed to "move quickly" if it wanted to make the most of its opportunities. Concurrently, he cautioned that "we do not as yet have sufficient information to assess the full implications of participation or to define the specific areas of our interest."[147] At the top of this memorandum, Thatcher wrote in pen that "the Germans have gotten farther than we have, this won't do."[148] The prime minister placed a political premium on being the first ally to secure a formal participation agreement, but Heseltine was not eager to strike a deal with the US on SDI.

For Britain, the political and industrial aspects of SDI were inseparable, and time was of the essence. Officials in Whitehall were still debating whether there should be a pan-European approach to SDI. UK Minister of State for Industry and Information Technology Geoffrey Pattie complained that the MoD was moving too slowly, and he identified Bonn as "far more active and supportive."[149] Since SDI was a long-term multi-billion-dollar research program into many types of advanced technologies, it did not need to come to fruition to be beneficial. Thatcher's chief scientific advisor, Robin Nicholson, argued that "whether or not SDI succeeds in its strategic aims, the very large US spend [sic] will produce technical advances in areas of importance to conventional defense and to civil industry."[150] Nicholson advocated a strictly bilateral arrangement with the US. He wrote in a memorandum to the prime minister, "We have a unique and hard won position of being the only country with a respected and trusted position on defense science and technology with both the US and Europe. We should exploit this position ruthlessly."[151] Nicholson believed that Heseltine's advocacy of a European approach was "feeble." Nicholas Owen, a policy advisor to Thatcher, maintained that because Britain already had unique and extensive special arrangements with the US in nuclear and intelligence technologies, the UK was especially well placed to benefit. A multilateral approach, he warned, could "make it harder for the USA to share technology." He similarly argued that Heseltine's "European approach is misconceived" and that British "participation is worth a high price and the Americans expect to pay one."[152]

Charles Powell continued to warn against Heseltine's European approach. He advised that the task was to "make up for lost time," since both West Germany and the Netherlands had already sent delegations of government officials and businessmen to discuss participating in SDI research.[153] Thatcher ultimately adopted Powell's recommendation that SDI and Eureka be handled as two separate issues. The former was more significant for UK security interests, and the latter would involve Britain but be less of a priority and take up fewer resources. Even with this bilateral framework in mind, London still needed to formalize the terms of its involvement. Powell doubted that the MoD was best equipped to handle negotiations with the US over SDI. Most problematic was the fact that Weinberger and Heseltine "cordially dislike[d] each other and tend[ed] to quarrel."[154]

Since SDI was so "close to the president's heart," SDI negotiations were about far more than intellectual property protections and contract amounts, they also had a direct bearing on the health of the Anglo-American partnership.[155] Thatcher understood that the president was wed to his nuclear abolitionist vision for strategic defense. It was in the British interest to support SDI research without encouraging Reagan's goal of a nuclear-free world that the prime minister thought was neither desirable nor possible. Participation in SDI would, Thatcher hoped, give Britain access to advanced technologies and also provide her with an opportunity to influence Reagan's decisions about SDI more directly. Heseltine's opposition to SDI and his personal issues with Weinberger were significant political liabilities. Consequently, Thatcher had to carefully manage dissenting opinions on SDI within Whitehall to see through to completion her agenda for British involvement in the program.

Heseltine met with Weinberger in late July 1985 to discuss UK involvement in SDI, including the creation of an SDI Participation Office in London.[156] This new institution would be a visible symbol of Britain's choice to embrace a bilateral agreement with the US and eschew a coordinated European approach to SDI. Of special importance to the MoD was that the UK could secure access to detailed information about SDI programs so that British companies could more effectively compete with American firms for contracts. The UK defense secretary explained how Britain would have to expend significant political capital to combat "vocal left-wing opposition" to SDI, which he believed justified Britain receiving a substantial percentage of SDI contracts.[157] Heseltine specifically proposed that the US set aside $1.5 billion worth of contracts for Britain over a five-year period. Notably, a few days prior to this meeting, Thatcher voiced concern about demanding a specific sum of money from the US.[158] Weinberger explained that due to US contracting laws, setting aside a specific percentage of the SDI budget for the UK would not be possible. The British defense secretary brushed these legalities aside and stressed that there had to be an equitable partnership with the UK due in large part to the political sensitivities of involvement.[159]

Powell suspected that Heseltine proposed such a large sum of money as a "wrecking bid." Historian of science Jon Agar agrees that Heseltine "probably was indeed offering a wrecking bid."[160] Heseltine, like Foreign Secretary Geoffrey Howe, had significant reservations about SDI. He was also unhappy with the prime minister's bilateral approach to SDI, which he believed was detrimental to the UK relationship with continental European powers. Regardless of Heseltine's intentions, Thatcher decided to push for a substantial portion of SDI's foreign research and development budget. The prime minister told Weinberger that Britain's "record as an ally, the history of our scientific and technical excellence entitled us to special consideration." Weinberger tried to explain yet again that the Reagan administration could not just set aside a specific percentage of the SDI budget due to US contracting laws. The prime minister was not, however, interested in these legal and procedural obstacles. Thatcher scoffed at Weinberger's offer of £10–£15 million in contracts as an "act of good faith," saying she was not keen on a "salami approach." It was incumbent on the White House to get Congress to see that "Britain was in a different category to other countries" and appropriate funds accordingly.[161] Later that same day, she reiterated to Vice President Bush that Britain "wanted to be full and worthy partners, not just small component makers."[162]

To make more substantive progress on securing concrete terms for Britain's integration into SDI research, the US and the UK established an Anglo-American working group devoted to this topic.[163] The MoD identified eighteen technical areas in which the UK was well placed to contribute to SDI, including optical computing, software security, lasers, signal processing, and materials science.[164] The MoD concluded that "absorption of additional sums for SDI work should not present industry with insurmountable difficulties."[165] The Department of Trade and Industry was, however, pessimistic about the ability of UK industry to exploit SDI technology. There was also the fundamental question of whether Britain had adequate resources to devote to SDI.[166]

Questions about such a large-scale British contribution to SDI were quickly overshadowed by more fundamental issues in the negotiations that the MoD described as "tough but cordial." The latter description was really an understatement. The MoD noted that the American participants in the Anglo-American working group were "clearly under instructions to define the smallest acceptable British contribution covering technologies of greatest interest to the US, on terms favorable to the US and with no commitment beyond the next year or so." The US did not want to make firm guarantees, in large part because Washington needed London's political support, while British technological contributions were primarily of peripheral concern.[167] US officials wanted to get the UK onboard as quickly as possible and worry about the technical arrangements at a later time, leaning on "existing bilateral agreements."[168] The key stumbling blocks for the MoD related to intellectual property rights, US technology transfer controls, and timely access to contract data.[169] The information exchange was particularly important so that Thatcher and her advisors had access to detailed and up-to-date information on where the US was heading with its SDI concept.

In the midst of these Anglo-American SDI negotiations, the French approached the British about collaboration on military space projects, satellite communications in particular. The French proposed that British Aerospace and Matra work together to develop a satellite based on Skynet 4. London was not, however, receptive. British officials concluded that the French strategy might be to develop a new satellite system "on the back of British technology," putting the UK at a disadvantage in the global satellite market. Most importantly, British defense leaders believed that closer collaboration with France might jeopardize "the Anglo–US space relationship, particularly at this sensitive stage of SDI discussions."[170]

The American–Soviet summit at Geneva scheduled for late November 1985 added urgency to the Anglo-American SDI negotiations. British officials noted that the US wanted to "conclude an agreement quickly . . . probably because they want our visible endorsement of SDI research before the United States–Soviet Summit on 19–20 November."[171] UK negotiators viewed Washington's desire to conclude an agreement quickly as leverage for securing favorable technology transfer arrangements, allowing Britain to use technologies generated through its involvement for other applications.

As the negotiations progressed, Thatcher and her advisors needed to define the priorities for UK involvement more specifically. The prime minister had to determine if she wanted to extract benefits primarily for British defense, for commercial interests, or for a mixture of both. Nicholas Owen, one of her policy advisors, advocated a complete focus on defense interests. He argued that "SDI is not important for civil work" and that "sending a man to the moon was a round-about way of designing a non-stick frying pan."[172] Owen was furthermore concerned about whether British industry could really deliver on $1.5 billion worth of SDI-related work over the following five years. If the UK could not meet its commitment, then delays in SDI research and development due to British firms might undermine American confidence in UK technical expertise.

As the negotiations entered their terminus, there were concerns on the part of British industrial leaders that they would not receive their fair share of SDI contracts. During the deliberations over the Trident submarine-launched ballistic system, UK trade union officials and defense contractors had come to believe that the government had not been forceful enough in protecting British interests, and they did not want to see the same mistake made again regarding SDI. Thatcher's government had to balance the desire to secure favorable terms with the need to conclude a memorandum of understanding (MOU) with Washington as quickly as possible because the MoD was asking British firms not to seek SDI contracts until a government-to-government agreement was finalized. The prime minister and her advisors believed that this was the only way to ensure British companies received "the best possible terms."[173] There were also sensitive arms control issues at play; the FCO and MoD continued to worry about the crucial distinctions between research and development and feared that any move toward the latter could undermine the ABM Treaty.[174]

In late October 1985, Heseltine informed the prime minister that the US had met the outstanding demands of the UK that stood in the way of moving forward with participation.[175] Weinberger had agreed to conditions that made it possible for British firms to compete for contracts effectively and to strengthen provisions for technology transfer. The US requested that the MOU be classified secret, which Heseltine believed reflected the degree to which the "American side feel they have moved to meet our concerns."[176] In reality, the security classification served the more practical purpose of allowing the Reagan administration to conceal from other potential SDI participants what kind of an agreement the UK was able to obtain. London and Washington modeled the SDI information exchange on the 1958 US–UK Mutual Defense Agreement.[177] Powell observed that the agreement was not "watertight . . . [but] is generally satisfactory."[178] Thatcher recognized that the agreement had few guarantees and acknowledged that "the scale of UK participation in practice is bound to remain uncertain under what is essentially a permissive arrangement."[179] Despite concerns about intellectual property rights and technology transfer, Thatcher pressed ahead with the MOU.

On December 6, 1985, the US and the UK signed the MOU.[180] For the US, the MOU was a political symbol rather than a substantive agreement for technological cooperation. While officially the UK supported only SDI research, not deployment of strategic defenses, Labour politicians alleged that Thatcher was endorsing "the principle and strategy behind Star Wars," which they identified as a quest for nuclear superiority.[181] The Soviet Union issued a demarche protesting the Anglo-American MOU, saying that it represented Britain "helping the US achieve military superiority" and that research would lead to development and testing.[182]

The Anglo-American MOU served as an added impetus for Bonn to establish a similar arrangement with the US regarding West German participation in SDI research. Official negotiations between Washington and Bonn only began around the time of the signing of the Anglo-American agreement. West German leaders wanted to obtain technological benefits from collaboration with the US on SDI but continued to worry about highly restrictive US policies on technology transfer.[183] Kohl also recognized that becoming involved in SDI research and development was politically sensitive in terms of both the German electorate and relations with the Soviet Union and other European states, especially France.[184] Even support for SDI research, but not deployment, created challenges for Bonn's *Ostpolitik*. The SPD criticized Kohl for "putting the relationship with France at risk over SDI" and presented Kohl as having an excessively Atlanticist orientation.[185]

The American–German negotiations had become difficult because Bonn wanted guarantees that the US would refrain from adopting a protectionist attitude toward SDI technologies. Despite concerns about technology being withheld by the US, West German Economics Minister Martin Bangemann and Weinberger signed the MOU in late March 1986.[186] The signing of the MOU resolved an interdepartmental scuffle between the German Ministries of Defense and Foreign Affairs; the former was supportive, and the latter was opposed. According to Wörner, "in the contest between the Foreign and Defense Ministries, [Foreign Minister] Genscher was the 'loser.'"[187]

Even after signing the MOU, West German officials remained skeptical about obtaining technological benefits. Wörner observed in a meeting with the French foreign and defense ministers that "the United States will surely reserve for itself the lion's share" of the SDI contracts.[188] Andreoni observes that having Bangemann, rather than Defense Minister Wörner, sign the MOU was "another sign that Bonn regarded its involvement in SDI research as a purely industrial and legal matter."[189] It would be more accurate to say that West German officials wanted it to *appear* that they were handling SDI as a primarily economic matter. In September 1986, Wörner informed Heseltine that Bonn was "encouraging firms to develop their technological links with the US in the SDI area, with a view not only to commercial objectives but to developing technologies which could be used for the improvement of conventional defense."[190] Wörner was especially interested in using SDI technologies for European air defense, harkening back to concepts associated with EDI.

The British were keen to know the terms of Bonn's MOU with the US. Although the terms of the American–German agreement were not public, a copy of the MOU was leaked to the press a month after it was signed. The West German federal prosecutor launched an investigation, but no one was formally charged.[191] This leak was fortuitous for Britain because it allowed a comparison of the two MOUs. UK officials "noted that in several important respects regarding ownership, use, disclosure and licensing of foreground rights the UK had secured more advantageous arrangements with the United States than those achieved by West Germany."[192] In reality, there was little guarantee in the British MOU that UK firms would indeed be able to use information derived from involvement in SDI projects.[193] Most significantly, a comparison of the two MOUs underlines the fact that although Britain did have unique intelligence and nuclear cooperative arrangements with the US, it had a limited ability to extract special benefits with regard to technology transfer in other areas. Even with the MOU, British industrial firms suspected that "a ring fence might be erected around SDI work, thus preventing possible exploitation of SDI technology through [other programs]."[194]

While negotiations were underway on West German SDI contracts, a terrorist attack reminded officials in Bonn of the political sensitivities attached to participation in the program. In early July 1986, a bomb detonated on the side of a road in Munich and demolished a limousine ferrying West German industrialist Karl Heinz Beckurt, a high-ranking executive with Siemens. The blast blew the car more than twenty feet from the road, killing both Beckurt and his chauffeur. Not long after this incident, the Red Army Faction, a West German terrorist organization, claimed responsibility and explained in a letter that Beckurt was targeted because Siemens was securing contracts with SDIO. The terrorists who orchestrated this attack hoped it would garner the support of the militant wing of the anti-nuclear movement in West Germany.[195] Although this incident did not affect Bonn's SDI policy, West German firms that secured SDI research contracts subsequently tried to keep a lower public profile regarding their partnership with SDIO.[196]

Discussions among West German, French, and British officials concerning involvement in SDI became increasingly superficial as they competed with each other for contracts, although Mitterrand still refused to negotiate a formal MOU with the US.[197] In early 1986, Mitterrand's approach to SDI became a significant issue in French domestic politics. March 1986 witnessed the coming to power of a coalition, including the Gaullist Rassemblement pour la République (RPR) and the center-right Union pour la Démocratie Française (UDF). This led to the appointment of Jacques Chirac, the head of the RPR, as prime minister. Chirac had been very critical of Mitterrand's outright rejection of participation in SDI, saying that such a course of action was not in the French national interest.[198] He pledged to correct this situation, explaining to a group of journalists that "France cannot refuse to associate herself to this great research called SDI . . . I will not let France remain on the margins of this great, unavoidable movement, which is irreversible and justified."[199] French firms competed for SDI contracts, but Paris still rejected an MOU with the US formalizing French involvement in the program. While Chirac was enthusiastic about potential technological benefits from participation, he was concerned about Reagan's anti-nuclear rhetoric, especially at the Reykjavik Summit in the fall of 1986.

Mitterrand's admonition that Europe must have autonomy in space took on even greater significance in 1986 after the space shuttle *Challenger* blew up shortly after launch, killing everyone onboard. This tragedy naturally affected the scheduled launchings of foreign payloads on the shuttle. With constrained American launch systems, European officials could not ignore the great demands that SDI components would place on the shuttle program, further marginalizing the space needs of Europe. In light of this new situation, Britain shifted its Skynet 4B satellite to Ariane, becoming the first shuttle customer to change over to its European competitor.[200]

The prospect of obtaining advanced technologies continued to attract foreign participation in SDI. In 1986, Israel and Italy became formally involved, followed by Japan in 1987.[201] MOUs served as political symbols more than anything else because they did not guarantee contracts or transfer of technology. A Department of Defense report to Congress on SDI stated that "an MOU is helpful, [but] it is not mandatory for participation. Companies in countries that have not signed an MOU have successfully competed for contracts."[202] Belgium, Canada, Denmark, France, and the Netherlands, all non-MOU countries, secured SDIO contracts of varying amounts:[203]

To the disappointment of Western European states, SDI did not materialize into the much hyped "technological Marshall Plan."[204] By the end of the 1980s, Britain had not obtained anything close to the original $1.5 billion that Michael Heseltine had requested in 1985. And British officials were not alone in their disappointment concerning SDI contracts. In general, European firms learned that SDI contracts would not be nearly as lucrative as they had originally hoped. Even though US officials had pledged that SDI-related technologies would have significant civil applications, many of the foreign contracts were in areas that could not be easily transferred to the commercial sector. Teltschik maintains that "West Germany did not really benefit industrially from participating [in SDI]."[205] By October 1989, Bonn had won approximately $70 million in SDI contracts—just $3 million less than Britain.[206] This situation once again underlines the fact that Britain's unique access to American defense and intelligence resources did not guarantee special status with regard to technology transfer. At nearly $185 million, Israel secured the largest number of contracts (in terms of monetary value), primarily due to its development of the Arrow missile defense system with American assistance.[207]

CONCLUSIONS

American military space activities in the 1980s, SDI especially, forced Western European politicians to directly confront the highly controversial subject of space militarization. When the motivations for Western European involvement in SDI are uncovered, we find that it was not so easy to disentangle the economic and military elements in Western European space policy. European debates over space militarization encompassed a broad range of economic, military, and political factors inextricably linked to European integration and transatlantic alliance dynamics. Rather than producing a wholly new problem in transatlantic affairs, SDI became quickly entangled in preexisting European discussions about autonomy in space and advanced technologies more broadly.

Perhaps most remarkable is that even with many prior difficult technology transfer experiences on joint projects with the US, Western European officials still held out hope that cooperation on SDI would somehow be different. They believed that Washington's political motivation to secure broader international support for SDI might indeed result in less restrictive policies on technology sharing. This situation created competition among European states for SDI contracts, further eroding the possibility of a common European position. On a practical level, Western European politicians and industrialists underestimated the difficulties of competing for SDI contracts alongside American firms that had long-established relationships with the Pentagon.

Most fundamentally, Western Europe's involvement in SDI forces us to reconsider its role in space militarization in the late Cold War. Tilmann Siebeneichner writes that through joint American–European projects that involved the shuttle, a dual-use system, Western Europeans contributed to "covert militarization" of the heavens.[208] In fact, Western European states *overtly* participated in space militarization. In the late 1970s and through the 1980s, France and Britain pursued military space projects dedicated to reconnaissance and communications. Notably, some of the same French officials who were advocates of space arms control later openly entertained a European space-based missile defense. The fear of being left behind technologically was clearly far more powerful than anxieties about a space arms race. Even though many European leaders distinguished between participating in

SDI research and supporting the *deployment* of space-based missile defenses, their countries' involvement in projects that could have been used to place weapons in space makes this distinction insignificant in a practical sense. To the disappointment of American officials, nothing close to a consensus in favor of space-based missile defense emerged among Western European allies, and SDI would remain a source of controversy in transatlantic relations well into the early 1990s.

5 OUT OF THE LABORATORY AND INTO SPACE

On the morning of September 5, 1986, a rocket roared off its pad at Cape Canaveral, Florida, carrying an SDI experiment into space. This project, called "Delta 180," successfully simulated the interception of a ballistic missile outside of the Earth's atmosphere. The Delta 180 test only reinforced Reagan's belief that SDI was indeed making significant progress. Delta 180's timing was critical; it took place only weeks before the 1986 superpower summit in Reykjavik, Iceland, where Ronald Reagan and Mikhail Gorbachev discussed eliminating all US and Soviet nuclear weapons. At Reykjavik, Reagan and Gorbachev found themselves at an impasse due to the Soviet demand that SDI be confined to the laboratory, which would have prohibited the testing of missile defense components in space. Reagan rejected Moscow's condition that SDI be kept in the laboratory, saying that this was tantamount to throwing it onto "the trash heap."[1]

Reagan and Gorbachev both wanted a nuclear-free world, but disagreement over SDI was a critical obstacle in their way. According to former US ambassador to the Soviet Union Jack Matlock, restriction of SDI to the laboratory "became in Reagan's mind nothing more than a backdoor way to destroy his dream."[2] But Reagan's intransigence concerning SDI during arms control negotiations was driven by more than his pursuit of an elusive dream; it was also shaped by the specific technological requirements of strategic defense system builders. In particular, Reagan wanted to keep open the option to deploy a strategic defense system in the near term using less-exotic technologies, such as kinetic interceptors on land and in space. In time, laser weapons could be added if the relevant technologies matured. Consequently, Reagan would not entertain *any* limits on technologies that might be relevant for strategic defenses. Even if a strategic defense system was a distant prospect, SDIO would need to integrate complex strategic defense technologies and test them in space in the near term. But even testing SDI technologies in space posed problems for the 1972 ABM Treaty, and preserving this treaty was an article of faith for both the Soviet Union and US allies in Europe. Fundamentally, the advancement of SDI required freedom of action in space.

Officials in Moscow believed that allowing the testing of missile defense components in space was tantamount to opening a Pandora's box of offensive military space capabilities. SDI's advanced software, sensors, and command and control systems could give the US military an even greater qualitative lead as well. Due to these considerations, Gorbachev concluded that he had to forestall a new phase in space militarization. Space arms control would prevent or retard, at the very least, Washington's ability to test and deploy military space systems with both ASAT and missile defense applications. For the Soviets, new military space technologies were the greatest threats to arms control, but for Reagan, they were the keys to a more peaceful future. Just as the Reagan administration used the prospect of high-technology cooperation to reduce European hostility to SDI, the president hoped that promising to "share the benefits" of SDI with Moscow would cause Gorbachev to acquiesce on strategic defense.[3]

"A SYSTEM OF SYSTEMS"

In mid to late 1984, the White House still did not know what specific form a strategic defense system might take, and work was proceeding on all relevant technologies.[4] While Abrahamson did not want to establish specific effectiveness criteria for a system, he maintained that "the technical approach being taken was structured by the President's objective of total effectiveness against BMs [ballistic missiles]." Such an expansive goal required a "system of systems": that is, multiple interlinked capabilities (e.g., space-, terrestrial-, and sea-based interceptors that were connected to tracking sensors on the ground and in space).[5] SDIO adopted a building-blocks approach in which a strategic defense system would be deployed in phases, with more exotic technologies, such as laser weapons, being added to much later iterations.

Developing the sensors necessary to track and target ballistic missiles was the costliest part of the SDI program, comprising about 35 percent of the overall budget.[6] SDIO also devoted significant attention to "space logistics": that is, new space launch vehicles that would lower the cost of placing systems into orbit. Abrahamson and his staff needed a heavy-lift vehicle for launching space-based platforms of up to 100 metric tons (220,000 pounds), along with a capability to service them in space.[7] Because of the energy requirements for all these space systems, new space-based power sources would be required too. SDIO only had a dedicated staff of about 100 military and civilian personnel. So, it depended on the US national laboratories, the military services, and partnerships with NASA for space logistics and on the Department of Energy for space power.[8] Although small in size, SDIO was primed to become a resource intensive space institution.

"SDI–ASAT ENTANGLEMENT"

As the Reagan administration prepared for the new round of arms control negotiations set to begin in March 1985, protecting SDI was at the forefront of the White House's objectives. While the "SDI Bible" that guided the US government's public diplomacy on strategic defense maintained that "SDI is not based on any single or preconceived notion of what an effective defense system would look like," administration officials recognized that space-based interceptors would indeed be essential for destroying ballistic missiles in their boost phase. Problematically for the White House, these same interceptors used for missile defense could also be used as ASATs.

Secretary of State George Shultz believed that the US "needed something to trade" to get the arms control negotiations moving and that ASATs were a good place to start. Kenneth Adelman, the head of the ACDA, proposed that the US and the Soviet Union, at the very least, establish "rules of behavior" in space, which could involve limits on ASAT testing.[9] The Pentagon carefully considered how any arms control measures could affect its military space plans. Because of the technological linkage between ASATs and SDI, Undersecretary of Defense for Policy Fred Iklé warned Caspar Weinberger in 1984 that "ASAT–SDI entanglement" was going to be a significant problem in the arms control negotiations.[10] Weinberger stressed to Robert McFarlane that "we have attempted to separate ASAT and SDI in public and Congressional fora, [but] it is clear that they are linked, both technically and from the perspective of arms control options . . . the distinctions between ASAT and SDI will not be capable of being maintained."[11] The Soviet Union, Western

European allies, and many SDI skeptics in Congress grasped the ASAT–SDI relationship and therefore sought to use ASAT arms control as a "stalking horse" to kill SDI.[12]

On the subject of US ASATs, Western European allies ranged from luke-warm to openly hostile.[13] McFarlane worried that Mitterrand's and Thatcher's public opposition to an arms race in space would generate even greater opposition to US military space plans.[14] In general, European allies supported "strict limits on dedicated ASAT systems" because of NATO's dependence on satellites for communications and reconnaissance. France, the UK, West Germany, and Italy favored a preemptive ban on high-altitude ASATs and believed that a space "rules of the road" agreement would indeed be useful.[15]

A ban on high-altitude ASATs seemed sensible, since they threatened satellites that were used for nuclear command and control, but Weinberger pointed out to McFarlane that SDIO was developing interceptors that would "have the inherent capability to attack high-altitude satellites." The secretary of defense also envisioned having ASATs to defend high-altitude satellites used for missile tracking and intelligence gathering.[16] In 1988, Moscow did indeed initiate research into a more advanced ASAT system designed to attack higher-altitude targets.[17] Since the Soviets were working on high-altitude intelligence satellites that could support military operations, having the ability to attack them in wartime would be useful from the Pentagon's perspective as well.[18]

In light of potential SDI system requirements and the Pentagon's military space agenda, Iklé and Weinberger conspired to "slow down the train speeding up toward extensive SDI limitations."[19] Weinberger turned his energies toward defeating Shultz's space arms control strategy by getting McFarlane, Casey, and the president on his side. The secretary of defense concluded that laying out for Reagan how space arms control would prevent SDI from advancing would convince the president to reject Shultz's proposal to negotiate with the Soviet Union on limiting ASATs. The SDI–ASAT linkage aside, Weinberger knew that Reagan harbored fears that the Soviets were "ahead of us in that dept. [military space capabilities]" and "want to freeze us into inferiority."[20]

Even though SDIO had not yet settled on what specific form a strategic defense system might take, there was clearly an understanding that futuristic technologies such as laser weapons might not mature quickly enough for near-term missile defense options. ASAT-related technologies (e.g., hit-to-kill interceptors) held the greatest promise for deploying a strategic defense system by the mid-1990s. Weinberger warned that limits on kinetic energy weapons would "eliminate our ability to develop the technology for *near-term* boost phase" defense systems (emphasis added).[21] In the public sphere, exotic missile defense technologies received the most attention because they conjured up images associated with works of science fiction. But existing capabilities offered the best hope for pursuing strategic defense in the next five to ten years. Even though Reagan had pledged that SDI would be a long-term research program, Weinberger clearly wanted to develop and demonstrate the technologies for a strategic defense system that could be deployed in the near future.

Reagan was initially receptive to the idea of ASAT arms control because he did not fully grasp its consequences for SDI. During a September 1984 meeting at the White House, Weinberger implored Reagan not to accept any ASAT constraints because of their implications for strategic defense research and development. Confused, Reagan said, "Concerning the ASAT thing, all theirs are ground-based. Ours are on a plane [in reference to the MHV jet-launched ASAT]. I don't know how limits on either ground-based or airborne ASAT systems interferes with SDI."[22] McFarlane then explained how key SDI technologies and ASATs were fundamentally one and the same. Limits on ASATs would hamper the development of SDI. Reagan did not like this answer because it undermined his view of SDI as a peace shield.

Brushing these technical details aside, the president said that it was imperative to show the Soviets that SDI was not threatening—that it was purely defensive. Admiral James Watkins, the chief of naval operations, cautioned that SDI should not be directly linked in the public eye with offensive space capabilities. He warned that "we must make certain that SDI is not made analogous to ASAT . . . there is a solid case for SDI, but we will always have problems in dealing with public opinion on space and ASAT. We must link research on SDI to making nuclear weapons obsolete." In response to Watkins' comments on ASATs, Reagan suggested that "we should first talk about getting rid of these offensive arms like this F-15 [air-launched] ASAT. We must make it clear that we are not seeking advantage, only defense."[23] Reagan characterized ASATs as "offensive weapons," whereas SDI was only a "nonnuclear defensive system."[24] Watkins explained to the president once again that there was an intimate technological connection between ASATs and SDI. Watkins conceded that the US ASAT program had serious limitations and could therefore "be given up, from a military point of view," but stressed the connections with SDI.[25]

Kenneth Adelman similarly told the president that "any SDI deployment would be an ASAT [sic]" and that it was therefore impossible to separate ASATs from SDI.[26] Despite these explanations about the SDI–ASAT linkage, Reagan asked if there was still some way to distinguish between offensive and defensive space systems. Ed Meese, an advisor to the president, stated bluntly that "the technology is the same; a treaty on ASAT testing could kill both ASAT and SDI."[27] Reagan nevertheless refused to abandon this distinction and would continue to say publicly, and during exchanges with the Soviet Union, that SDI was only a defensive system that could in no way threaten Soviet interests.

While Shultz believed that the US could, and should, at least discuss limits on the so-called space-strike arms (e.g., SDI and ASATs), Weinberger and Casey remained vehemently opposed. When reflecting on SDI years after his tenure as defense secretary, Weinberger said that he had feared that the Soviets, or the State Department, might have been able to persuade the president to go along with "some sort of chimerical arms control agreement."[28] But there was no chance of Reagan trading away SDI because he believed that it was the ultimate insurance policy in case an arms control agreement failed. It was the president's vision of SDI as a means of delivering the world from the threat of nuclear annihilation that made the preservation of the program a primary objective of his arms control strategy. And he did not want to place any limits on SDI-related technologies that could make the job of SDI system designers more difficult in the future.

The Pentagon's success in convincing Reagan that ASAT limits would be detrimental to SDI was reflected in the president's instructions for the first round of the Nuclear and Space Talks set to begin in Geneva on March 12, 1985. Reagan directed that "we will protect the promise offered by the US ASAT and SDI research program to . . . provide a basis for a more stable deterrent at some future time."[29] Notably, this set of instructions explicitly identified the ASAT–SDI relationship, saying the "promise offered by the US *ASAT and SDI*" (emphasis added).[30] For Reagan, preserving ASATs was important not primarily because of their deterrent potential and military utility, as stated publicly, but rather due to their relationship with missile defense.

In early January 1985, Shultz and Gromyko had met in Geneva to develop a new framework for strategic arms negotiations.[31] At that time, Gromyko emphatically stated that the superpowers needed to "prevent the militarization of space" through banning space attack weapons of all kinds.[32] The notion that space militarization could be prevented was nonsensical, since space had long been used for military purposes. Rather than demilitarization, the US officially sought to reach an agreement concerning restrictions on those systems and military activities in space that "could diminish stability."[33] There was, however, much interpretive flexibility associated with the term "stability."[34] In reality, the White House would not accept any proposal that limited its freedom of action in space.

Gromyko and Shultz were able to agree on three arms negotiations forums: strategic nuclear weapons (START), intermediate-range nuclear forces (INF), and space and defensive arms, collectively known as the Nuclear and Space Talks.[35] McFarlane praised Shultz to Reagan, saying that an agreement had been reached because he had an "iron-ass Secretary of State."[36] The president was pleased and hopeful, but this was cautious optimism. American officials noted that "the stalemate between the two countries on space weaponry" would continue to pose a problem.[37] A Soviet spokesman similarly cautioned that the resumption of talks did not signal a new détente but was rather a "small crack in the East–West ice."[38] Even if it was a small opening, this was a ray of light after an especially tense period in superpower relations.

CONTINUITY AND CHANGE UNDER GORBACHEV

Konstantin Chernenko's death on the day before the Nuclear and Space Talks were set to begin (March 12, 1985) added a whole new level of complexity to the arms control dialogue. US officials could only speculate whether Mikhail Gorbachev, Chernenko's successor, would be a more cooperative partner for Reagan. Despite the Soviet Union having a new leader with fresh ideas, many US officials were hesitant to see Gorbachev as inaugurating a new era in American–Soviet relations. John Lenczowski of the NSC staff described Gorbachev as "a quintessential Communist Party man."[39] Even if he sought reform, members of the Reagan administration believed that Gorbachev would be constrained by more conservative elements in the Soviet bureaucracy. A CIA report entitled "Gorbachev, the New Broom" observed that he had been "the most activist Soviet leader since Khrushchev." Analysts believed, however, that his objective was not radical reform but rather an "attack on corruption and inefficiency."[40] Significantly, Gorbachev did shake up the Kremlin by bringing in people loyal to him; in July 1985, he replaced Gromyko with Eduard Shevardnadze, who would breathe new life into Soviet diplomacy.

Even though Gorbachev wanted fresh ideas brought into the Soviet bureaucracy, he still had to contend with hardliners, especially in the Ministry of Defense. One area in which there was marked continuity, at least initially, was SDI. When Gorbachev met with Vice President George H. W. Bush in Moscow after Chernenko's funeral, he stressed the need to prevent the arms race from moving into space.[41] SDI remained a source of confusion and concern in the Soviet Union due in no small part to the fact that the specific goals of the program were still unclear.

There were Soviet officials who most certainly believed that SDI was intended to give the US a decisive nuclear first-strike capability and to control space, but these views were based on speculation and perhaps mirror imaging in some cases. KGB documents from 1985 reveal that Soviet intelligence sought to understand better not only the technical aspects of SDI research and development but also the political objectives underpinning the program. The Center in Moscow wanted its *rezidenturi* (intelligence stations) to gather information to determine if the Reagan administration planned to "use SDI to place military-political pressure on the USSR" and to secure concessions from the Soviet Union on "heavy ICBMs."[42] Soviet intelligence officials also sought information about how different groups within the Pentagon and State Department viewed SDI and the prospects for the US observing the 1972 ABM Treaty.[43] Soviet intelligence officers recognized that there were divergent views on SDI within the US national security apparatus, just as there were different perspectives on the program within the defense and foreign policy communities in the Soviet Union. Even though Moscow placed a premium on information concerning SDI, the KGB was limited in its ability to penetrate the program and based much of its intelligence on US press reports and public statements.[44]

Evaluating the technological feasibility of American space-based missile defense was not a straightforward task either. Andropov had commissioned a technical study on space-based missile defense shortly after Reagan made his March 1983 speech. After several months, the group in charge of the investigation concluded that the American effort would fail to produce a leak-proof defense system over the next fifteen to twenty years, but Soviet officials were still troubled by the "thrust of the program." Additionally, SDI could lead to "a whole line of new technologies in the United States that were not completely devoted to defending against a strike from the Soviet Union."[45] In other words, SDI could give the US an advantage in advanced technologies with many defense applications.

By the early 1980s, Soviet defense planners were devoting more attention to what they termed *razvedyvatel'no-udarnye kompleksy* ("reconnaissance strike complexes").[46] This concept involved using advanced technologies for automated command, control, communications, and intelligence systems to enable precision strikes against targets deep behind enemy lines.[47] Key technologies associated with SDI research and development (e.g., command and control systems and advanced sensors) were critical for a reconnaissance strike complex. On an even broader scale, Soviet leaders worried about a widening technology gap between East and West due to predominantly American and Japanese advances in computing and electronics that would have far-reaching political and economic consequences.[48] In 1983, Soviet General Nikolai Ogarkov, chief of the Soviet General Staff, told Leslie Gelb of the *New York Times* that modern military power was based not on sheer numbers but rather on computers. He said that "computers are everywhere in America. Here, we don't even have computers in every office of the Defense Ministry."[49] Even if SDI did not produce a functioning missile defense system, Soviet officials realized that it could indeed lead to advances in software and hardware that would expand America's qualitative military edge, in addition to economic competitiveness, over the Soviet Union.[50]

There were certainly Soviet scientists and officials, especially in the Ministry of Foreign Affairs, who were skeptical that SDI would produce revolutionary technologies that would upset the strategic balance. According to Simon Miles, "while most in the MID [Ministry of Foreign Affairs] saw SDI as a fool's errand, the military took it deadly seriously."[51] Pavel Podvig observes that there were officials in the Soviet military-industrial complex who wanted to use SDI as a means of securing more funding for large-scale missile defense and space projects that had been underway since the 1970s.[52] Jonathan Haslam points out that, at least in early 1985, Gorbachev was swayed more by the Ministry of Defense's alarmist view of SDI, strengthening Moscow's push for space arms control.[53] Regardless of conflicting assessments on SDI, curbing the arms race was essential for the economic reforms that Gorbachev had in mind.

In evaluating Gorbachev's policy on SDI, the psychological dimensions of the program should not be dismissed as insignificant. For the two years prior to Gorbachev becoming general secretary, Soviet officials had characterized space-based missile defense as a grave threat to Moscow's security. It is likely that the Soviet Union was concerned about the psychological consequences of US space-based missile defense system, regardless of its effectiveness. In 1977, Zbigniew Brzezinski had warned Jimmy Carter that a Soviet laser weapon in space, even if it did not alter the military balance, could have a greater psychological impact on Americans than Sputnik.[54] The CIA similarly observed in 1983 that "the psychological effect of the first test of a space-based laser in a weapon-related mode would be greater than the actual military significance of such a weapon in its initial applications."[55] These anxieties about public perceptions of space weapons were not unique to the American context. Vladimir Shcherbitsky, a member of the Politburo, said to Reagan during a March 1985 meeting at the White House, "The prospect of space weapons is particularly frightening. People would feel that destruction is poised above their heads. To have weapons on earth and under the water is one thing, but something which is poised in space above your head all the time is enough to drive people crazy."[56] Many American and Soviet leaders remembered the psychological impact of Sputnik in 1957. One could only imagine the public reaction to an orbital weapon system. Such a situation could not be ignored by Soviet officials; there would be substantial pressure, especially from the defense establishment, for a military response.

"OTHER PHYSICAL PRINCIPLES": SDI AND THE ABM TREATY

Whether the US would ultimately deploy a space-based missile defense system, and when that decision might take place, were distant issues. Reagan acknowledged that it would be left up to a future president and Congress to make the decision to deploy strategic defenses.[57] The pressing matter was the testing of missile defense technologies, especially in space. US officials offered conflicting assessments of when the Pentagon would need to move forward with a testing schedule that required modifications to the ABM Treaty. Abrahamson saw 1994 "as the ABM Treaty critical year" in which a decision would have to be made about moving beyond the confines of the treaty.[58] Paul Nitze, by contrast, predicted that it would be "many, many years, perhaps well into the next century" before SDI testing would begin.[59]

The ABM Treaty permitted the testing of missile defense systems and components only if they were fixed on land and at specific sites.[60] For example, the HOE in 1984 that involved using a kinetic interceptor to destroy a missile in flight was permitted by the treaty. As far as the US and the Soviet Union were concerned, missile defense components referred to missiles, launchers, and radars.[61] Despite the definitions agreed upon by the US and the Soviet Union at the time of the ABM Treaty's signing, there was still disagreement regarding what constituted a missile defense component and where the boundary between the laboratory and field testing was located.

Testing SDI technologies was important not only to make progress toward developing a strategic defense system, but also for building momentum in favor of the program. SDIO, with the backing of the White House, planned a series of experiments called Significant Technical Achievements in Research (STAR).[62] The tests were intended to "provide visible and convincing validation of technology levels needed to proceed to deployment of effective defenses."[63] SDIO planned to begin STAR demonstrations in 1989 that were compliant with the ABM Treaty.[64]

Soviet and Western European concerns over the ABM Treaty were exacerbated by Abrahamson's announcement in February 1985 that testing in space of tracking and targeting systems had been brought forward to 1987 and that the US could use active ballistic missile targets during experiments.[65] The Pentagon specified that it had ten tests planned that were allowable under the ABM Treaty. The US government had deemed that laboratory testing, field tests of non-ABM components, and field tests of fixed land-based ABM components were all permissible.[66] European allies were, however, uncomfortable with any tests that could even be perceived as a violation of the ABM Treaty. British diplomats, in particular, worried about an expedited SDI testing schedule. UK officials were not wholly convinced by American assurances that upcoming SDI-related demonstrations would not be problematic for the ABM Treaty. The UK FCO warned the prime minister that "piecemeal activities relevant to the grey areas of the Treaty could risk undoing its central provisions much sooner than otherwise be expected."[67]

For the Reagan administration, the ABM Treaty was not a source of stability but rather a means of preventing the US from obtaining a technology-based advantage. Reagan believed that the US had "compromised [its] clear technological lead in the anti-ballistic missile system, the ABM, for the sake of a deal" with the Soviet Union.[68] Richard Perle, a senior official in the Department of Defense, said publicly that the treaty "was a mistake in 1972 and the sooner we face up to the implications of recognizing that mistake the better."[69] Caspar Weinberger had already expressed a willingness to amend the treaty to move forward with SDI.[70] These administration views were well known by the European allies, and they feared that the White House might unilaterally abrogate the treaty with little notice.

Transatlantic tensions over the ABM Treaty heightened when the White House started publicly highlighting alleged Soviet violations of the treaty. At a meeting of the NATO NPG on March 26 and 27, 1985, Weinberger tried to convince his counterparts to issue a condemnatory statement about alleged Soviet violations of the ABM Treaty, particularly the Krasnoyarsk radar. The Soviets maintained that that the radar was only used for tracking space objects, but senior administration officials were adamant that it was also designed for missile defense battle management. A 1984 NIE observed that "the Soviets continue construction of large phased-array radars that, to varying degrees, could provide . . . [missile defense] battle management support. A sixth such radar was detected under construction in 1983 near Krasnoyarsk."[71] Multiple media outlets reported in March 1985 that an intelligence estimate entitled "Implications of a New Soviet Phased-Array Radar" had concluded that the radar was "not well designed" to be used in support of an anti-ballistic missile mission.[72] This intelligence report is still not declassified, but a 1987 memorandum from the ACDA concluded that "there are uncertainties and differences of view about their [radars] suitability to support a widespread [missile defense system] deployment."[73]

Despite the unknowns about the Krasnoyarsk radar, Weinberger wanted NATO to issue a communique saying that the alliance "deplored the new phased array radar under construction at Krasnoyarsk in violation of the ABM Treaty." In a memorandum to the prime minister, Michael Heseltine pushed back on Weinberger's certainty about the Krasnoyarsk radar, writing that "our expert advice is that it is not possible on the current evidence to determine whether or not the radar contravenes the ABM Treaty." Heseltine also worried that characterizing the Krasnoyarsk radar as a treaty violation could put the US and the UK in a difficult position if the Soviet Union alleged, in retaliation, that the construction of a new early warning radar at RAF Fylingdales in the UK constituted a breach of the ABM Treaty.[74] In reality, Fylingdales and Krasnoyarsk were not at all similar. The ABM Treaty specifically dictated that missile warning radars be located in border areas; the Krasnoyarsk radar was located nearly two thousand miles from a border.

Heseltine became convinced that Weinberger's insistence on an alliance-wide condemnatory statement regarding the Krasnoyarsk radar indicated that the Reagan administration might be laying the groundwork for pulling out of the ABM Treaty. He wrote to the prime minister that "the ground is being prepared . . . for the argument to be used at the appropriate point that the regime stabilized by the Treaty has effectively broken down because of Soviet behavior and that there is therefore no reason for the US Administration to feel any obligation on its own part to uphold its provisions."[75] Charles Powell, Thatcher's most trusted advisor, warned her that "Weinberger is now, like Shultz, canvassing the possibility of US withdrawal from the ABM Treaty."[76] To the chagrin of Weinberger, the March 1985 NATO NPG communique said only that "we noted with concern the extensive and long-standing efforts in the strategic defense field by the Soviet Union which already deploys the world's only ABM and anti-satellite systems."[77] The Krasnoyarsk radar was not mentioned at all. Notably, in 1989, Soviet officials would say that the radar was indeed a violation of the ABM Treaty and take measures to be in full compliance.[78]

Eyes around the world turned to the heavens in June 1985 when SDIO conducted its first laser experiment in space using the space shuttle *Discovery*. This activity was designed to comply fully with the ABM Treaty and served as an opportunity to show the world that SDI was more than a concept on paper. The mission involved a mirror reflecting a laser beam aimed from an Air Force facility in Maui, Hawaii. Engineers hoped to validate their ability to keep the laser pointed at the orbiting mirror.[79] Precise tracking and pointing of the laser beam would be necessary for destroying a ballistic missile in flight. The experiment became politicized when it failed on the first attempt. Ground controllers sent their commands for the shuttle's position in nautical miles instead of feet, which caused the spacecraft to be in the wrong orientation for the execution of the experiment. SDI skeptics immediately seized on this failure. John Pike of the Federation of American Scientists asked, "If they can't do a simple experiment like this, what are they going to do in combat?"[80] Abrahamson was clearly frustrated by such comments and said that SDI detractors were "seizing on ridiculous things to try to criticize the program" and that this was only a "small procedural error."[81]

The experiment's second attempt was successful. While a tape recorder played Tchaikovsky's *1812 Overture* in the shuttle's cockpit, a laser beam "painted" a blue-green light on the nose of *Discovery* for least two and a half minutes.[82] Being able to point a laser beam accurately was important, but creating a laser with enough power that could actually destroy ballistic missiles would be an even bigger challenge. Regardless, this first SDI shuttle experiment suggested that future such efforts would be heavily politicized in Congress and in the media. It created a significant public relations challenge for SDIO because highly complex experiments representing a small portion of SDI research could be seen as representative of the program's overall potential for success or failure.

As the Reagan administration planned future SDI tests, it created significant controversy by announcing a new interpretation of the ABM Treaty in the fall of 1985. McFarlane presented a "broad" interpretation of the ABM Treaty that would allow the testing of new futuristic technologies. In the treaty, the US and the Soviet Union had included a provision called "Agreed Statement D" that specified if missile defense components "based on other physical principles" were created in the future, "specific limitations on such systems and their components would be subject to discussion."[83] Proponents of the broad definition argued that because SDI included exotic technologies (e.g., laser weapons) that were not yet operational when the treaty was signed, SDI entailed capabilities based on other physical principles and could be exempt from the treaty's testing restrictions. Many legal scholars contested this new interpretation.[84] To calm anxieties in the wake of McFarlane's statement, Reagan signed NSDD-192 that directed the Department of Defense to continue using a more restrictive treaty interpretation, but maintained that the broader interpretation was indeed correct and could be used if deemed necessary in the future.[85] Competing interpretations of the ABM Treaty would remain a controversial issue for the remainder of Reagan's presidency.

MUSCLE FLEXING IN SPACE

To address the fundamental disagreements over space weapons and nuclear arms in the American–Soviet strategic arms dialogue, in May 1985, both countries agreed to a superpower summit in Geneva scheduled for November of that same year. This would be the first time that an American president had met with his Soviet counterpart in six years. Neither the US nor the Soviet Union anticipated a substantial breakthrough in relations to take place. According to Simon Miles, Reagan saw Geneva as the beginning of "a lengthy process of negotiation from which he expected to emerge as victor. The summit was not meant to yield tangible results."[86] Gorbachev's objective was to lay the groundwork for improving American–Soviet relations, but SDI was the key obstacle in his way.[87]

As summit preparations got underway, the Soviet Union considered various options for militarily responding to SDI. In July 1985, the Central Committee and the Council of Ministers approved two umbrella programs of ground- and space-based systems.[88] The first program, called "D-20," involved research and development related to ground-based missile defenses. The second effort, called "SK-1000," focused on space-based missile defenses and ASATs (both ground and space-based).[89] Key aspects of these various efforts had started in the 1970s. ASAT concepts included space mines, a ground-based ASAT, and weapons based on "other physical principles," for example lasers.[90] The Soviet Ministry of Defense would ultimately settle on countermeasures, including ASATs, designed to negate critical elements of the SDI space-based infrastructure.[91] The vulnerability of SDI components would therefore become a key political, as well as a technical, issue. The Reagan administration would need to convince lawmakers that a strategic defense system would be resilient to attack.

A little more than a week after the US and the Soviet Union agreed to a summit, the Senate approved the first destructive test of the US MHV ASAT against an object in space. Congress had earlier placed restrictions on ASAT testing with the hope that they would help kick-start negotiations with the Soviet Union on strategic weapons. In May 1985, Senator John Kerry (D-MA) had proposed a new testing ban as a "signal of our good faith and willingness to restrain the arms race and keep it out of space."[92] Senator John Warner (R-VA) opposed Kerry and introduced a bill that would allow a three-stage test, with the final one being against a US target in space. He said that a limit on US ASAT testing would erode the American negotiating position at the upcoming Geneva Summit.[93] John Steinbrunner, director of foreign policy studies at the Brookings Institution, called the test a "gun to the head approach" that would drive the superpowers into a military space race that would "leave both sides worse off." He maintained that "the whole strategic relationship of the US and Soviet Union depends on getting some kind of anti-satellite agreement." To the Soviets, the US ASAT test appeared to be muscle flexing ahead of Geneva.

On September 13, 1985, an F-15 fighter jet took off from Vandenberg Air Force Base in California and fired off an ASM-135 missile that successfully destroyed an American satellite in low Earth orbit. The administration wanted to complete the test before October because it anticipated that Congress would impose a new ASAT testing ban. The thirty-pound MHV interceptor collided with its target, a US satellite, at a closing velocity of 15,000 miles per hour. The collision produced debris that would remain in orbit for many years to come, and it was increasingly apparent that debris from ASAT tests posed a long-term threat to spacecraft operations.[94] In a win for pro-space arms control advocates, shortly after the MHV test Congress imposed a moratorium on testing ASATs against objects in space.[95] Consequently, this September test would be the only destructive demonstration of the MHV interceptor.

Unsurprisingly, Moscow reacted very negatively to the ASAT test; it came shortly after the Soviet Union proposed a United Nations–sponsored international conference intended to "prevent the military use of outer space."[96] Alexander Druzhinin, a political observer in Moscow, reported that this test represented the practical implementation of the "so-called 'Star Wars' program." He said that "hitherto, representatives of Reagan's administration patently attempted to mislead the public, [and] have presented the matter as if this program is restricted to harmless and completely safe scientific research developmental work."[97] Soviet Foreign Minister Eduard Shevardnadze resolutely condemned the test.[98] Notably, the Soviets had already conducted more than twenty ASAT tests in space. And it must be remembered that Andropov removed the Soviet ASAT from active service in August 1983, hoping to make progress on space arms control.[99] Presumably due to this US ASAT test in September 1985, the Soviets reactivated their own ASAT program that same year.[100]

The US ASAT program, and this test specifically, served multiple purposes. Ahead of talks with the Soviets, the ASAT demonstration showed the Soviets that the US possessed a sophisticated weapon for attacking satellites. During the ASAT negotiations in the late 1970s, the Soviets had little incentive to accept limits on their own ASAT program, since the US did not yet have an operational capability.[101] For hardliners such as Weinberger and Perle, the ASAT test signaled US resolve not to abandon its offensive space capabilities. The successful collision of MHV with a satellite was also connected to SDI research and development. According to Richard Matlock, a former senior official at the MDA, "the Air Force's air-launched anti-satellite demonstration in September 1985 . . . bolstered our confidence and convinced the Soviets that we could apply the kinetic intercept technology in a layered missile defense architecture."[102] In other words, ASAT technologies could be used for missile defense and vice versa.

A "SPLENDIDLY NAÏVE NOTION": SHARING SDI WITH THE SOVIET UNION

After Gorbachev's black limousine pulled up to the Fleur d'Eau in Geneva on November 19, 1985, the Soviet leader stepped out bundled up in a topcoat, scarf, and hat. Despite the frigid temperature, Reagan greeted his Soviet counterpart in an elegant suit with no overcoat. The president looked energetic when standing next to Gorbachev, who appeared like a caricature of a Bolshevik. The summit theatrics had officially begun. This meeting, which was the first time Reagan and Gorbachev had been in the same room, provided the two leaders an opportunity to get to know one another. Neither Reagan nor Gorbachev anticipated a significant breakthrough in American– Soviet relations to occur over the two-day summit, but they were both enthusiastic about making progress in the strategic arms dialogue.

It did not take long for the subject of space weapons to arise. Gorbachev warned that SDI would lead to an arms race in space. The Soviet leader claimed that "any shield can be pierced, so SDI cannot save us." He further pledged that the Soviet Union would "build up in order to smash your [space] shield."[103] While no plans had been solidified, the general secretary clearly had been briefed by the Ministry of Defense and scientific advisors on the various countermeasures that Moscow could pursue should the US move forward with deployment of strategic defenses.

Gorbachev went on to condemn US military space activities, disingenuously saying that "Soviet [space] research is for peaceful purposes."[104] He and Reagan both knew that the vast majority of Soviet expenditures on its space program were for military aims.[105] The Soviet Union was investing in reconnaissance satellites, early warning satellites, manned military space missions, and ASATs (among others).[106] Gorbachev's objections to the "militarization of space" overlooked the fact that American and Soviet space activities had been militarized since the beginning of the Space Age. Nevertheless, opposing the "militarization of space" held sway in public diplomacy because so few people knew the extent of secretive superpower military and intelligence space programs.

Reagan tried to convince Gorbachev that SDI was only defensive in nature.[107] The president emphatically claimed that "SDI was not a weapon system or a plan for conducting a war in space." Gorbachev expressed his concern that the US could use SDI technologies to destroy satellites, ballistic missiles, and targets on land. In response, Reagan once again emphasized that SDI was only defensive and thus not part of an arms race. The president focused on his grand vision for SDI and ignored its technical connections to ASATs that had offensive applications.

Reagan was clearly caught off guard by Gorbachev's assertion that SDI technologies would be used to strike targets on the ground from space. Peter Westwick has described how the potential for space strikes against terrestrial objects was a key source of Soviet anxiety about SDI.[108] Such a proposition was not, however, realistic in the near term. The power output levels for directed-energy weapons prevented them from being used to destroy objects on Earth from space. The general secretary's point about SDI having the potential to be used offensively was, nevertheless, a legitimate concern. This exchange between Gorbachev and Reagan on the offensive and defensive attributes of SDI underlines Colin Gray's observation that "political perspective determines judgment on the offensive or defensive character of weaponry and military posture."[109] Reagan was sincere when he pledged that SDI would not be used offensively, but Gorbachev was correct in his assertion that SDI technologies had the potential to be used offensively. Even if the president was uninterested in the offensive applications of SDI technologies, the Soviet Union could not be certain about future American administrations.

Reagan attempted to alleviate Gorbachev's concerns about the US using SDI to achieve a strategic advantage by promising to share its strategic defense research with the Soviet Union. Additionally, the president wanted both countries to open their laboratories where defense research was taking place as a trust-building measure.[110] This was not a new idea for Reagan; within a few days of his March 23, 1983, speech that would lead to SDI, he said that if strategic defenses could be developed, they would be internationalized.[111] Frances FitzGerald describes how the president's advisors were generally opposed to the idea of sharing.[112] Kenneth Adelman characterized the president's desire to share SDI as a "splendidly naïve notion [that] only Reagan could have believed, much less conceived."[113] It was certainly an unexpected proposal from a president who was trying to crack down on technology transfer to the Eastern Bloc. Nevertheless, not all of Reagan's advisors believed it was a ridiculous idea. In November 1985, Casey sent a letter to Weinberger that outlined options for cooperating with the Soviets on SDI. He identified "open labs" and "joint control of deployed [strategic defense] systems" as potential frameworks for cooperating with the Soviet Union. NSC staffers presented the 1975 American–Soviet Apollo–Soyuz space mission as a replicable model for SDI.[114] Both countries could cooperate, according to these officials, without compromising sensitive technologies. But SDI involved technologies of much greater sensitivity than the Apollo–Soyuz Test Project.

Some US officials involved in studies about sharing SDI with the Soviet Union believed it was possible to have joint US–Soviet control of a strategic defense system without providing the Soviets direct access to sensitive technologies.[115] However, few details were provided concerning how this cooperative framework would be established. The very idea of sharing SDI research with the Soviets confounded US allies. Casey acknowledged that "we will have to deal with the apparent disconnect between SDI technology sharing [with the Soviets] and [strict] COCOM controls, which could create problems with the allies."[116] American allies, especially in Europe, had long since viewed US technology transfer policies as highly restrictive and would therefore be upset by a US willingness even to consider providing Moscow access to SDI.[117] At Geneva, Gorbachev outrightly dismissed Reagan's pledge to share, saying that "the US did not share its most advanced technology even with its allies."[118]

As the Geneva Summit was ending, it was apparent to all participants that the space part of the Nuclear and Space Talks was the primary impediment to progress on arms control. Simon Miles observes that Reagan would not relent on SDI because he saw it as a catalyst for the eventual elimination of nuclear weapons and as a means of exerting pressure on the Soviet Union.[119] While the latter was certainly a key consideration for Reagan's advisors (and Thatcher), the president's nuclear abolitionism was the primary motivation for not compromising on SDI. Gorbachev conceded that on a human level, he could understand that the "idea of strategic defense had captivated the President's imagination," but as a political leader, "he could not possibly agree with the President with regard to this concept." On a practical level, divergent interpretations of the ABM Treaty were clearly going to be critical obstacles to making progress. Gorbachev believed that the treaty needed to be strengthened, whereas Reagan saw it as a threat to SDI. In the near term, Paul Nitze stressed that the parties needed an agreement on what was permitted and prohibited with regard to research, development, and testing, which would be easier said than done.[120]

After the summit, the two leaders exchanged letters, in which the military uses of space loomed large. The president assured Gorbachev that there was "no development underway to create space-based offensive weapons."[121] The general secretary was not convinced and provided a very technical description of how space weapons could be used to destroy reconnaissance, navigation, and communications satellites and thus "blind" the Soviet Union before an attack. Gorbachev pointed out that space weapons could be used "defensively and offensively," and that even if Reagan did not intend to use SDI offensively, the future was uncertain, and it was not at all clear what another president might do.[122] Both Reagan and Gorbachev wanted the same end: the elimination of nuclear weapons. The problem was that they had diametrically opposed views on the means for achieving their goal.

WHERE IS THE LABORATORY BOUNDARY?

In late January 1986, a space disaster struck that had significant implications for SDI. As the world watched on live television, the shuttle *Challenger* disintegrated seventy-three seconds after launch, killing all crewmembers (including a schoolteacher). Less than two months later, a Titan-D rocket carrying a US reconnaissance satellite blew up shortly after launch. These events called into question the reliability of US launch systems—a point that Thatcher would raise in a meeting with Abrahamson in the summer of 1986.[123] An Air Force spokesman had said that SDI deployment could require "about 600 space launches during a three-year period, more than one rocket liftoff every two days."[124] SDIO reported to Congress that the success of the entire program was heavily dependent on major advances in space transportation and logistics.[125] The halting of shuttle launches would have an immediate impact on SDI experiments, delaying one by a year and another by two years.[126] The *Challenger* tragedy took with it all hopes that the shuttle would provide ready and cost-effective access to space and once again highlighted the fact that space logistics posed a significant challenge for both the testing and deployment of space-based missile defense technologies.

Controversy over limits on SDI development and testing became a major issue in Congress in the summer of 1986. The Reagan administration fought for an SDIO budget of $4.8 billion for FY87, which would have been an increase of $2 billion from the previous year; in the end, Congress appropriated $3.2 billion.[127] SDIO planned a major demonstration of hit-to-kill technologies that would simulate a missile intercept in space for September 1986; this was called the "Delta 180" vector sum experiment.[128] The primary goal of Delta 180 was to intercept a target vehicle in powered flight in space safely. The Johns Hopkins University Applied Physics Laboratory (APL) led the effort for SDIO. To move swiftly, APL used off-the-shelf capabilities; the Delta 180 experiment went from the drawing board to launch in less than eighteen months (for comparison, the NRO built a satellite in this same time period that took five years from initial approval to launch). Delta 180 planning began in 1985; the decision to use expendable launch vehicles, rather than the shuttle, proved especially prudent after the *Challenger* disaster. The experiment had to be compliant with the ABM Treaty, which ruled out the use of ICBM components. On September 5, 1986, a Delta rocket carrying the experiment payloads launched from Cape Canaveral. The second and third rocket stages (the former was the target and the latter served as the seeker) separated, maneuvered, and then the third-stage payload module successfully intercepted its target.[129]

From the standpoint of the Reagan administration, the experiment was both a technological and political success. SDIO demonstrated that it could rapidly develop and launch space capabilities. Creating a functioning strategic defense system was, however, an even greater task that would require much more than an intercept in a curated test environment. The timing of the Delta 180 experiment was very important as well. It occurred only weeks before the next superpower summit scheduled for early October in Reykjavik, Iceland. General Abrahamson said that his goal was to place "gun camera film on the table at Reykjavik."[130] While the success of the Delta 180 experiment in no way guaranteed that a strategic defense system would be feasible in the near term, it gave Reagan confidence going into the summit that SDI was indeed making progress and strengthened his belief in the program's potential to lead to a nuclear-free world.

Just as in the lead-up to Geneva, US officials anticipated that SDI would be the primary obstacle to progress on nuclear arms control in Reykjavik. An NIE from September 1986 reinforced the belief of administration officials that SDI was a form of leverage in negotiations. Intelligence analysts maintained that "the main Soviet motive for considering and negotiating about large nuclear force reductions at present is to undermine the US Strategic

Defense Initiative."[131] To this end, Moscow sought to strengthen the ABM Treaty. Shevardnadze told Shultz that "outer space and the ABM Treaty is the cornerstone [of an agreement], and we need it first."[132] The Soviet Union specifically wanted a US pledge not to withdraw from the treaty for at least fifteen years and to have research "confined to laboratories."[133] But Soviet officials did not really define what "laboratory" meant.

Technological (and political) uncertainty were key factors in the Soviet approach to arms negotiations. Even if a highly sophisticated strategic defense was not possible in the near term, Soviet officials worried about reducing nuclear forces, while the US continued to research, develop, and test advanced missile defense and space technologies. For the Soviets, SDI was a Pandora's box that would catalyze the American lead in high-technology arenas. Cutting off SDI development before any breakthroughs was a critical task for Moscow.[134] More specifically, Soviet officials built their strategy around the central goal of preventing SDI from moving into a space-based testing phase.

Reagan and Gorbachev met at Hofdi House near downtown Reykjavik on October 11, 1986. In their opening exchange, Gorbachev proposed reducing strategic arms by 50 percent and eliminating all American and Soviet medium-range nuclear weapons in Europe. Gorbachev excluded British and French nuclear forces, which had been a point of contention with London and Paris. Regarding the ABM Treaty, there needed to be a guaranteed period of non-withdrawal; he suggested ten years as a compromise followed by three to five years of negotiations on how to proceed from that point. As an addendum, Gorbachev wanted a prohibition on ASATs because allowing them would "open a channel for development of ABM weapons."[135] Shultz told Reagan that Gorbachev "was laying gifts at our feet."[136]

The ideological difference between the two leaders on the ABM Treaty was the key problem. Reagan aimed to create new provisions in the ABM Treaty that would allow tests in space. He promised that Soviet representatives would be permitted to observe experiments and that an eventual system would be shared. The president argued that in the future there could be a Hitler-like madman who sought nuclear weapons, but "if both countries had such a defense system, we would not need to be concerned about what others might do and we could rid the world of strategic nuclear arms."[137] (In reality, a space-based missile defense capability could not defend against all nuclear delivery systems). Gorbachev did not find the president's arguments compelling and echoed Shevardnadze's earlier comments that the uncertainty about where SDI technologies might lead made them destabilizing.

In their afternoon meeting on October 11, the two leaders became locked in familiar arguments about SDI. Gorbachev was adamant that SDI could be used offensively to launch a nuclear first strike. He further stressed that if they eliminated all ballistic missiles, then there would be no need for SDI in the first place. Reagan once again pledged that SDI would be shared with the Soviet Union and identified it as the "greatest opportunity for peace of the 20th century." To illustrate his point that SDI was a security blanket for the world, he asked Gorbachev to consider Qaddafi. According to Reagan, "if [Qaddafi] had [nuclear weapons] he would certainly have used them," and SDI would be the ultimate protection against such a scenario.[138] Regarding the ABM Treaty, Reagan stressed that it would not be eliminated. Rather, he wanted to allow both countries to continue developing strategic defenses and to share the benefits if these measures prove feasible.[139] But if both countries possessed strategic defenses on the scale that Reagan envisioned, then the ABM Treaty would have no purpose.

Gorbachev warned Reagan that if the US moved forward with strategic defense deployment, Moscow would be forced to respond, but that it would do so asymmetrically. Even though Gorbachev did not go into detail, he was clearly referencing ASATs, faster burn ICBMs, and decoys that could be used to defeat missile defenses. Reagan returned to the idea that "both sides should go ahead [with strategic defense] and if the Soviets do better, they can give us theirs." The general secretary once again emphasized that the Soviet Union would not build a space-based missile defense system but would instead focus on cheaper countermeasures to American space-based missile defenses. He moreover challenged Reagan's sincerity about sharing the benefits of SDI research, famously saying that the US refused to share milk factory technologies with the Soviet Union.[140]

In their final evening of meetings together on October 12, all agreement concerning nuclear weapons was contingent on Reagan and Gorbachev finding common ground on the limits of SDI-related laboratory research. On nuclear weapons, they made the most substantial breakthroughs in the history of the Cold War. They agreed that *all* ballistic missiles should be eliminated by 1996. Reagan then said it would be "fine with him if we eliminated all nuclear weapons" and Gorbachev replied, "We can do that. We can eliminate them."[141] The leaders of the two superpowers had just declared that they could get rid of their *entire* nuclear arsenals. Then, this all fell apart over one word: laboratory. The general secretary could not return to Moscow without a prohibition on weapons testing in space, which Reagan refused. Consequently, there was no deal to be had. Both leaders rose, shook hands, and departed.

Even though they walked away without an arms control agreement, Reykjavik was not a complete failure. Elizabeth Charles observes that "Reagan and Gorbachev broke the ice that had been forming in the American and Soviet relationship since Geneva almost a year earlier. The two leaders made great strides toward arms reductions and the elimination of nuclear weapons."[142] According to polls taken after the summit, a majority of people (68 percent) believed that Reagan was right not to accept limits on SDI.[143] Even if he had agreed to the constraints demanded by Gorbachev, it is highly unlikely that the US Senate would have ever ratified a treaty eliminating all American nuclear weapons; many Republicans were very critical of Reagan for even considering it. Thatcher was appalled at the proposal to get rid of all nuclear weapons. She told Reagan over the phone that this would be "tantamount to surrender" and that Britain "has no intention of giving up its independent nuclear deterrent."[144] National Security Advisor Admiral John Poindexter strongly urged Reagan to "step back from any discussion of eliminating all nuclear weapons in 10 years" and that he should make no further public comment endorsing the idea of the total elimination of all nuclear weapons.[145] But the elimination of nuclear weapons through space-based missile defense had become an article of faith for the president. There was no turning back now.

A FIRST-PHASE STRATEGIC DEFENSE SYSTEM

Never before had the meaning of "laboratory" become such a critical issue in superpower politics. Importantly, however, the Soviets had not specifically articulated at Reykjavik what confining SDI to the laboratory entailed.[146] To clarify the Soviet position on laboratory testing, Shevardnadze called upon Roald Sagdeev, a prominent Soviet scientist and advisor to Gorbachev, to explain "what is a laboratory in this context? Is it a small room in the basement tinkering with something, or what is it?" Sagdeev pointed out that it would be difficult to hold the US to only testing on Earth, since the Soviet Union had conducted experiments on its orbital space laboratory.[147]

Marshall Sergey Akhromeyev would admit to Henry Kissinger in early 1987 that it was difficult to delineate space experiments used for military and purely scientific purposes but that "we need to seek such distinctions."[148] Shortly after the meeting with Shevardnadze, Sagdeev traveled to New York and gave a press conference. The *Washington Post* reported that "Soviet scientist says 'modest' SDI testing is compatible with ABM pact."[149] It appeared that cracks in the Soviet position on laboratory testing might be forming, but there was no substantive change in reality.

Gromyko reminded his colleagues in the Politburo that SDI was about much more than missile defense. Satellites were becoming increasingly important for modern warfare through functions such as near real-time intelligence gathering, precision targeting, and command and control; all of these areas significantly overlapped with the SDI research agenda. SDI would serve as a catalyst for digital systems that would later be associated with the so-called revolution in military affairs.[150] Satellites in space would become critical information nodes for warfighting on Earth, and the Soviet Union could not keep pace with American developments in this arena. Gromyko identified limits on space-based testing of SDI technologies as the only way forward. He asked, "If we abolish nuclear weapons, then what? . . . Where are the guarantees that [the Americans] would not overtake us in the space race?"[151] Gorbachev believed that time was still on the Soviet Union's side, since deployment of a space-based missile defense system was a distant prospect.[152]

As pressure on the Pentagon to make progress intensified, SDIO sketched out a concept for an initial phase of an SDI system that might be ready for a deployment decision in the early 1990s. In late December 1986, Weinberger briefed Reagan on a new plan for a strategic defense system that would involve a constellation of orbiting kinetic interceptors designed to destroy ICBMs in their boost phase of flight.[153] Notably, more exotic technologies such as laser weapons were not included because they would not be mature enough for deployment anytime in the near future. In September 1987, Weinberger would approve this concept called "Strategic Defense System (SDS) Phase I."[154]

Just as the Soviets anticipated, SDS would also have offensive space applications. A classified report to Congress on the SDS Phase I detailed how "in the event of conflict, the SDS would contribute to the support of all military operations through protection of space-based assets while denying freedom of action to the enemy": that is, destroying adversary satellites.[155] Kenneth Adelman later observed that "this [space control role] is just what Gorbachev said was the real reason for SDI, which Regan ferociously denied. [It's] good this wasn't leaked."[156] In reality, this classified report only acknowledged what experts had been saying publicly for some time: space-based missile defenses would be used offensively in wartime to attack space systems. Fundamentally, space offense and defense were inseparable.

The archival record in the former Soviet Union reveals that immediately after Reykjavik, senior officials in Moscow took a keen interest in work being done to counter the space-based elements of a deployed SDI system. On October 14, the Politburo asked the Ministry of Defense to present proposals for "hastening work on countermeasures for a deployed American multi-echelon missile defense, especially its space components."[157] These countermeasures were exactly what Gorbachev was referring to when he earlier told Reagan that the Soviet Union would pursue an asymmetric (and cost-effective) response to any deployed space-based missile defense capability. Even before this, the CIA reported that the Soviet Union would invest in improved ASAT systems, in addition to other measures, to respond to a US strategic defense system.[158] But blocking testing of US space weapons remained Gorbachev's primary goal.

SDIO was now concentrating on developing a system that, according to Abrahamson, would evolve over time in response to the Soviet threat environment. The first iteration would focus on kinetic interceptors, but he explained that later stages could add directed-energy capabilities as those technologies matured.[159] On March 15, 1987, he personally presented the SDS Phase I in detail to Thatcher.[160] Abrahamson described how interceptors would be housed in space "garages," initially three hundred would be needed, although he confirmed that this number might later rise to more than a thousand.[161] But the general admitted that the US did not "currently have any adequate launch capability." The logistical requirements for getting so many SDI components into space were clearly going to be a huge problem. Abrahamson was nevertheless adamant that phased deployment would be feasible beginning in the mid-1990s—a position that senior officials at the UK MoD rejected. To the irritation of MoD scientists, Thatcher took Abrahamson's word over theirs.[162] Problematically for US defense planners, the space components of SDS Phase I were vulnerable to attack from Soviet ASATs. The garages housing interceptors would be sitting ducks in space for Soviet ASATs.[163] SDS Phase I, despite its shortcomings, signaled that the White House wanted to deploy strategic defenses as soon as possible, with the idea that the system's problems would be worked out later.

By late January 1987, conflicting messages were coming out of the Reagan administration regarding the deployment timeline for a strategic defense system. Weinberger told a gathering of the National Space Foundation in Colorado Springs that "we are now seeing opportunities for earlier deployment of the first phase of strategic defense than we previously thought possible."[164] The defense secretary explained that while the first phase would not be able to destroy all incoming Soviet missiles, it would upset Moscow's strategic calculus because Soviet defense planners would not know how many of their missiles would make it through US strategic defenses. In contrast to Weinberger, Admiral William Crowe, the chairman of the joint chiefs, reported to the Senate Armed Services Committee that "it will be quite some time before a [deployment] decision can be made."[165]

Despite Crowe's statement, some European allies were concerned that the US was indeed expediting SDI deployment and that it could further upset the strategic arms dialogue with Moscow; Italian Prime Minister Bettino Craxi shared these very concerns with Thatcher in February 1987. The British prime minister conceded that there had been "a lot of careless talk, [but] there was no question of a deployment of an SDI system for many years." Thatcher added the caveat that "she did not share the worries of some people about the SDI. It was vital that the West should always be in the forefront of new defense technology."[166] Her views on SDI and the ABM Treaty were evolving. She told Nitze and Perle on February 25 that "if the [SDI] study now being undertaken by the United State Department of Defense demonstrated that feasibility could only be established on the basis of the broader interpretation of the ABM Treaty, she would wish to see the program go ahead on that basis."[167] She nevertheless cautioned against any move to the broader interpretation of the ABM Treaty until it was absolutely necessary.[168]

After Reykjavik, Moscow was still holding to its unilateral moratorium on ASAT testing against targets in space. In February 1987, Soviet officials went a step further and ordered that "everything that could resemble tests of space-based weapon systems" be cancelled.[169] Despite this position, in May 1987, the Politburo approved the launch of the Skif-DM laser weapons system, but it failed to reach orbit.[170] After this, the MoD lost all remaining political support for space-based weapons. Pavel Podvig observes that "a successful 'Skif-DM' mission would have complicated the efforts to limit development of space-based weapons systems" and that by "1987 the Soviet political leadership considered this program an impediment to . . . an arms control agreement."[171]

Around the same time that the Soviet Union cancelled testing of its space-based weapons, Reagan signed a national security directive pressing "for the elimination of the [ASAT] testing moratorium at the earliest opportunity" in support of his goal to deploy an ASAT as soon as practicable.[172] In a section of the document that was previously secret, and remains partially redacted, a restructured ASAT program is mentioned that would be "jointly funded by the Air Force and the Strategic Defense Initiative Organization."[173] Developing ASATs jointly between the Air Force and SDIO made sense from a funding and program management standpoint, but overtly connecting missile defense and ASATs was still politically sensitive. Although it was not publicly acknowledged, SDIO managers believed that ASATs would be necessary to carry out space control operations to defend the space-based components of a strategic defense system.

THE SPACE ROADBLOCK BEGINS TO CRUMBLE

In American–Soviet exchanges after Reykjavik, there was still substantial disagreement about what the ABM Treaty did and did not permit with regard to testing in space. Both Reagan and Gorbachev wanted to secure agreements on INF and START, but divergent views on space were causing progress to grind to a halt. A decisive turning point in the arms control dialogue was a February 1987 Politburo meeting in which Gorbachev decided to de-link SDI from negotiations on INF. Gorbachev made this decision not because of pressure from SDI, but rather because he recognized that doing so was the only path toward an INF Treaty. Moreover, curbing the arms race was vital for his economic agenda.[174] In addition to the political benefits of de-linkage, technical factors were also clearly important for Gorbachev's decision. Because of advice from Yevgeny Velikhov, and other members of Gorbachev's scientific advisory team, the general secretary "felt more secure knowing that with countermeasures the Soviet Union could render SDI useless."[175]

Moscow and Washington were making substantial progress in curbing the arms race, but the military uses of space would remain a serious stumbling block in the START negotiations. If Gorbachev had confidence in Soviet countermeasures to SDI, then why did he insist on maintaining linkage between SDI and START? The answer to this question lies in the fact that even a partially functioning space-based missile defense system could have had a negative impact on Soviet strategic forces.[176] Building SDI countermeasures (e.g., more ASATs and increased numbers of ICBMs) would also have been costly, and Gorbachev wanted to reduce defense spending across the board. Additionally, officials in Moscow were still concerned about SDI leading to more advanced technologies that would give the US an even greater qualitative lead over the Soviet Union. Shevardnadze was therefore insistent that both sides agree to a comprehensive ban on space weapons technologies.[177] Reagan was not, however, giving any ground on SDI, even though he truly wanted a START agreement.

Frank Carlucci, who became the president's national security advisor in December 1986, implored Reagan to consider making at least some concessions on the ABM Treaty, such as a guaranteed non-withdrawal period. Pushing back on Abrahamson's optimistic assessments, Carlucci said that he "would bet every penny . . . that SDI would not be deployed until 1996 or 1997 at the earliest." Reagan would not, however, entertain any concessions on SDI and maintained that the Soviets wanted to "kill SDI so that they can build their own systems."[178] As in earlier periods, US intelligence did not provide a definitive assessment regarding the status of Soviet research into strategic defense. A July 1987 NIE asserted that there was "strong evidence of Soviet efforts to develop high-energy lasers for air defense, antisatellite (ASAT) and ballistic missile defense (BMD) applications." It further stated that "the Soviets also appear to be considering space-based lasers for BMD." Even though analysts disagreed about the feasibility for deployment before the year 2000, these kinds of intelligence reports only reinforced Reagan's belief that SDI remained a necessary program. For Reagan, SDI was at times a cooperative tool of peace, while at others, it served as a guarantee that Moscow would not secure an advantage with a "red shield" in space.[179]

One of the key problems in the START negotiations toward the end of Reagan's second term was determining whether field testing of weapons in space would be permitted. As the Soviet Union once again pushed for a complete ban on "space-strike arms," Reagan claimed that "the space threat posed by the Soviet Union is growing more serious as time goes on" and that a US ASAT would "help preserve the security of the nation and our men and women in uniform."[180] Even some lawmakers in the Democratic Party believed that the US should move forward with an ASAT effort. Representative George Darden (D-GA) said during a heated debate on the House floor that "there is no reason to eliminate the entire ASAT program" and that the "prospective SDI system may have promising ASAT capabilities."[181] This (public) advocacy of more fully integrating SDI with the US ASAT effort further undermined Reagan's prior statements that SDI was only a defensive capability and a tool of peace. Reagan called on Congress to remove its restrictions on testing the US ASAT against objects in space. The Air Force was planning three ASAT tests in space for 1988. Congressional leaders, however, preserved the moratorium on ASAT testing, which led to the cancellation of the MHV ASAT program in 1988.[182]

To resolve American–Soviet differences over missile defense testing in space, the White House proposed that the US and the Soviet Union agree to a Defense and Space Treaty, separate from START, which would allow specific kinds of missile defense tests in space.[183] Most importantly, Reagan wanted to convince Moscow to drop its linkage of START with an agreement on the ABM Treaty. In the immediate term, Adelman proposed that Reagan agree to prohibit space-based tests against targets on Earth as an act of good faith. Even though the Pentagon had no space-to-Earth tests planned, Weinberger fought against Adelman's position, saying that the need might arise and that "we ought not to give up any flexibility now."[184] Reagan sided with Weinberger and continued to reject *any* limits on military activities in space, which only played into Soviet fears.

In an attempt to allay Moscow's concerns, US negotiators developed a "predictability package" that would include Soviet observations of American space-based tests, US and Soviet personnel visiting each other's defense laboratories, and exchanges of data on missile defense research.[185] Washington and Moscow could not, however, resolve outstanding issues in START negotiations, including missile defense testing in space, before Reagan left the White House. Freedom of action in space remained at the forefront of the US arms control negotiating strategy.

CONCLUSIONS

After Mikhail Gorbachev replaced Konstantin Chernenko, he breathed new life into the Soviet bureaucracy. Ronald Reagan finally had a Soviet counterpart who shared his desire to make progress on arms control. Even though the president and the general secretary had a common goal of a nuclear-free world, their views on the means of securing it could not have been more different. Reagan viewed space-based missile defense as the insurance policy against nuclear threats and the foundation of prospective arms control treaties. In stark contrast, Gorbachev maintained that nuclear reductions were only possible after first curbing the further militarization of space.

Even though the Reagan administration recognized that a comprehensive strategic defense system would not come to fruition for many years, probably not until the next century, SDIO needed to be able to test critical missile defense components in space long before a deployment decision was reached. The ABM Treaty was the key obstacle in the way of more expansive testing, and it was a symbol, for Reagan and many of his advisors, of a weight holding down US technological competitiveness. In the arms control negotiations, Reagan had to carefully balance his desire to achieve deep cuts in nuclear weapons with his need to secure Soviet acquiescence to ABM Treaty modifications.

American, Soviet, and British declassified documents provide new insights into how SDI system design considerations shaped arms control developments in the second half of the 1980s. Notably, the research and development requirements for a strategic defense capability was an especially critical factor in the US decision to eschew any form of space arms control. In the end, Gorbachev's pragmatic decision to de-link SDI from INF negotiations paved the way for the first landmark arms control agreement of the Cold War's final days. The general secretary's confidence in making this decision was partly based on reports by the Soviet Ministry of Defense that the Soviet Union would indeed be able to exploit vulnerabilities of a space-based missile defense capability if the US ever deployed one. Fundamentally, specific technological considerations, rather than abstract ideas about SDI alone, shaped key American and Soviet decisions with regard to arms control.

As Reagan approached the end of his presidency, he had an arms control agreement (the INF Treaty) that eliminated an entire class of nuclear weapons. Concurrently, the Pentagon solidified a first-phase concept for strategic defense. Even though the Department of Defense had demonstrated hit-to-kill interceptor technologies, integrating all of the components necessary for an effective space-based missile defense system was still a distant prospect. SDI's future was not only uncertain from a technological standpoint. As the US and the Soviet Union improved their relationship, SDI system designers would have to create new justifications for space-based missile defense in a rapidly changing security environment.

6 SDI AND THE NEW WORLD ORDER

More than ever before, space is the "High Ground" that we must occupy.
—National Military Strategy of the United States (1992)

On August 2, 1990, at 2:00am local time, Iraqi military forces crossed over the border with their oil-rich neighbor Kuwait and quickly seized control of the country. Approximately a week later, the US began deploying military units in Saudi Arabia as part of Operation Desert Shield. On November 29, 1990, the United Nations Security Council authorized the use of force against Iraq if Saddam Hussein did not remove his three hundred thousand troops from Kuwait. As expected, Saddam refused, and on January 16, 1991, the US and its coalition partners from thirty-two nations commenced Operation Desert Storm to liberate Kuwait. After nearly six weeks of aerial bombardment of Iraqi targets, the US led a multinational ground offensive that resulted in the majority of Iraq's armed forces in Kuwait either retreating or surrendering. This short conflict witnessed the introduction of some of the most technologically advanced military hardware in the US arsenal, including so-called smart weapons guided by GPS satellites. The chief of staff of the US Air Force, General Merrill McPeak, said that space capabilities had been so effectively integrated into combat operations that Operation Desert Storm constituted "the first space war."[1]

Saddam's decision to invade Kuwait would have significant consequences for the future of SDI. Prior to Operation Desert Storm, many members of Congress were questioning the need for a large-scale strategic defense effort in light of improving relations with the Soviet Union. Shortly after Desert Storm began, Iraqi forces fired multiple Scud missiles at Saudi Arabia and Israel. The media depicted US Patriot tactical missile defense units as having successfully intercepted the majority of Iraqi Scuds.[2] Consequently, there was a surge in congressional support, on each side of the political aisle, for missile defense. Concerns over the proliferation of ballistic missiles to states such as Libya and Iran, coupled with the alleged successes of Patriot, served to validate at least some US investment in missile defense. However, deployment of missile defense in space remained a significant point of divergence among lawmakers and defense officials.

Early in his presidency, George H. W. Bush spoke of a rapidly changing global geopolitical situation and a "new world order" that could emerge; "a new era—freer from the threat of terror, stronger in the pursuit of justice and more secure in the quest for peace."[3] Missile defense advocates promoted SDI as a key element of security in this new world order. It would provide protection against the proliferation of missile technologies across the Middle East, in North Africa, and other areas. At the same time, US national security and defense strategy documents identified space as an even more critical arena for the US.[4] SDI technologies deployed in space could be valuable tools for establishing American hegemony in the cosmos.

The fate of SDI was not primarily a technological question. Rather, lawmakers had to decide whether it was in the US interest to deploy space weapons as the Soviet threat diminished and budgets were increasingly constrained. Technological problems with SDI research and development in the late 1980s and early 1990s complicated the efforts of the program's advocates to push forward with the deployment of strategic defenses in space. But considerations related to delicate American–Soviet (and later American–Russian) relations, arms control, and the changing geopolitical landscape were as important as questions surrounding the technological feasibility of a deployed strategic defense system.

AN UNCERTAIN FUTURE: SPACE, STRATEGIC DEFENSE, AND THE REAGAN–BUSH TRANSITION

In the final two years of Reagan's presidency, some national security experts were concerned that the US was not keeping pace with the Soviet Union's military space program.[5] Senior administration officials suspected that

American leadership in space was "being perceived by the domestic and foreign public as less than fully credible."[6] Weinberger believed that it was time to address "the perception that exists in some quarters that the Soviet Union has surpassed the United States in space capabilities" by formulating an updated national space policy.[7] Weinberger alleged that this anxiety about Moscow having gained an advantage in space was driven by "recent Soviet [space] successes." The secretary of defense was alluding to Moscow's development of the SL-16 medium-lift and the SL-17 heavy-lift rockets that could carry large payloads into higher orbits.[8] A 1987 intelligence report predicted that the SL-17 would provide options for orbiting large components for space weapons.[9] The *Washington Times* reported in May 1987 that due to the Soviet Union's "new rocket," the Soviet Union could secure a "commanding lead in 'star wars' space-based military systems."[10]

In addition to Soviet space developments, SDI was also a key factor in the push to revise the US national space policy. Since SDI had reached a stage where more testing in space was being planned, it would soon make greater demands on US space infrastructure.[11] To deploy SDI components, the US needed a new launch capability with heavier payload capacity, which would be an expensive undertaking. Initial concepts for what became known as the "Advanced Launch System" (ALS) were estimated to cost somewhere between $8 and $14 billion.[12] ALS highlighted, once again, that SDI's greatest cost were associated with infrastructure for launching and servicing space components. Political controversy over weapons in space overshadowed the reality that a deployed strategic defense system would be one of the largest space programs ever conceived.

Before SDI could get off the ground, it would have to survive at least one new presidency and multiple budget cycles. In the final year of Reagan's tenure as president, the fight over SDI became especially contentious between the executive and legislative branches. In early August 1988, Reagan, against the advice of Secretary of Defense Frank Carlucci and National Security Advisor Colin Powell, vetoed the FY1989 National Defense Authorization Act (NDAA), in large part, due to its constraints on SDI. It included a nearly 20 percent cut in SDI funding, in addition to restrictions on funding for a space interceptor.[13] The president said that "the way this bill restricts our proposed space-based interceptors would cripple the very concept of a space shield against nuclear attack. And I will not abide this, particularly in view of the technical progress that SDI is making. They say this bill would take the 'stars' out of Star Wars. With my veto today, I'm putting back the 'I'—initiative—in SDI."[14] Ultimately, Congress dropped the funding constraints for a space interceptor, but kept the SDI budget at approximately $3.7 billion. This was the first time since the establishment of SDI that the program's budget remained nearly identical to the prior year.

To build momentum for SDI, advocates of the program looked for ways in which strategic defense technologies could be immediately applied and demonstrated. ASATs were one such area for using SDI technologies, which caught the media's attention. The *New York Times* reported in the fall of 1988 that "exotic weapons being developed by the Pentagon to shoot down enemy missiles are now being promoted for a new role that is less taxing and more controversial: the destruction of enemy satellites in space."[15] Lasers and the Exoatmospheric Reentry-Vehicle Interceptor Subsystem (ERIS), a ground-based kinetic interceptor for the Strategic Defense System, were both potential candidates for ASATs.[16] British diplomats remarked in a cable back to London that there had "been a resurgence of interest in anti-satellite weapons in the dog days of the Reagan administration."[17] For nearly a year, the FCO had been following efforts to repurpose SDI's Mid-Infrared Advanced Chemical Laser (MIRACL) for an offensive ASAT role.[18]

Reagan approved an updated national space policy in February 1988 that identified space control—that is, being able to deny an adversary the use of space through military force—as an official Department of Defense mission.[19] Carlucci emphasized the urgent need for an operational ASAT system, claiming that the absence of a US ASAT capability was the "single most vulnerable point" in the nation's defenses.[20] To remedy this situation, the Reagan administration's final defense authorization included a request for funds to develop ASATs further.[21] In an effort to garner the support of all the military services, the Department of Defense established an ASAT Joint Program Office, rather than having one service lead the entire effort as had been done with the air-launched MHV.[22]

The resurgence in ASAT interest was due to multiple factors. Key officials in the administration certainly viewed the lack of an operational ASAT as a military deficiency. Also important was the fact that using SDI technologies for ASATs served as another avenue for furthering strategic defense research. Charles Monfort, the Washington director of the Union of Concerned Scientists, wrote in a 1988 *Bulletin of the Atomic Scientists* article that the "Pentagon is reviving antisatellite weapons as a cheaper, less controversial way than the Strategic Defense Initiative to work on defensive technologies."[23] He pointed out that "the line between antisatellite weapons and strategic defense weapons has become so blurred that it may be impossible to distinguish between the two . . . as defense budgets tighten, it will be difficult for the Pentagon to find the tens of billions of dollars that would be needed for full-scale deployment of strategic defenses." In reality, there had never been a clear boundary between missile defense and ASATs; the earliest US ASATs were modified missile defense capabilities. But Monfort was correct that attempting to garner greater support for ASATs was another way to keep up the momentum for research into strategic defenses. Notably, openly discussing the use of missile defense technologies for ASATs was a break with the past practice of distancing SDI from offensive space weapons. Nevertheless, it would be up to the Bush administration to decide how to proceed on both SDI and ASATs, and his military space agenda was anything but clear.

Vice President George H. W. Bush accepted the Republican nomination for the presidency on August 18, 1988. In his speech, he pledged to "modernize and preserve our technological edge, and that includes strategic defense."[24] Bush was, however, circumspect regarding deployment of a missile defense system. In August 1988, he had said that "a 'full deployment' of the system would be 'very expensive,'" and that any decisions about it would depend on further research."[25] In response to Bush's comments on SDI, the conservative political commentator William F. Buckley Jr. wrote an op-ed for the *Washington Post* questioning Bush's commitment to the program. He said that the vice president had been "stepfatherly in his treatment of SDI" during his party nomination acceptance speech.[26] Bush was surprised by the Buckley piece, and he subsequently sought to reassure conservatives that he supported missile defense. Yet, he remained cautious in discussing deployment and would not fully commit to it at this stage.[27] The announcement that veteran national security expert Brent Scowcroft, an SDI skeptic and ardent supporter of the ABM Treaty, would be Bush's national security advisor only fueled speculation that the incoming president would let SDI wither away.[28] In stark contrast with Reagan, neither Bush nor his running mate Dan Quayle (who wanted a near-term space-based missile defense deployment) expressed the idea that strategic defenses could lead to a nuclear-free world. The nuclear abolitionism that was one of the driving forces behind SDI was now gone.

British officials in the FCO and the MoD saw Bush's election as an opportunity to intervene early and convince the new president not to move forward with the deployment of a strategic defense capability. On November 11, 1988 (right after the election), the FCO and MoD advised Thatcher to tell Bush that the UK government was "opposed to either cooperative or unilateral deployment of an SDI system" and wanted to "shift our own policy to one of active discouragement of SDI." Next to these recommendations, Thatcher wrote "NO" in large letters and double underlined it.[29] Thatcher shared the view of Charles Powell, her trusted advisor, that SDI could "complicate the calculations of an attack[er] and therefore add to deterrence." It could also serve as a guard against nuclear proliferation from "some maverick power such as Libya or North Korea." Powell, like Thatcher, maintained that technological progress was unstoppable; he argued that it was not prudent to "stop SDI: the march of technology will simply engulf us."[30] Amid the renewed debates in the MoD and FCO concerning SDI, Thatcher informed her cabinet that "there were good arguments for eventual deployment of strategic defenses" especially because such defenses could upset the Soviet strategic calculus, thereby enhancing deterrence.[31] She not only refused to try to prevent SDI from moving forward, but also actively encouraged Bush to "remain committed to SDI" during their first meeting in June 1989.[32]

Within weeks of the inauguration, Bush outlined his administration's priorities and pledged to "vigorously pursue the Strategic Defense Initiative."[33] Just as during the campaign, the president was cautious regarding deployment. Approximately five months after entering the White House, Vice President Quayle said that the administration was indeed committed to the development and *deployment* of a strategic defense system and decried the "MAD mullahs" opposed to it.[34] It was, however, going to take a lot more than vocal support from the vice president to overcome the technological and political obstacles to deploying a space-based missile defense system.

FROM SMART ROCKS TO BRILLIANT PEBBLES: STRATEGIC DEFENSE TAKES SHAPE

In 1987, it was becoming increasingly clear to the Department of Defense that the space-based interceptor for the SDS Phase I was going to be a problem. A "garage" in space housing multiple interceptors would be vulnerable to Soviet ASATs, and its size would render launching it into space cost prohibitive. Even with the miniaturization efforts of SDIO, the "garage" concept was still too costly given the fiscal environment. Due to these problems, Lowell

Wood of Lawrence Livermore National Laboratory had come to believe that small autonomous interceptors could solve the vulnerability and cost problems associated with a space-based interceptor system.[35] Most beneficially for SDIO, these autonomous interceptors could be built using off-the-shelf technologies and would therefore potentially allow a deployment decision in the early 1990s. Wood coined the name "Brilliant Pebbles," which was based on Daniel Graham's "smart rocks" idea that involved highly capable sensors guiding small missiles (i.e., the rocks) in space to destroy incoming ballistic missiles.[36] Wood briefed Abrahamson on the concept in early 1987, and the general was immediately enthusiastic.

In the winter of 1986, SDI had formally entered into the defense acquisition process, under the purview of the Defense Acquisition Board (DAB), which meant that the program would be subject to more scrutiny regarding key system decisions.[37] While Brilliant Pebbles made the most sense to Abrahamson, it would be up to the DAB to make formal recommendations to the secretary of defense on how to proceed. A major DAB review of SDS Phase I led to DAB Milestone 1 approval of work directed toward "development and deployment coupled with endorsement of program efforts directed to subsequent SDS phases." DAB Milestone 1 approval meant only that the program was moving into the demonstration and validation stage of work. Notably, the feasibility of critical subsystems had not yet been established. Defense experts anticipated that it could take at least another four to five years to gather enough data to make an informed decision about the full-scale engineering development of a strategic defense system and its requisite subsystems. The Department of Defense awarded Martin Marietta a contract (worth approximately $500 million) to establish a US National Test Bed—that is, a gigantic simulation facility—to be based in Colorado Springs. This computing facility would carry out the complex simulations for the modeling and design of the SDS Phase I. But the accuracy of the computer simulations would become a politically contentious issue, as lawmakers had to decide in the coming years how to proceed with SDI funding.

At this stage, the joint chiefs of staff issued their effectiveness criteria for the first phase of a strategic defense system. They required that it "be capable of destroying 50% of the first wave of SS-18s [strategic Soviet ICBMs] and 30% of [all] other systems."[38] The strategy being pursued was to "enhance deterrence by ensuring that the structure of a full-scale attack could be disrupted."[39] In other words, defense planners in Moscow would not know how many of their nuclear delivery vehicles would survive, and they would thereby be deterred from launching a strike. Thatcher had been convinced that this was indeed a sound deterrent strategy, although the FCO and MoD did not share her view.[40]

Although the joint chiefs of staff had settled on effectiveness criteria for a strategic defense system, there were still more questions than answers concerning technological feasibility. In SDI's early years, Reagan had said that the program would be a long-term effort, and it was becoming increasingly clear that it might take thirty years or more to develop the technologies to deploy a comprehensive defense. British officials observed that integrating all of the components, even just for a phase 1 system, would constitute "the largest and most complex undertaking of its kind ever attempted."[41] Additionally, insufficient space launch infrastructure was once again identified as a key logistical problem. In the public sphere, the debate over SDI was oftentimes framed as a scientific problem, when in reality the challenges associated with systems engineering and maintaining such a large infrastructure in space were the most significant issues faced by the program's leadership. Louis Marquet, who served as the deputy director for technology within SDIO, said that "interceptors to destroy the warheads are not the problem. We know we can build those. The key [problem] is sensors" to track ballistic missile components in various phases of flight.[42] British technical experts in the MoD warned that these "problems are likely to take longer to solve than the US are willing to admit."[43]

The Defense Science Board (DSB) published a report on the status of SDI in May 1988. It recommended a phased deployment plan "rather than a single major action" to meet the joint chiefs' requirements.[44] The DSB placed great emphasis on the further development of effective sensors and command and control capabilities. The report stated: "We believe it would be better to think about ballistic missile defenses as first of all a surveillance system together with its associate processing and communications, whose purpose is to determine the actual characteristics of an attack, to find the boosters against the background and to find the RVs amid the decoys, chaff, nuclear effects, and other countermeasures and to determine where they are and where they are going."[45] Robert Everett, who led the DSB review of SDI, explained in Senate testimony that "instead of thinking of it [SDI Phase I] as a collection of weapons," program managers should "think of it as a central sensor processing system into which you then add those reaction devices, weapons . . . as they make sense."[46] The *New York Times* said that Everett's group emphasized "the brains, rather than the brawn, behind the Star Wars system."[47] Again, space infrastructure was the key issue.

Also in May 1988, the OTA published its own report entitled "SDI: Technology, Survivability, and Software." Its findings can best be described as a mixture of qualified optimism and significant uncertainty. The report's authors concluded that "after 30 years of BMD research, including the first few years of the Strategic Defense Initiative (SDI), defense scientists and engineers have produced impressive technical achievements, but questions remain about the feasibility of meeting the goals of the SDI."[48] OTA pointed out that a key SDIO deficiency had been devoting "little analysis of any kind of space-based threats to BMD system survivability" and that "SDIO analyses assume that US BMD technologies will remain superior to Soviet technologies."[49] To negate the space components of SDI, OTA assessed that the Soviet Union could indeed produce more effective ASATs. To counter this threat, the US would have to secure "control of certain sectors of space." In other words, protecting a strategic defense system required more robust preparation for combat operations in space. OTA also identified software dependability as a significant challenge and a point of vulnerability for a deployed system.[50]

Deployment was a long-term issue, but SDIO soon needed to confront the legal complexities concerning the expansion of missile defense testing in space. There was still disagreement within the Department of Defense about how long SDIO could go on without moving beyond the more restrictive interpretation of the ABM Treaty. Abrahamson informed Thatcher in January 1989 that there had been much progress with the Brilliant Pebbles concept and that it would greatly reduce the cost of the overall system and the strain on launch capacity. He predicted that testing would begin in two years' time.[51] This meeting with Thatcher was Abrahamson's last as head of the SDI program. (Even though Reagan had nominated him for a fourth star, which would have added even more political clout to SDIO, Congress would not approve it.) At this January 1989 meeting, Abrahamson introduced the prime minister to his successor, Lieutenant General George Monahan, who lacked Abrahamson's charm, charisma, and zeal for SDI. Thatcher told Monahan that she hoped he would continue Abrahamson's practice of traveling to London to provide SDI updates. Few people recognized just how significant the relationship between Thatcher and Abrahamson was for UK policy on SDI. It was largely because of their interactions that the prime minister came to believe in SDI technologies and even supported the deployment of strategic defenses against the advice of her government ministers.[52]

In his end of tour report right before retiring, Abrahamson strongly endorsed Brilliant Pebbles. According to Baucom, Abrahamson was convinced that Brilliant Pebbles "could be operational in five years at a cost of less than $25 billion"; this was a huge saving when compared to earlier estimates for a first-phase system that far exceeded $100 billion.[53] As a point of comparison, a Nimitz class nuclear-powered aircraft carrier commissioned in the late 1980s cost approximately $2.5 billion (1988 dollars). Even with the potential cost savings, the Pentagon's adoption of Brilliant Pebbles was not guaranteed.

Early in Bush's presidency, Secretary of Defense Richard Cheney signaled that he would be a firm advocate of space weapons. In April 1989, Cheney said that developing a "satellite-killing weapon [was] 'most important,' arguing that the Soviets have an edge in the technology and could knock out US surveillance and communication satellites in a war."[54] Pushing back on Cheney's space agenda, in July 1989, Senator John Kerry (D-MA) introduced a bill entitled the "Satellite Security Act" that was intended to restrict funding for ASAT development and testing. He argued that the US was more dependent on space systems than the Soviet Union and therefore had more to lose from an ASAT competition. Kerry pointed out that the Soviet Union had opened its laser test facility at Sary Shagan to US scientists, journalists, and Congressmen in the summer of 1989, revealing that Soviet research into laser weapons was not as far along as previously believed.[55] Most fundamentally, the senator argued that "this new information raises the question of whether past assessments of the Soviet laser program have significantly overestimated or exaggerated the military capability of the lasers themselves and of the program overall."[56] If the intelligence was wrong, it would "undermine the very foundation of the rationale for the billions we have spent on the Strategic Defense Initiative and the current crash [ASAT] program." Only a year prior, Representative George Brown Jr. (D-CA) had introduced a bill to ban the testing of ASATs indefinitely. Neither Brown nor Kerry was successful in their efforts.[57]

Clearly, there was reluctance among many lawmakers to push forward with the testing of ASATs. Nevertheless, some opponents to ASAT testing maintained that it was necessary to keep an ASAT research and development program to be able to deploy them quickly should they become necessary.

Brown said that his bill was defeated, at least in part, because there was a "high and rapidly increasing priority being given to the development of a warfighting capability in space."[58] Around this same time, Pentagon representatives were publicly discussing the vital role of ASATs in US defense strategy. In 1989, General Crosbie Saint, the commander-in-chief of US Army Europe, said that he saw "an effective ASAT system as the key to the control of space and also conceivably as the key to victory."[59] While Crosbie overstated the value of ASATs, having the ability to destroy Soviet ocean surveillance satellites certainly made sense from the standpoint of defense planners who were worried about keeping sea lines of communication open in wartime. These views were very much in line with the ASAT arguments made by the Buchsbaum panel in the mid-1970s.[60]

For the FY1990 defense budget, the Department of Defense requested an increase in funding for ASATs. The Army established the Kinetic Energy ASAT (KE ASAT) program in 1989, which was designed to use a ground-launched missile to destroy satellites in low Earth orbit. British diplomats observed that Cheney "is beginning to think and talk like a conservative in his growing affection for defense-related space projects (at a time when he is talking of slashing the defense budget as a whole)."[61] Cheney said that the US needed an "ASAT capability both to counter the Soviet targeting satellites and to deter the use of the existing Soviet ASAT system."[62] He further opposed restrictions on ASAT testing because of the negative impact they would have on moving to an operational stage and due to their potential consequences for SDI testing in space.[63] It is important to note that Cheney was Ford's chief of staff when the president decided in January 1977 to establish a new ASAT program. Cheney was therefore intimately aware of the military arguments in favor of having an ASAT capability. Scowcroft, who was Ford's deputy national security advisor (1973–1975), had publicly criticized ASATs as destabilizing before Bush appointed him as his national security advisor.[64] Disagreements between Scowcroft and Cheney on military space programs would extend to SDI as well.

One of Cheney's early tasks was advising the president on how the administration should plan SDI's future. Even though Bush had said in February 1989 that he intended to proceed with SDI, it did not appear to be a top priority. Quayle, on the other hand, regularly shared his enthusiasm for moving forward with deployment of a strategic defense system. In a speech to the Navy League's Sea-Air-Space Exposition in March 1989 he said that "one area where events are moving especially swiftly, and which I'd like to discuss with you this morning, is space. Nowhere else are our scientific, technical, commercial, defense, and foreign policy goals more closely intertwined . . . space is vital to our international competitiveness, to our continued economic growth, and, indeed, to our very survival as a nation."[65] He saw strategic defense in space as a necessary step toward significantly reducing nuclear threats to the US. Quayle's comments also echoed High Frontier's message about the need to protect US interests in space through military power, which would lead to even greater economic prosperity as the US developed its commercial space resources.

Cheney realized that a decision was going to have to be made in the near future regarding the space-based infrastructure of a strategic defense system, especially the interceptor. In April 1989, the secretary of defense indicated that he saw Brilliant Pebbles as the most cost-effective option. Cheney had already announced that he would be cutting defense spending, including for the SDI budget, and expressed hope that Brilliant Pebbles could "save a lot of money." Earlier in the year, he was signaling that SDI would move forward but in a reduced form. Cheney said that Reagan had "oversold" the program and that its focus would be on enhancing deterrence through partial deployment. The Department of Defense moreover presented SDI as a safeguard against the proliferation of missile technologies and nuclear weapons.[66]

In the late 1980s, US and allied national security officials were growing even more concerned about the proliferation of ballistic missiles and nuclear fissile materials to rogue regimes. Even Senator Sam Nunn (D-GA), who was a vocal critic of SDI, had expressed support for a more limited defense system— that is, a smaller land-based capability—that could provide protection against small-scale ballistic missile attacks.[67] Space-based interceptors were, however, vital for destroying missiles in their boost phase. SDI supporters did not want to see the space layer reduced in any way, while critics viewed it as still having the potential to start a new arms race.

Bush wanted to find a way to fit SDI into a more constrained budgetary environment without upsetting key elements within the GOP. It would have been politically problematic to abandon SDI altogether because there was bipartisan consensus that having a research program, as a hedge against a Soviet breakout in missile defense was indeed important, and SDI had become an article of faith for many vocal conservatives. Additionally, there were people, such as Cheney and Thatcher, who believed that a limited space-based missile defense system could upset the Soviet calculus just enough to enhance deterrence. Cheney also saw a partial strategic defense system as a means of protecting US ICBM silos against more accurate Soviet ballistic missiles, in addition to guarding against missile proliferation to "rogue" states.[68]

Quayle described Reagan's belief that SDI could lead to a nuclear-free world as "political jargon."[69] Marlin Fitzwater, the White House press secretary, tried to walk back from the vice president's remarks, but even ardent supporters of SDI, such as Daniel Graham, came to Quayle's defense. Graham said that SDI's goal was never a perfect defense. Richard Perle, a former senior defense official in the Reagan administration, described Quayle's comments as "long overdue," and predicted that they might help garner greater support for SDI because he was offering a less ideological portrayal of it.[70]

The vice president explained that that the emphasis would now be on developing a realistic military strategy for strategic defense.[71] In other words, Reagan's vision for SDI was not all or nothing; there were variations of his dream for eliminating the threat of nuclear annihilation through advanced space technologies. Even some firm believers in the necessity of nuclear deterrence had concluded that a partial missile shield could reduce the vulnerability of the US and its allies. A *Newsweek* article said in 1988 that "if the previous question was whether SDI was feasible, Americans must now ask if it is desirable."[72] Answering this question required going beyond the realm of purely technological considerations and evaluating US defense needs in a changing geopolitical environment. Cheney had come to believe that Brilliant Pebbles offered the best hope for moving the program forward in a cost-effective manner that could enhance US security interests. Nevertheless, there were still fundamental systems engineering and political issues that had to be addressed. Congress was unlikely to continue to fund SDI, especially space interceptors, unless there was sufficient indication that a deployed system was indeed feasible and there were guarantees that the program would not derail arms control negotiations.

With all of the technical and political uncertainty concerning SDI, Bush decided to proceed cautiously. In June 1989, the president signed NSD-14, entitled "ICBM Modernization and the Strategic Defense Initiative," which established that the goals of the SDI program remained "sound" and that research and development of advanced technologies necessary for "strategic defenses" should continue to be a major US response to the "Soviet challenge." In this R&D effort, "particular emphasis" was to be placed on promising concepts for effective boost-phase defenses, for example "Brilliant Pebbles."[73] The NSD also directed Cheney to have an independent review of the SDI program. The secretary of defense selected Ambassador Henry Cooper, who had served as a negotiator for the Nuclear and Space Talks, was a known SDI enthusiast, and would succeed Monahan as the head of SDIO. His examination of the SDI program would be completed in early March 1990 and end with a strong endorsement of Brilliant Pebbles.[74]

In the summer of 1989, the Department of Defense had two outside bodies, the JASONs and the DSB, conduct additional technical feasibility studies of various strategic defense concepts, including Brilliant Pebbles. The JASONs are a group of America's leading scientists who work under the aegis of the MITRE Corporation and advise the Department of Defense and the intelligence community on complex issues.[75] The JASON study focused on the feasibility of Brilliant Pebbles' component technologies and the battle management command and control system that would be used along with it.[76] The study participants also examined other interceptor concepts as points of comparison. Dr. John Cornwall, a physicist from Cornell University and the leader of the JASON Brilliant Pebbles review, presented his findings to Monahan in August 1989. Cornwall noted that Brilliant Pebbles was largely based on existing technologies, and the bottom line was that there were no technological "showstoppers."[77] Monahan also invited the UK MoD to conduct its own independent review of Brilliant Pebbles, and that study concluded that "there are no 'showstoppers' to making Brilliant Pebbles work but there are a number of technical risks which lead us to believe they will cost more and take longer to get into service." The MoD was especially concerned about the US being able to create effective software to make the entire system function, especially the autonomous interceptors.[78] Notably, the head of the JASON study was a physicist, and Slayton has pointed out that physicists oftentimes overlooked the software and information technologies requirements for a strategic defense system.[79]

Having Brilliant Pebbles operate as autonomous, or semi-autonomous, interceptors with an on-board suite of target-tracking sensors would greatly reduce the cost of the overall program because other space-based tracking systems could be eliminated. At least for a first-phase system, the smaller constellation size would significantly lower the strain on US launch systems. The report was not, however, completely laudatory; the JASONs noted that "there are several problems which must be addressed," including "serious countermeasures threats" such as ASATs.[80] Technical experts and seasoned intelligence analysts agreed that the Soviet Union could develop countermeasures, but there was disagreement on how effective they would be, and there was no consensus regarding Moscow's willingness to spend the money required to develop and deploy them.[81]

At the same time that the JASONs were completing their study, the DSB examined Brilliant Pebbles and submitted its report to the Department of Defense in January 1989. According to Baucom, "the DSB concluded that BP [Brilliant Pebbles] faced some technical problems that would have to be overcome, but found no fundamental flaws with the concept." Because Baucom was the official Department of Defense historian of SDI, he had full access to the two studies, but they have not been declassified and thus cannot be scrutinized. British MoD documents suggest that the reports were shared, either in part or in whole, with UK representatives. One MoD official noted in January 1990 that "it was equally obvious from the Defense Science Board and JASONS reports on Brilliant Pebbles that Lowell Wood's claims as to the maturity of the [Brilliant Pebbles] project were somewhat exaggerated."[82] This UK analysis was, however, vague and does not provide specific details about the feasibility of Brilliant Pebbles beyond the aforementioned UK review that said there were no "showstoppers." It is noteworthy that the UK report used the same phrase, "[no] showstoppers"; it is not clear if the British officials were influenced in any way by the American evaluations taking place. At the very least, UK officials were highly skeptical of the early 1990s deployment timeline as presented by their American counterparts.[83] Tests of Brilliant Pebbles hardware would raise more questions about the feasibility of a near-term deployment of an effective space interceptor.

Even though there was widespread support for Brilliant Pebbles among SDI promoters, it was not the only space-based interceptor being considered. A UK MoD memorandum noted that the DSB had recommended in September 1989 that "the [space-based interceptor] concepts should be allowed to compete for the next two years." The Air Force was promoting a "gun rack" interceptor, a modified version of the missiles housed in a space "garage," which was included in an SDIO space infrastructure study at the behest of the Air Force and in the face of opposition from both the Department of Defense and SDIO.[84] The competing interceptor concepts, driven in part by interdepartmental rivalry, provide an important reminder of Donald Mackenzie's observation that weapon systems are not the product of a natural trajectory of technological development. Rather, bureaucratic forces, like parochial military service interests, can be overwhelmingly influential.[85] Organizational interests aside, it was still not clear in the fall of 1989 how significant a priority military space programs, SDI in particular, were for Bush.[86]

To coordinate space policy more effectively across the civil, commercial, and national security sectors, Bush established a National Space Council chaired by the vice president. Quayle directed a "fast-paced" review of existing US space policy, acknowledging that it was "not expected to result in basic modifications to existing policies."[87] With a rapidly changing geopolitical environment, it was too soon to make any serious changes to the American space agenda. In November 1989, Bush approved his first national space policy, which was nearly identical to Reagan's final space policy. Within the document's national security section, the White House directed the Pentagon to "ensure [US] freedom of action in space," which required "an integrated combination of antisatellite, survivability, and surveillance capabilities." The president pledged that an ASAT would be developed and deployed "with initial operations capability at the earliest possible date." Regarding SDI, the Department of Defense planned to incorporate the requirements of space-based missile defense into its national security space program.[88]

For FY1990, the administration requested $4.6 billion for SDI, which was only marginally more than the previous year's amount.[89] Cheney was concerned that Congress might substantially reduce the program's funding, which would have pushed the deployment timeline out even more. Consequently, the secretary of defense tried to drum up support for SDI, predicting that it "could be the single most important military bequest this generation could make to the future."[90] He threatened to recommend that the president veto the FY1990 NDAA if Congress did not provide sufficient funding for SDI. Ultimately, Congress appropriated $4 billion for FY1990, and Bush signed the NDAA into law on November 29, 1989.[91] Research could move forward, but the financial belt would tighten even more.

Financial and technological considerations were not the only factors shaping SDI's future. In late September 1989, Secretary of State James Baker and Soviet Foreign Minister Eduard Shevardnadze met to discuss the ongoing arms control negotiations.[92] At this meeting, the Soviet Union announced that it would drop its requirement that an agreement be reached on the ABM Treaty before signing onto START.[93] This shift did not mean that all of the Kremlin's concerns about SDI had evaporated; senior Soviet officials still preferred formal limits on *kosmos-zemlya* ("space-to-Earth") weapons and

ASATs.[94] There was also concern about the US moving ahead with advanced strategic defense technologies while strategic nuclear weapons stockpiles were reduced.[95] The KGB predicted that "the United States would not be able to deploy a highly effective strategic defense system in the near term."[96] But just as in the US, there were divergent Soviet ideas about system "effectiveness." In any case, since Washington and Moscow could not agree to specific terms regarding missile defense testing and deployment, the Kremlin sidestepped these complexities by moving forward with START, but added the caveat that any violation of the ABM Treaty could result in the abrogation of START.[97] Opponents of SDI would point toward progress on START as evidence that there was no compelling justification for continuing to invest in strategic defense.[98]

Monahan convinced Cheney in the first half of January 1990 to include Brilliant Pebbles in a phase 1 strategic defense system.[99] Shortly thereafter, on February 7 Bush visited Lawrence Livermore National Laboratory where much of the Brilliant Pebbles work was being done; General Monahan accompanied him. During that visit, the president expressed enthusiasm for Brilliant Pebbles, saying that "if the technology I've seen today proves feasible—and I'm told it looks very promising—no war planner could be confident of the consequences of a ballistic missile attack."[100] He stressed that strategic defenses were more important than ever. Within a matter of days, Lowell Wood and Monahan presented the Brilliant Pebbles development plan at a press briefing. They explained that the first phase would include 4,614 Brilliant Pebbles in orbit at a cost of $1.1 to $1.4 million per "singlet": that is, per interceptor. The price tag for launching these systems into space would be between $2 and $3 billion. In an effort to lower costs, the interceptors would have a more limited number of tracking satellites to assist them in detecting and targeting ballistic missiles. Therefore, the total price tag of a phase 1 plan was estimated to be $55 billion.[101]

Wood explained that each Brilliant Pebble would be housed in a "life jacket," which "serves a variety of functions from survivability to in-space functionality."[102] In peacetime, the Brilliant Pebbles would function as highly sophisticated sensor platforms that could detect and track a variety of missile threats. What made them "brilliant" were the data processing capabilities that could be housed on such a small satellite, which resulted from great strides in miniaturization over the previous decade. During crisis and conflict, US Space Command, the designated operator of a strategic defense system, would then "convert some or all of them into weapons platforms, understood to be in the interdiction mode."[103] Importantly, combining sensors and weapons on the same satellite would have eliminated the boundary between "passive" and armed space systems. One of the most innovative features of Brilliant Pebbles was their propulsion systems that had a very high thrust-to-weight ratio. Wood said that this new propulsion capability was the key to Brilliant Pebbles' agility, and that it if were commanded to do so, it "could fly a minimum energy trajectory to the planet Mars." Notably, rocket engineers would later propose using propulsion (and sensor) technologies derived from Brilliant Pebbles for planetary science missions.[104]

Among the key arguments in favor of Brilliant Pebbles was its cost efficiency and survivability as compared to other space-based interceptor concepts. However, there were a range of countermeasures the Soviets could choose from: creating faster-burn ICBMs, building more ICBMs, and fielding a larger number of ASATs. Monahan and Wood assured the public that Brilliant Pebbles would be hardened against ASATs.[105] Wood's confidence in the survivability of Brilliant Pebbles was undeterred by a March 1990 Lawrence Livermore report claiming that Brilliant Pebbles would indeed be vulnerable to Soviet ASATs. Because the new phase 1 concept would have fewer space-based sensors that could track both ballistic missiles and potential threats (e.g., ASATs) to the strategic defense system, Brilliant Pebbles sensors would have greater difficulty in detecting and responding to ASATs.[106] The author of the Lawrence Livermore study concluded that "although the BP is small and relatively inexpensive compared to other space-based concepts, it does not appear feasible to meet this [cost-effective] requirement when the [Soviet] ASAT carries a small non-nuclear homing device."[107] Brilliant Pebbles would therefore "not be economically competitive" unless some relatively inexpensive countermeasures to defeat the ASATs could be found.[108] To address the vulnerability of Brilliant Pebbles to ASAT attacks, the Department of Defense would have to be ready to launch replenishment interceptors into orbit, have more effective countermeasures (e.g., maneuvering satellites) and/or have ASATs to defend the space-based interceptors.

GETTING WITH THE TIMES: SDI IN A CHANGING GEOPOLITICAL ENVIRONMENT

The beginning of the 1990s was a time of some of the most significant geopolitical changes during the Cold War. Shortly after the fall of the Berlin Wall in 1989, West German Chancellor Helmut Kohl declared that his goal was the "unity of our nation."[109] German unification quickly took center stage in European security. Tim Sayle observes that in 1989, "all eyes were on Germany, both the great prize in the Cold War and the potential battleground in a hot war."[110] In August, Scowcroft warned the president that "managing our relations with Germany is likely to be the most serious geopolitical challenge our country faces over the next decade."[111] The integration of a unified Germany into NATO would be an especially difficult task. After the Soviet Union announced its de-linkage of START and SDI in September 1989, it appeared that the superpowers could indeed make serious progress toward signing a treaty to limit strategic nuclear arms. SDI's future in this changing political environment was uncertain to say the least.

Cheney told the House Armed Services Committee in February 1990 that he wanted to move forward with strategic defense deployment.[112] SDI officials and supporters used the "no showstoppers" from the JASON report as evidence that there was, according to Monahan, consensus among American scientists "that says indeed, [strategic defense] can be done." William Happer, a Princeton physicist who chaired the JASON steering group, cast doubt on the wholly positive characterization of the JASON report, saying that "I think that's overstating the case by quite a bit. I don't remember anywhere where we expressed confidence the problems [with Brilliant Pebbles] could be overcome. Maybe they can and maybe they can't."[113]

SDIO had to develop a strategy quickly for building greater momentum for strategic defense when it appeared that Congressional and public interest were waning. In March 1990, Ambassador Cooper completed his independent review of SDI and submitted it to Cheney.[114] He called for greater funding for Brilliant Pebbles, a strengthened commitment to the high-endoatmospheric defense interceptor (HEDI) for ground-launched mid-course interception, and a reduction in the number of tracking satellites. Cooper also sought to create a "national debate" on missile defense deployment that would make it a prominent issue in the next presidential election. Nevertheless, even the staunchest SDI supporters, like Cooper, recognized that the program would need to modify its approach to strategic defense to prevent further funding cuts and continue working toward a mid-1990s deployment goal.

With cost cutting in mind, in May 1990, the OSD began a six-month study of a concept called Global Protection Against Limited Strikes (GPALS), a scaled-down strategic defense system with land- and space-based interceptors. Approximately two years earlier, Senator Sam Nunn (D-GA) had introduced the idea of an Accidental Launch Protection System (ALPS). He envisioned a missile defense capability that would provide some protection against accidental launches rather than full protection from a large-scale nuclear attack. Congressman Les Aspin (D-WI) agreed that "there is great longing 'out there' for protecting against incoming missiles and that ALPS may be the answer."[115] A limited terrestrial-based defense served as a political middle ground to appease missile defense supporters while still preserving the ABM Treaty. But SDI boosters would not give up space weapons because they were the essential element for going after missiles in their initial phase of flight and would contribute to the Pentagon's space control capabilities.[116]

Cooper believed that GPALS would be more easily sold to space-based missile defense skeptics who saw some value in protection from limited missile threats and accidental launches. He shared with Thatcher that "he thought the best way to make progress with the overall program was to change the focus from concentration on a massive Soviet missile attack . . . to the risk of missile attack in any part of the world [especially rogue states]."[117] Cooper was echoing comments from Cheney, who had already started using "emerging third world threats" as a justification for continuing to fund strategic defense. During a March 1990 speech to the American Preparedness Association in Washington, DC, the secretary of defense had warned that by the year 2000, fifteen "developing countries will produce or own their own ballistic missiles."[118] Cheney believed that these emerging missile threats should not be dismissed as insignificant and provided a compelling reason to continue supporting Brilliant Pebbles.

To the relief of SDIO, in early June 1990, the DAB approved moving forward with Brilliant Pebbles.[119] Within a few weeks of this, Monahan retired and was succeeded by Henry Cooper. In choosing Cooper to replace Monahan, Cheney had "turned the clock back to the Abrahamson era, with a known champion of SDI as Director."[120] SDI boosters, like the Heritage Foundation, emphatically supported Cooper taking the SDIO helm.

Now, SDIO turned its attention to demonstrating the feasibility of space-based interceptors. In August 1990, SDIO conducted its first test of Brilliant Pebbles in space. Its objectives were to (1) demonstrate the ability to acquire stars, navigate, and stabilize the interceptor using the attitude control system; (2) demonstrate the ability of the interceptor to detect, acquire, and track an accelerating target's rocket plume; (3) gather data with infrared and ultraviolet sensors; and (4) demonstrate basic hardware performance versus design requirements in a realistic environment.[121] This preliminary flight test of Brilliant Pebbles was largely a failure because a bolt released prematurely eighty-one seconds after launch, which prevented transmission and recording of performance information from the interceptor. This test failure did not indicate a fundamental flaw with the Brilliant Pebbles system, but it did not help Cooper secure more funding for FY1991. The second test would not take place until the following year. For FY1991, SDIO only received $2.9 billion out of the more than $4 billion the president had requested. Without a significant change, SDI's future looked grim.

STRATEGIC DEFENSE IN THE NEW WORLD ORDER

In his 1991 State of the Union speech, Bush declared that the "end of the Cold War has been a victory for all humanity."[122] Germany was united, and the Soviet Union had voted for the UN resolution that empowered states to use "all necessary means" to remove Iraq from Kuwait. This was a very different world from the one he had known as Reagan's vice president for eight years. These significant changes in the international system were the inspiration for Bush's idea that states should embrace a "new world order, where brutality will go unrewarded and aggression will meet collective resistance."[123] Multilateralism was a key element in this emerging political system. But where did this leave strategic defense? The president announced that the US would officially reorient its strategic defense effort toward "providing protection from limited ballistic missile strikes, whatever their source" through the GPALS concept.[124]

GPALS was designed to provide a defense against a limited number of ballistic missiles from any place on Earth. It included two space-based systems: Brilliant Pebbles interceptors and Brilliant Eyes surveillance satellites. Additionally, there would be ground- and/or sea-based defenses for targeting ballistic missiles in their mid-course and terminal phases of flight. GPALS required a significant reduction in interceptors, bringing the total number down to between five hundred and a thousand.[125] The space-based infrastructure for *tracking* ballistic missiles would receive bipartisan support, but Brilliant Pebbles remained a source of political controversy.

The first Gulf War created a new wave of support for missile defense and provided SDIO with what it saw as living proof of the need for a global missile defense capability albeit on a reduced scale. British diplomats observed that the Gulf War had been instrumental in "creating a favorable political climate [for missile defense]." London believed that the US regarded GPALS as "more readily 'saleable' to allies, the Congress, and perhaps even the Russians."[126] Cooper and other SDI boosters certainly hoped that a reorientation toward GPALS would build momentum in favor of including Brilliant Pebbles in a first-phase strategic defense deployment.

To the chagrin of Cooper, due to tighter budgets and warming relations with the Soviet Union, even some Republican senators, such as John Warner (R-VA), William Cohen (R-ME), and Richard Lugar (R-IN), suggested that the US should focus on land-based interceptors for the first iteration of a strategic defense capability.[127] Predictably, this proposal was vehemently rejected by SDI boosters. The space interceptors were not only essential for attacking ballistic missiles in their boost phase of flight but they also served as "the symbol of Star Wars legitimacy."[128] Sociologist Graham Spinardi has compellingly argued that there was a battle among missile defense supporters for "epistemic authority" based on competing knowledge claims about ground- and space-based missile defense. He writes that for space-based missile defense advocates, ground-focused missile defense technologies were the "wrong approach *in principle*" (emphasis in original) because a space-based system is the only way to perform boost phase interception. Mid-course defenses are susceptible to simple countermeasures.

Spinardi does not, however, go far enough in explaining why space interceptors were so important for SDI boosters. Space offered more than an ideal vantage point for going after ballistic missiles in their boost phase. As already noted, for many missile defense promoters, space-based interceptors were just one part of a broader agenda to expand American military and economic power in space.[129] Daniel Graham's organization dedicated to promoting missile defense was called "High Frontier" because of the founder's belief that conquering space could provide the US unprecedented defense and economic advantages.[130] Westwick details the connections between space colonization groups, such as the L5 Society, and the embrace of missile defense in the 1970s.[131] Reagan often invoked the idea of space as a frontier that could be conquered. During a 1988 speech at Johnson Space Center, Reagan had said that it is "mankind's manifest destiny to bring our humanity into space; to colonize this galaxy; and as a nation, we have the power to determine if America will lead or follow. I say that America must lead."[132]

For SDI purists, there could be no compromise on any fundamental aspect of the space infrastructure for strategic defense. Thomas Rona, a defense official in the Reagan administration, had warned that "interim options focused or aimed exclusively at ground [defenses] . . . are unnecessary and possibly detrimental to the long-term SDI goals." But in the 1990s, space-based missile defense visionaries had to face geopolitical and fiscal conditions that were increasingly hostile to any expansive military space program, especially one whose deployment could upend progress in arms control. A GAO report stressed that SDIO had to contend with the changing Soviet threat.[133] This factor, even more so than any technical challenges, was driving the scaling down of US strategic defense concepts.

The first Gulf War not only highlighted the potential value of a more limited missile defense, but also shined a spotlight on the importance of military satellites for tactical combat operations. In many ways, Operation Desert Storm was the coming to fruition of the 1976 Buchsbaum panel's prediction that "this trend toward effective integration of space assets into military combat operations will continue and that real-time space capabilities will become increasingly important—even essential to the effective use of military forces."[134] Combat operations in Kuwait also appeared to validate Secretary of the Air Force Edward Aldridge's claim in 1988 that "spacepower will be as decisive in future combat as airpower is today."[135] (This statement begs the question of how decisive airpower had been in reality.) In this environment, space control was more important than ever according to military space promoters. General Thomas Moorman, who would go on to become the Air Force Vice Chief of Staff in 1994, said: "The ability of the US to maintain the initiative and to sustain surprise by masking its military actions would have been much more difficult if Saddam Hussein—or a future adversary—had his own space reconnaissance assets. This prospect argues for an ASAT system to assure that, just as US forces achieved control of the air and the battlefield, we can control space as well (i.e., achieve space superiority)."[136]

Even before the US-led coalition began to remove Iraqi forces from Kuwait, more US national security leaders drew connections between spacepower and maintaining military preeminence in all domains of operation. In May 1990, Deputy Assistant Secretary of Defense for Strategic Defense, Space and Verification Policy Douglas Graham said that "space control has become as important to the USA as sea control capabilities are to the exercise of maritime strategy and air power is to land and air warfare." He further declared that "our global responsibilities remain vital to US national security, and space will be an increasingly critical link to our forces and friends overseas."[137] In 1990, Chairman of the Joint Chiefs Colin Powell justified a new US ASAT program during Congressional testimony by asserting that it was needed to deter the Soviet Union from using its ASATs in wartime.[138] The Soviet Union, moreover, was no longer the only spacefaring nation to consider. More players, including China, Israel, Pakistan, and India, had entered the space arena.[139] This reality now had to factor into US military space strategy.

As the next presidential election approached, Cooper wanted to do everything in his power to shore up support for Brilliant Pebbles. In April 1991, SDIO conducted its second flight test of Brilliant Pebbles. This test occurred around the time that the White House requested an increase in funding for Brilliant Pebbles, with 2001 as the new target year for deployment. After the test, SDIO officials, including Henry Cooper, said that the test had been "90% successful."[140] The following year, however, GAO published a report claiming that SDIO had significantly mischaracterized the results of the experiment. Several of the test's goals were not achieved. Most importantly, the interceptor failed to demonstrate that it could acquire and track its target.[141] Cooper stood by his statement that the test was a success and said that the claim of "90 percent success is a generic kind of a statement. I don't believe anybody was thinking about a quantitative assessment when they said 90 percent . . . It probably shouldn't have been given out."[142] Spinardi points out that there is significant disagreement over whether tests are an accurate reflection of how a system will perform in an operational setting. Consequently, test results are easily politicized and endlessly debated. While Brilliant Pebbles' testing problems most certainly helped the case of SDI critics, technological problems were secondary to the political considerations of America's new security realities with the Cold War now over.

As lawmakers in the House and Senate were mulling over the FY1992 NDAA, there was clearly bipartisan support for theater and tactical missile defense systems. Brilliant Pebbles, on the other hand, was still a significant point of contention. The House proposed eliminating all funding for it, while the Senate wanted $625 million. Congressman Les Aspin said that Congress should not fund Brilliant Pebbles because "we don't have the money to explore every system, particularly ones we are doubtful of."[143]

Ultimately, Congress appropriated $4.2 billion for SDIO for FY1992, which was a $1.2 billion increase from the previous year.[144] Additionally, Congress passed the Missile Defense Act of 1991, which provided funding for Brilliant Pebbles but removed it from a phase 1 national missile defense plan.[145]

The Missile Defense Act served as a compromise that allowed research on space-based interceptors to move ahead but mandated a treaty-compliant missile defense deployment at only one site.[146] Eliminating Brilliant Pebbles funding completely could have resulted in a presidential veto. While Congress was clearly unsupportive of space-based interceptors, it did mandate that a missile defense deployment would utilize space-based sensors. Even if space weapons were under threat, moving forward with "passive" space infrastructure for missile defense had received bipartisan support. The Missile Defense Act went into effect on December 5, 1991; three days later, the Soviet Union officially ceased to exist.

TAKING THE STARS OUT OF STAR WARS: ELIMINATING SPACE WEAPONS FROM SDI

The March 1992 National Military Strategy, the first one published after the Soviet Union collapsed, revealed how significantly public US government descriptions of space had changed since the late 1950s. This strategy document identified space as "the 'High Ground' that we must occupy," which seemed to signal a US commitment to securing space supremacy.[147] The 1992 National Security Strategy was more reserved, describing ASATs as necessary for "active defense" but not suggesting that military dominance in space was the end goal. With the Cold War over, the US was returning to a more restrained approach to space, despite policy language that suggested otherwise. This shift would have important implications for Brilliant Pebbles.

With the Soviet Union gone, it appeared that there was no longer a substantial threat to the US military and intelligence satellites that were key components of America's system for strategic warning and for warfighting. Even though the language used about space in US policy documents had become increasingly militarized, especially since the mid-1970s, there was clearly nothing close to a consensus that space weapons were necessary for furthering US interests. Even amid ever-growing American concerns about Soviet military space and missile defense capabilities in the late 1970s and throughout the 1980s, Congress remained uneasy about funding the development of space weapons and permitting their testing. In addition to domestic political factors, transatlantic allies were still generally opposed to space weapons. James C. Moltz has pointed out that there was also more awareness about the damaging effects of weapons testing on the space environment. ASAT tests could create debris that could stay in orbit for weeks, months, or years and threaten the safe operation of satellites.[148]

The budget deliberations and their results for FY1992 reveal that even though national security officials agreed that the US needed to protect its interests in space, military leaders and lawmakers did not see maintaining ASATs as a high priority. Despite the statements from senior uniformed officers about maintaining freedom of action in space and conducting combat operations reaching into the cosmos, more traditional platforms such as bombers, missiles, ships, and tanks prevailed over ASATs. The US Army proposed eliminating the KE ASAT program and was only stopped by White House intervention. Despite the fact that the number of spacefaring nations was predicted to grow, the threat to American military and intelligence satellites appeared to be a ghost of the past.

Even with the Cold War over, the preservation of the ABM Treaty remained a hot political issue in the US, among the transatlantic allies, and in post-Soviet Russia. In January 1992, Russian President Boris Yeltsin announced, "Today in our military doctrine we no longer consider the US as being our potential opponent. And we want to be allies. And if a global system of protection from outer space is thus set up, and the joint exploitation [sic], there would be no need for nuclear weapons in submarines, based on land, and so on."[149] In this new political atmosphere, Bush announced that it was worth discussing cooperation on missile defense with Russia. But just as cooperation with the Soviet Union on missile defense had been a sensitive subject among US allies, so too would cooperating with the Russian Federation.

SDI advocates lost a key ally when Conservative John Major replaced Margaret Thatcher in November 1990 after she had been pushed out of office by her party colleagues. To the relief of the FCO and MoD, Major was no SDI enthusiast. Shortly after taking office, Major informed his staff that he "did not want to follow the practice of his predecessor [of meeting with the director of SDIO]."[150] The British were concerned about reports that Washington might decide to share missile defense–related information with Russia. Even more importantly, Major's advisors were uncomfortable with the prospect of significant modifications to the ABM Treaty because of their implications for the UK nuclear deterrent. London did, however, ultimately "support limited amendments to the ABM Treaty and the sharing of early warning data with Russia" largely to appease the Bush administration.[151]

In early 1992, administration officials informed European allies that the US was committed to a GPALS system.[152] Stephen Wall, a senior British diplomat who served as private secretary to Major on foreign policy matters, complained to the prime minister that the US had not consulted the UK ahead of time about working with Russia on missile defense. He said that because the US was the world's only superpower now, Britain would likely "see more of the American tendency to think up policies which suit them and launch them without notice."[153]

Even though Britain did not want to allow differences on missile defense to become a source of tension with the US, Major did voice concerns over GPALS in a letter to Bush. The prime minister wrote that he did not oppose "some extension of ballistic missile defense," but cautioned that there was no consensus in the alliance on space-based weapons and cooperation with Russia in strategic defense technologies. Despite the fact that the Soviet Union was gone, Major warned that Russia could decide to invest more in missile defense in the future to the detriment of alliance interests and that "a future Russian government may not be as comfortable a partner for the West."[154] The British were not unique in their views on GPALS deployment; an FCO report observed that "France is strongly opposed to space-based deployments, and Germany [is] decidedly cool. Japan, interestingly, has also submitted a long and skeptical questionnaire."[155] The French hyperbolically warned that "US policy on GPALS casts doubt on the future of [NATO]."[156]

It became apparent to British officials during a May 1992 lunch with Secretary of Defense Dick Cheney, Deputy Secretary of State Lawrence Eagleburger, and National Security Advisor Brent Scowcroft that there were significant differences of opinion in the Bush administration regarding how the US should proceed with its strategic defense program. Cheney acknowledged that there was "little support in Congress for space-based systems. But the Pentagon will pursue their ambitions to try to get the ABM Treaty modified."[157] In a cable to FCO headquarters, British Ambassador to the US Robin Renwick wrote in a section marked "please protect," indicating its sensitivity, that Eagleburger had confided in him that "he felt that the American ideas on this subject [GPALS] had been as ill-thought through as the German ideas on the

Franco-German Corps." Eagleburger encouraged the UK to continue "vigorously" expressing its concerns about GPALS and the ABM Treaty. Scowcroft
conceded that "the initial US approach [to cooperating with Yeltsin on missile defense] had been ill-conceived. He had tried to help get this genie back
into the bottle."[158] Scowcroft and Eagleburger did not hide their aspiration to
rein in Cheney's missile defense and space ambitions. Eagleburger had earlier
told Renwick that the British "were absolutely right to have made a major
fuss about the president's message" regarding missile defense cooperation
with Russia, and that "they [Americans] had not worked that [cooperation]
out themselves."[159] The deputy secretary of state was clandestinely trying to
use British opposition to missile defense deployment and cooperation with
Russia as a means of undermining Cheney's objectives for GPALS and ABM
Treaty modification.

The differences between Cheney and his colleagues in the administration over strategic defense could easily conceal the fact that they represented
a much larger divergence of opinion over the future of American national
security. In March 1992, a Pentagon policy document outlining a post–Cold
War vision of US defense was leaked to the press.[160] Paul Wolfowitz, the
Undersecretary of Defense for Policy, was its primary author; it subsequently
became known as the "Wolfowitz Doctrine." A core idea of this new Defense
Planning Guidance was that "a collective response will not always be timely
and, in the absence of US leadership, may not gel . . . [we cannot] allow our
critical interests to depend solely on international mechanisms that can be
blocked by countries whose interests may be very different from our own."[161]
The US, therefore, had to retain the capabilities to "act independently, as necessary, to protect our critical interests."[162] Missile defense had to be developed
with the "right combination of nuclear deterrent forces" to address threats
wherever they might arise. (By contrast, the British maintained that it was
"[not] clear how offensive and defensive strategic systems can co-exist in a
stable balance."[163]) Wolfowitz also proposed using missile defense to provide
"extended protection," such as extended nuclear deterrence, to allies. His
emphasis on being able to act unilaterally was at odds with the multilateralism and collective action envisioned by Bush's "new world order."

The political environment of the early 1990s prompted the embrace of
a more limited strategic defense, with Brilliant Pebbles excised from a first-
phase deployment. Politics, more so than technological considerations, were
driving the GPALS system design. Despite Brilliant Pebbles being restricted

from involvement in a first-phase deployment, Henry Cooper continued to pursue it vigorously. Baucom writes that Cooper's commitment to Brilliant Pebbles would place "his agency on a collision course with congressional Democrats, many of whom were committed to arms control and staunch opponents of SDI."[164] Senator Nunn accused Cooper of pursuing Brilliant Pebbles as his top priority, even though it would not contribute to a 1996 deployment of a limited defense system. In response to this congressional scrutiny of his development plan, Cooper cut $2 billion from the funding profile of the space interceptor effort, which resulted in a slippage of thirty months in the Brilliant Pebbles program.[165]

Key figures in Congress were signaling to the Department of Defense that SDIO was going to be facing more budget cuts for the FY1993 NDAA. Senator Carl Levin (D-MI), who served on the Senate Armed Services Committee, said that "space-based sensors are something we should be continuing research on but space-based interceptors like Brilliant Pebbles should be explored for a follow-up system, not funded as the crash course program."[166] Once again, Congress was drawing a line between "passive" space-based infrastructure for missile warning and tracking and space-based weapons. The former was acceptable, but the latter was too politically sensitive and carried with it too many technological risks and uncertainties.

SDIO's efforts to save Brilliant Pebbles were not helped by the failure of a third, and final, flight test on October 22, 1992. This test was supposed to demonstrate the ability of a Brilliant Pebbles interceptor to come to within ten meters of its target. Fifty-five seconds into the launch, however, range safety personnel had to destroy the carrier rocket because of a booster problem. Once again, Brilliant Pebbles failed a test due to problems with the launch system. The nuances of this Brilliant Pebbles failure would be lost on both lawmakers and the general public. In terms of publicity, it didn't look good for a program that had not proven itself technologically, was facing growing opposition in Congress, and had no support from key allies.[167]

The FY1993 NDAA included a modification to the 1991 Missile Defense Act that made it explicitly clear that a deployed missile defense system would comply with the ABM Treaty. Furthermore, it cut funding for space-based interceptors to no more than $300 million and characterized them as being part of a research program only. With this new legislation, SDIO's aspirations for a near-term deployment of missile defense interceptors in space quickly evaporated. During the 1992 presidential campaign, Bush remained committed to strategic defense and criticized then-Governor Bill Clinton for threatening to cut SDI funding.[168] After Clinton misspoke about Patriot missile defense operations during the first Gulf War, Bush said at a rally that the Arkansas governor "might be a Rhodes scholar, but he's no rocket scientist."[169] But Clinton would have the last word on SDI. In November 1992, Clinton won the election, which rang the death knell for Brilliant Pebbles. Shortly after winning, he nominated Aspin, one of the most vocal critics of SDI, as secretary of defense.

The new administration renamed SDIO the "Ballistic Missile Defense Organization" to dissociate it from Reagan and his space-based missile defense agenda. In April 1993, Clinton signed Presidential Review Directive/NSC-31 on the future of American ballistic missile defense and the ABM Treaty.[170] While Aspin directed that FY1994 funding for missile defense be maintained at the FY1993 nominal level, he instructed the Pentagon to focus on theater missile defense, with national missile defense—that is, strategic defense—as a secondary priority. Brilliant Pebbles funding was to "be reduced to support a technology base program. Brilliant Eyes [tracking satellites] development should be slowed" until a thorough review of its role in a reduced-scale national missile defense architecture could take place.[171] Even this more limited missile defense was pushed out to the twenty-first century. Notably, the 1995 National Military Strategy stripped out all language from the 1992 version that referred to the cosmos as the "high ground" that had to be occupied.[172] Ground- and sea-based missile defense development would continue, but national security space infrastructure would be reoriented toward support functions alone.

CONCLUSIONS

The transition to a post–Cold War world raised many questions about the future of US military space programs. Anxieties about the Soviet Union were quickly eclipsed by fears about rogue regimes in the Middle East, North Africa, and Asia. The geopolitical transformations taking place provided both opportunities and challenges for SDI advocates. On the one hand, the first Gulf War seemed to offer living proof that missile defense was necessary, but on the other, it did not change the minds of opponents regarding space-based missile defense. Nothing close to a majority of lawmakers supported space weapons, both to control access to the cosmos and to carry out missile defense operations. Transatlantic allies also wanted to see a return to the status quo of nuclear deterrence and the preservation of the ABM Treaty.

Even though SDIO had not demonstrated that Brilliant Pebbles was indeed ready for deployment, technological difficulties were not the primary reason why the program did not move forward. When we open the "black box" of the strategic defense systems envisaged by the Bush administration, we discover that it was the domestic and international political forces of the period that were the most powerful influences on missile defense system design. Cooper, like Abrahamson, was trying to sell a revised version of SDI to meet a threat that was feared by congressmen on both sides of the aisle and European allies alike. Systems engineers had to remake the infrastructure to fit a changing security framework while still trying to maintain the goal of SDI purists to have interceptors in space.

It is impossible to know whether Brilliant Pebbles would have advanced had the Soviet Union not dissolved in 1991. If legislators had continued to fund space-based interceptors, the Department of Defense could have indeed deployed them, but how effective they would have been is an open question. Divergent views on strategic defense system effectiveness plagued SDIO from the beginning. Support for SDI in Congress was waning even before the Cold War officially ended, but the collapse of the Soviet Union fundamentally called into question the need for such an expansive investment into space-based defense. Scud missiles in the Persian Gulf and fears about Iran, Libya, and North Korea could not save SDI from the budgetary knife. Consequently, space would be free of orbital weapons, at least for the time being.

SDI RECONSIDERED

When the dust of the Soviet Union's collapse had settled, space-based missile defense quickly faded. The closest that Brilliant Pebbles ever came to deployment was the 1994 joint NASA–Ballistic Missile Defense Organization (SDIO's successor) *Clementine* probe that used missile defense technologies for a planetary science mission.[1] Strategic defense advocates saw *Clementine* as irrefutable proof that Brilliant Pebbles was more technologically mature than critics had alleged. Nevertheless, whether the Department of Defense could have integrated all of the components required to have an effective space-based missile defense capability remains an open question.

Much of the historical inquiry concerning SDI has focused on Reagan's nuclear abolitionism and whether the pursuit of space-based missile defense played a role in the demise of the Soviet Union.[2] And the fact that Reagan's strategic defense dream never came to fruition makes it easy to dismiss SDI as only a science-fiction fantasy. Due to this situation, scholars have neglected to explain the rationale for particular missile defense technological choices, along with their political consequences. Moving SDI technologies to center stage reveals that this controversial program cannot be understood as a Reagan phenomenon alone and that it continues to shape the space security environment at the present time.

Certainly, SDI would not have emerged without Ronald Reagan and his conviction that technology could provide a solution to the nuclear dilemma. But shifting the frame of reference to SDI's technologies underscores that the program grew out of what Peter Westwick terms the "remilitarization of space" in the second half of the 1970s when both the US and the Soviet

Union intensified development of ASATs.[3] Even though American ASATs and space-based missile defense would not be operationally deployed, Reagan's commitment to SDI left an indelible mark on the arms control framework that emerged out of the Nuclear and Space Talks in the 1980s and early 1990s. Scholars have argued that Reagan's intransigence on SDI was an impediment to the signing of INF and START, but have tended to overlook its consequences for the space component of the Nuclear and Space Talks.[4] In the space forum, Reagan's unwillingness to accept any constraints on technologies that could be used for missile defense and ASATs eliminated the possibility for a space arms control agreement. Missile defense aside, key advisors to the president recognized that the US had an overwhelming advantage in space technologies and therefore viewed *any* constraints on military space activities as detrimental to US defense interests. Consequently, unprecedented reductions in the US and Russian nuclear arsenals would not be accompanied by a new treaty limiting military activities in the cosmos.

A NEW PHASE IN SPACE MILITARIZATION

Reagan would take space policy to new heights, and he leaned into a more competitive view of the cosmos that had begun to circulate in key corners of the US national security establishment in the decade prior to his presidency. In the 1970s, technological factors, coupled with the erosion of détente, were driving American officials to reconsider US space policy. The advent of satellites that could rapidly transmit data to military forces meant that space could be exploited for both tactical and strategic advantages in wartime. Publicly, reconnaissance satellites were presented as tools for monitoring arms control agreements and creating transparency. Secretly, these systems were increasingly multiuse platforms that enabled functions such as precision targeting of enemy forces. Even though these satellites were only "passive" sensors in space, publicly acknowledging their emerging warfighting role would have created difficulties for American public diplomacy that stressed US commitment to the peaceful use of outer space. In reality, however, there had long been significant interpretative flexibility concerning what indeed constituted "peaceful" activities in space. Since satellites were fast becoming critical information nodes in both American and Soviet warfighting strategies, US officials concluded that having the ability to destroy Soviet military support satellites was vital.

The same advances in electronics and sensor technologies that enabled a "tactical shift" in the use of satellites also created the opportunity for non-nuclear ASATs that could more precisely attack enemy satellites. In particular, the 1970s witnessed marked strides in hit-to-kill capabilities that could more efficiently destroy targets in low Earth orbit. These same technologies would have missile defense applications as well. It must be reemphasized that missile defense and space warfare technologies are inseparable. Efforts to divorce the former from the latter have been driven primarily by political considerations not technical realities.

Since the Soviet Union was developing nonnuclear ASATs even before the US, it could easily be concluded that the more militarized US space policy emerging in the late mid to late 1970s was largely a response to Soviet actions. US ASAT development, in particular, appears from afar to be part of an action–reaction phenomenon. The declassified record, however, provides a much different picture. Advisors to Gerald Ford acknowledged that the US ASAT program would in no way deter the Soviet Union from using its own ASATs in wartime. Having the ability to destroy Soviet military support satellites, rather than matching Soviet ASAT capabilities, was the primary rationale for Ford's ASAT decision. Most fundamentally, well before Reagan came into office, US restraint in space was eroding.

Reagan established SDI at a time when the US was preparing for the transformation of space into a potential battlefield, but the president did not view SDI solely as a means of militarily dominating space. The Reagan administration's attempt to distinguish SDI from its larger military space agenda created a significant public diplomacy challenge. We find that consistently, going back to 1983, advisors to the president cautioned that many Americans and European allies would be uncomfortable with the prospect of placing weapons in space. Space weapons were vital for being able to destroy ballistic missiles in their boost phase of flight, but the widespread hostility toward them created a political liability for Reagan's SDI vision. Consequently, the White House tried to deflect attention away from the technological realities of strategic defense and focus instead on the president's aim to use the program to reduce the threat of nuclear annihilation.

Since SDI signaled, for both advocates and critics of space-based missile defense, that the heavens could become a battlefield, it captivated the popular imagination. Astroculture—the ways in which people make meaning of outer space—had a decisive impact on the politics of strategic defense technologies.[5] For better or worse, many people looked to popular culture to make sense of the highly technical and arcane terminology of military space technologies. Technical details aside, the Reagan administration's presentation of SDI as both a peaceful endeavor *and* a response to highly destabilizing Soviet strategic defense and military space buildup created confusion about the nature of the program. As the Cold War was ending, however, SDI program managers more openly acknowledged, and even advocated, the offensive applications of SDI technologies. But SDI never fully escaped the tension between its public framing as a peace initiative and the reality that it involved the development of space weapons that could be used offensively.

The fact that the US government had generally eschewed any substantive public discussion of its military space efforts prior to the 1980s created an even more fundamental challenge for the Reagan administration's public diplomacy concerning SDI. The White House had to find a way to talk about this closed-off arena of American statecraft without compromising sensitive national security space programs. In general, Reagan discussed SDI in the abstract, but the debate on SDI moved sensitive topics such as the vulnerability of space-based systems into a more public position. The NRO, in particular, was uncomfortable with any open analysis of national security space matters and made a concerted effort to ensure that the lid of secrecy would not be lifted too high.[6]

SDI AS AN INTERNATIONAL PHENOMENON

Since Reagan made the preservation of SDI the central objective of his arms control agenda, he had to assuage European and Soviet concerns about deploying space weapons. To allay these anxieties, the president pledged that SDI would be an international project benefiting both the Soviet Union and American allies. Although this idea was politically controversial and technologically doubtful, Reagan was indeed sincere in his desire to share SDI technologies with the Soviet Union. However, by the early 1990s, "sharing SDI" with Moscow had morphed into more limited cooperative proposals, including exchanging missile defense data and reciprocal American–Soviet visits to strategic defense research facilities.

Although US and Soviet space activities were at the center of the space security dialogue in the 1980s, detailing the involvement of Europeans in

SDI research reframes strategic defense in this period as an international phenomenon that transcends the bipolar superpower dynamic. The US needed European support for SDI to demonstrate that there was alliance-wide consensus behind the program. And US officials recognized that the Europeans saw SDI as a catalyst for advanced technologies that could have significant implications for industrial competitiveness, as well as defense, and hoped that the prospect of technology transfer would convince European politicians to endorse the program.

The prospect of European participation in SDI research and development quickly became enmeshed in the politics of European integration. Western European states had to choose between bilateral arrangements with the US and a common European approach. To the chagrin of French officials, the former prevailed. Rather than European involvement leading to greater transatlantic unity, a competition ensued within Europe to obtain lucrative SDI contracts. In the end, the memoranda of understanding that the US signed with Britain, West Germany, and other allies to guide their participation in SDI were largely symbolic. The US offered few guarantees regarding technology transfer and made no promises about awarding specific contracts to its European partners.

Although reducing European hostility toward SDI was a driving factor behind the US invitation to allow allies into the SDI program, it would be a mistake to conclude that the US saw European participation as purely performative. Internal SDIO documents reveal that US officials wanted to exploit the expertise of European firms in specific areas. Moreover, British defense scientists provided their American counterparts an extra set of expert eyes to check their assumptions and conclusions on technical matters. To the disappointment of European participants, projects awarded to European companies, laboratories, and universities were generally small in scale. Not only did American defense contractors and laboratories have more resources than their European colleagues, but they also had established relationships with the Pentagon and years of experience in navigating the byzantine US defense contracting process. In certain instances, secrecy restrictions also proved to be a limiting factor for allies.

Despite the fact that Europeans were generally dissatisfied with their SDI contracts, they cumulatively obtained hundreds of millions of dollars' worth of work. Consequently, the "geography" of SDI research and development was international. This book has primarily focused on Western European participants in SDI, but the nature of Japanese and Israeli work on SDI contracts deserves more attention.[7] There is also more work to be done in investigating how SDI impacted foreign defense firms, university research establishments, and industrial organizations. How other countries such as India and China perceived SDI and the ways in which it might have influenced their own defense programs and economic interests are also important areas for further consideration. In sum, we are only just beginning to pivot away from a primarily American–Soviet view of SDI specifically and the militarization of space more broadly to explore these subjects as truly global issues.

INVISIBLE INFRASTRUCTURE

When we begin to look deeper into Paglen's "other night sky," we find that military and intelligence satellites were not disparate devices in space but rather individual parts of a much larger US space infrastructure. Although US space policy documents distinguished between military, intelligence, and civil space programs, their boundaries were fluid and porous. The shuttle, for example, was designed to deliver both national security and civil payloads into space. Although it was not intended to be a weapon, the Soviets viewed it as a weapons-*capable* platform and therefore wanted it included in any prospective ASAT agreement. As already noted, the nature of US national security space infrastructure was changing in the 1970s due to satellites becoming more integrated with US combat forces. Nevertheless, US national security space infrastructure was still primarily passive: that is, focused on gathering and transmitting data. In stark contrast, a strategic defense system on the scale envisioned by Reagan would have required a massive expansion of American space infrastructure, both passive satellites for tracking missiles and interceptors with offensive and defensive applications.

When space-based missile defense is examined as a technological system, we find that the obstacles to transforming it into a reality were primarily engineering, rather than scientific, in nature. A substantial percentage of SDI's budget was devoted to building sensors and command and control capabilities that could be used to track ballistic missiles in various phases of flight and then seamlessly transfer targeting data to an interceptor. Along with this hardware, SDIO had to create software of an unprecedented degree of complexity and sophistication to link all of these elements together.

The SDI space renaissance that Abrahamson so vividly described was totally dependent on the creation of new space logistics systems that could efficiently launch and service a large number of missile defense components in orbit. Cost of launch had to be brought down significantly to make strategic defense "cost effective at the margin."[8] To the chagrin of SDI advocates, the space shuttle did not materialize into a cost-effective space delivery asset. The *Challenger* disaster forced SDIO planners to consider more carefully other strategies for creating a cost-effective space-based defense infrastructure. In particular, strategic defense system builders pushed for smaller, more agile space systems that would be cheaper to deploy. As budgets became tighter with the ending of the Cold War, infrastructural constraints even more substantially influenced the strategic defense system concepts pursued by the Department of Defense.

The vulnerability of the space-based missile defense systems emerged as a politically charged issue when it became clear that they would be susceptible to Soviet ASAT barrages. Consequently, SDIO had to convince European allies and US lawmakers that a strategic defense capability would be resilient in the face of attacks. They had to sell SDI as a continuous technological revolution that would always be one step ahead of Soviet countermeasures. While the Department of Defense did not point to it publicly, the Pentagon recognized that it would need the ability to carry out "space control" missions to protect and defend a deployed space-based missile defense system. And directly linking space-based defense with space warfare ran contrary to the White House's blueprint for "selling" SDI to the public.

Soviet defense planners viewed space infrastructure as the weak underbelly of any prospective strategic defense system.[9] Although Reagan expressed concerns about a Soviet SDI equivalent program, Gorbachev stressed that the Soviet Union would invest in only asymmetric countermeasures (e.g., ASATs) to undermine any American space-based missile defense system, rather than deploying its own strategic defense capability.[10] By the late 1980s Soviet defense experts had concluded that countermeasures would be sufficient for addressing the threats posed by any near-term strategic defense system.[11] This viewpoint contributed to the Kremlin's willingness to move forward with START by dropping the treaty's linkage with a prohibition on the development and testing of strategic defense technologies.

POLITICALLY CHARGED ARTIFACTS

Employing a technological perspective reveals that we cannot fully understand the political consequences of SDI without analyzing the rationale for
system decisions. Even though there was very little certainty in the early to
mid-1980s about what technologies would be included in a strategic defense
system, technological considerations shaped significant US foreign policy
decisions that impacted transatlantic and superpower relations. A key question for US officials as they prepared for the Nuclear and Space Talks was
whether they would be willing to entertain limits on military space technologies that could be used for both ASATs and strategic defense. Pentagon
officials recognized early on that exotic technologies, such as laser weapons,
might not mature before well into the next century, which meant that near-
term missile defense options using existing technologies had to be protected
from arms control limits.

In particular, hit-to-kill ASATs were largely indistinguishable from
kinetic missile defense interceptors that had the potential to form the basis
of a first-phase strategic defense. Due to this relationship, SDI critics hoped
to use ASAT arms control as a means of limiting strategic defense development. Even though Reagan had identified the MHV ASAT as essential for
deterring the Soviet Union and for controlling access to space, he was ready
to trade it away for an agreement limiting "offensive space weapons" until
he recognized that it would negatively impact SDI. Due to this ASAT–SDI
connection, the US consistently rejected any limits on ASATs.

Opening the "black box" of SDI technologies also reveals how concepts
such as "system effectiveness" and "system vulnerability" were not only
technical issues but also fundamentally political. Some defense strategists
maintained that a strategic defense system only had to be effective enough
to introduce more uncertainty in the Soviet strategic calculus, thereby convincing Moscow that a large-scale nuclear strike would be unsuccessful. This
position was based on assumptions about Soviet defense strategy and could
not be answered by purely technical studies of strategic defense components.

EVOLVING SYSTEM FOR CHANGING TIMES

Since the US never actually deployed a space-based missile defense system,
SDI might appear to be a technological failure. But such a position implies

that SDI was a specific program, when in reality it was an umbrella for multiple research efforts. This fact was lost on many people in the 1980s, and still causes confusion today, because many officials from the US, Western Europe, and the Soviet Union spoke about SDI as if it were one artifact. SDI was not in fact a technological failure. Krige reminds us that "a technological system is a complex, fluid, evolving thing, and there is no one best way to bring it to fruition: in an important sense it is always a work in progress."[12] The scope and objectives of SDI changed over time primarily due to transformations in the international security environment. When the Reagan administration first tasked SDIO with developing the technologies for a comprehensive strategic defense, Gorbachev was not yet the leader of the Soviet Union. In the early 1980s, tensions between Washington and Moscow were running high, and anxieties about nuclear war were accelerating. By the end of Reagan's presidency, the US and the Soviet Union had signed a landmark arms control treaty eliminating an entire class of nuclear weapons. While walking on Red Square in June 1988, a reporter asked Reagan if he still considered the Soviet Union to be an "evil empire" as he had said in 1983. The president smiled and said, "No, you are talking about another time, another era."[13]

SDI research and development did not exist in a vacuum, and system designers were affected by changes in US foreign relations and domestic American politics. As the Cold War ended, strategic defense advocates had to find a way to keep their program relevant for new security circumstances. The Bush administration attempted to refocus SDI to address emerging ballistic missile threats from so-called rogue nations. The first Gulf War even led to a brief surge in missile defense support on both sides of the political aisle. But there was still significant opposition to space-based missile defense. The failure of Brilliant Pebbles flight tests did not help the case of SDIO advocates, but the more significant reality was that Scud missile attacks from Iraq and missile proliferation to places such as Libya and North Korea were not enough, in the eyes of lawmakers, to justify the continuation of large-scale funding for technologies that were designed for a threat that appeared to be dissipating.

In an attempt to distance missile defense from Reagan, the Clinton administration changed the name of SDIO to the Ballistic Missile Defense Organization. Missile defense development was not going away; it was just refocused to land- and sea-based systems. Even though the US was no longer pursuing space-based interceptors as the basis of a strategic defense system, the Department of Defense continued to work on space-based sensors that could aid ground- and sea-based interceptors in hitting their targets. Consequently, space-based infrastructure was still an important part of US missile defense systems. SDIO also led to the creation of a permanent infrastructure for developing technologies with both missile defense and space warfare applications.

SANCTUARY OR BATTLEFIELD?

Although it has been largely forgotten, there was real anxiety in the 1980s about space being transformed into a warzone. Fear about conflict in space had been building in the 1970s with the resumption of Soviet ASAT testing and the US establishing its own new ASAT program, but SDI placed space militarization in a more prominent position in the public sphere. In the wake of the Apollo program, the US had turned away from bolder space exploration initiatives, creating uncertainty about humanity's future in the cosmos. In the public eye, SDI signaled that space had evolved from a domain of superpower competition, serving as a proxy for war, to a potential battleground.

While some people feared the consequences of space-based weaponry, others embraced them due to the belief that they could secure strategic advantages for the US and lead to an improved security framework that did not depend upon the threat of nuclear annihilation. People debated whether space could indeed be preserved as a sanctuary from war, despite the fact that neither superpower had ever truly treated space as a sanctuary. It is almost certain that had the US and the Soviet Union gone to war, they would have employed ASATs to destroy each other's satellites. These military realities aside, there was still the aspiration that space could somehow be kept untouched by the horrors of war. Critics of this position maintained that war in space was preferable to conflict on Earth, but this argument incorrectly presupposed that space could be, in effect, cordoned off.

In the decade prior to SDI, American and Soviet officials had already begun to discuss more rigorous constraints on space militarization. There were (and remain) few limits on military activities in space beyond the 1967 Outer Space Treaty and the now defunct 1972 ABM Treaty that forbade deployment of space-based missile defense. In the 1970s, with the advent of more advanced ASATs and military satellites, along with the introduction of semi-reusable spacecraft (e.g., the shuttle), a new space race appeared to be likely. The negotiations over SDI in the Nuclear and Space Talks pertained not only to nuclear stability but also to the boundaries of military space activities. The space component of the Nuclear and Space Talks was, moreover, a continuation of arms control negotiations of the late 1970s that aimed to limit emerging space technologies.

Arms control negotiations between the US and the Soviet Union shaped the nature of spaceflight up through the present writing. The Nuclear and Space Talks of the 1980s led to nuclear reductions but failed to establish agreements on norms of behavior in space. Senior officials in the US State Department and the ACDA advocated negotiating "rules of the road" for space. Ken Adelman, the head of the ACDA, specifically argued that the US needed "to make the world safer through the controlled use of space, that negotiations along these lines were better than doing nothing." He maintained that "rules of behavior" were required to guide superpower activities in space and to promote stability.[14] But his advice was not heeded, and progress was not made. These issues, left unresolved, would not reemerge as urgent diplomatic subjects until the twenty-first century when more countries began fielding ASAT capabilities.[15]

As the archival record on space militarization opens more widely, we discover a variety of space arms control measures that were contemplated but which never left the diplomatic drawing board. American officials recognized that more advanced ASATs designed to attack higher-altitude satellites used for nuclear command and control might have dangerous implications for nuclear stability. But American and Soviet negotiators had great difficulty in finding common ground on a framework for constraining space technologies, many of which had multiple applications. The entanglement of missile defense and ASAT technologies was (and remains today) one of the most challenging aspects of attempting to constrain nonnuclear space weapons.

Space arms control, like SDI, was an international affair. Certainly, Moscow and Washington were the two main players. But an international group that included diplomats, scientists, defense intellectuals, and even the Vatican participated in the debate about space arms control as well. Nevertheless, the high wall of secrecy surrounding US and Soviet military space programs created challenges for non-US actors who became involved in dialogues on space militarization. Among the Western European allies, Britain had privileged knowledge, although still limited in key respects, about US military and intelligence space capabilities, while many NATO allies were kept largely in the dark. American military space strategy was, in many ways, even more closed off than its nuclear policy. At NATO, there was at least a consultative mechanism for nuclear issues through the NPG; nothing remotely similar existed for space issues. All of NATO was dependent on US military and intelligence space infrastructure, but only with the advent of more capable ASATs in the 1970s and the prospect of space-based missile defense in the 1980s and early1990s did NATO member states begin to pay greater attention to stability in space. In the end, US allies in NATO had a very limited ability to shape American military space policy, which at times contributed to preexisting tensions in the alliance.

The Reagan administration refused to accept any limits on space activities out of fear that doing so might negatively impact SDI development at some point. Concurrently, the joint chiefs and some senior intelligence officials worried that space arms control might open up other US military space activities to greater scrutiny, thereby placing "black" space programs in jeopardy. Admiral Crowe candidly said that "we don't want the Soviets crawling all over our space vehicles."[16] Consequently, the US would continue to prioritize freedom of action in space until the Clinton administration realigned US missile defense priorities, but there was no longer any impetus for a treaty that formally constrained American and Russian military space capabilities.

It is impossible to know what might have happened if the Reagan or Bush (41) administrations decided to accept at least some limits on military space programs, for example on high-altitude ASATs. Despite Soviet interest in space arms control, it is not at all certain that Moscow and Washington could have indeed found common ground concerning *verifiable* limits. In any case, the world emerged out of the Cold War with substantial nuclear arms reductions, but no new space arms control treaty. Although SDI would be brought down to earth, American research into missile defense technologies that also had space warfare applications would remain firmly in place.

A SENSE OF DÉJÀ VU

If someone were to walk into one of the many Washington, DC, watering holes where national security intellectuals congregate to debate US foreign policy and ask what was significant about January 2007, the defense experts present would likely point to the so-called surge in the Iraq War. The surge, which involved a temporary increase of twenty thousand US troops in Iraq, was intended to get the Iraqi insurgency under control, and it officially began on January 10, 2007. One day after the surge commenced, at an altitude of approximately 530 miles, China prepared for a very different kind of war than the ones raging below in Iraq and Afghanistan. At 5:28pm (Eastern Standard Time), China used a modified ballistic missile to obliterate one of its defunct weather satellites, generating more than two thousand pieces of debris.[1] Liu Jianchao, a spokesman for the Chinese Ministry of Foreign Affairs, quickly reassured the international community that there was "no need to feel threatened about this [ASAT test]" and pledged that Beijing would "not participate in any kind of arms race in outer space."[2] Defense and space policy experts in Washington were not calmed by Liu's words. Theresa Hitchens, a space policy expert, alarmingly described the Chinese test as a potential "shot across the bow."[3]

Only one year later, the US used a modified SM-3 missile defense interceptor to eviscerate a defunct US satellite as part of Operation Burnt Frost.[4] Officially, the US destroyed the satellite because it had toxic hydrazine onboard, and there was concern it might impact land and contaminate the surrounding area.[5] Some observers, nevertheless, interpreted Burnt Frost as a response to the Chinese ASAT activity. Russia accused the US of using this operation as a cover to test an ASAT capability.[6] Within a short period of time, headlines appeared, describing the US and China as preparing for war in outer space.[7]

Not even ten years into the twenty-first century, insecurity in space—a ghost of the Cold War—had returned. However, not all observers have been surprised by this development. The Chinese ASAT test seemed to confirm the warning of a space commission, headed by Donald Rumsfeld, that the US could face a "space Pearl Harbor."[8] This study, officially called the "Commission to Assess United States National Security Space Management and Organization," released its findings on January 11, 2001 (six years to the exact day before the Chinese ASAT test). Rumsfeld and his colleagues highlighted US dependence on space infrastructure for economic and national security activities and implored the US government to "take seriously the possibility of an attack on US space systems."[9] To prevent a "space Pearl Harbor," the space commission advocated greater investment in resilient space systems, along with the development of "superior" capabilities for "power projection in, from, and through space" in order to "negate the hostile use of space against US interests."[10]

Studying the origins and evolution of SDI does not provide easy solutions to current space security challenges, but it is essential for understanding how we arrived at the present state of affairs. Examining the declassified record reveals, moreover, that policymakers today are concerned with many of the same space security issues that their predecessors raised in the 1970s and 1980s: Can arms control prevent an arms race in outer space? What is the military utility of kinetic and non-kinetic ASATs? Can missile defense enhance deterrence? How do states establish a deterrence framework involving space systems? What is the optimal strategy for reducing the vulnerability of space systems? We are, moreover, still grappling with the question of what *should* the limits of military activities in space be.

The US was forever changed by the terrorist attacks on September 11, 2001 (9/11). Less than five months after 9/11, President George W. Bush introduced the term "axis of evil" during his State of the Union address to describe "rogue regimes," such as North Korea and Iran, that supported terrorism and sought weapons of mass destruction.[11] Due to concerns about missile proliferation, missile defense took on even greater importance for Bush. In his final year in office, Bill Clinton had decided not to move forward with a national missile defense system because he concluded that while the requisite technologies were "promising," they were "not yet proven."[12] Bush, on the other hand, wanted to push forward with a national missile defense system as quickly as possible. To remove one of his primary obstacles, the president withdrew the US from the 1972 ABM Treaty in June 2002.

While on the campaign trail in 2008, then-presidential candidate Barack Obama described missile defense technologies as "unproven" and pledged to cut MDA funding, along with promising not to "weaponize outer space."[13] Despite these statements, he ultimately pushed forward with theater-level missile defense systems and the Ground-Based Midcourse Defense designed to defend the American homeland from ICBMs. The Pentagon would ultimately deploy forty-four interceptors spread across Alaska and California, supported by radars on land and at sea and sensors in space, to detect and target ballistic missiles more accurately.[14]

Obama's national security team increasingly worried about China's growing military power, including its space and counter-space systems (i.e., weapons for interfering with and/or destroying satellites). Consequently, Obama's space policy characterized the space domain as "increasingly congested, contested, and competitive."[15] The 2011 unclassified World Threat Assessment published by the Office of the Director of National Intelligence pointed toward growing insecurity in space, but stopped short of identifying specific countries and weapons.[16] The final World Threat Assessment of the Obama administration, published in 2016, specifically highlighted both Russian and Chinese investment in counter-space technologies (e.g., ASATs) and explained that the Russian "senior leadership probably views countering the US space advantage as a critical component of warfighting."[17] Why were Moscow and Beijing investing in counter-space technologies? They had closely watched US combat forces, enabled by space systems, rapidly defeat the Iraqi military in the First Gulf War and viewed US space dependence as an Achilles heel. Consequently, Chinese and Russian defense experts saw space weapons as a potential way to level the playing field with the US in wartime.

Chinese and Russian defense publications reveal that SDI remains an influential factor in Beijing's and Moscow's view of recent US military space capabilities and strategy. A RAND Corporation study observed that "SDI is the most referenced US military activity in the space domain within the literature [RAND] reviewed, both in terms of raw count and in span of relevancy to PRC [People's Republic of China] perceptions."[18] From the standpoint of Chinese military experts, SDI represented the US embrace of "space superiority" and the view that space is indeed a warfighting domain.[19] Similarly,

Russian defense literature identifies SDI as the foundation of the alleged space superiority objectives of subsequent US presidential administrations.[20] Fundamentally, SDI continues to play an important role in both Russian and Chinese narratives about space militarization.

The Trump administration placed a spotlight on national security space policy to a level of intensity not witnessed since the Reagan era. Trump revived the National Space Council, created the Space Force as an independent military service, and reestablished US Space Command that had been dissolved in 2002.[21] Revealing scant awareness of US military space programs, critics alleged that Trump raised the danger of "militarizing space."[22] In reality, the Space Force primarily entailed moving preexisting US Air Force Space Command units, along with space capabilities from the Army and Navy, into the new service. Notably, the NRO and the MDA remain independent organizations. It is too early to judge whether the Space Force will indeed make US military space operations more efficient. Any actual combat actions involving space would, however, be carried out by Space Command, one of the eleven US combatant commands. Despite disagreements over the new service, the creation of the Space Force has raised public awareness about the importance of national security space operations. In light of the proliferation of counter-space capabilities, General Jay Raymond, the first head of the US Space Force, identified space as "a warfighting domain just like air, land, and sea" and added that he "couldn't have said that five or six years ago."[23] Raymond's words are reminiscent of some of the very same language used by the Buchsbaum panel in the Ford administration.[24]

It appeared that SDI might even be revived when Trump announced in 2019 that the Pentagon would invest in "a space-based missile defense layer" with the goal of being able to detect and destroy missiles launched from anywhere on Earth.[25] The president declared that space is ultimately "going to be a very big part of our defense and offense."[26] Senator Edward Markey (D-MA) called the White House's plans "a Star Wars sequel" and decried building a "wall in space" as an answer to America's security problems.[27] Defense officials confirmed that the Pentagon would study the feasibility of space-based weapons, including lasers, but acknowledged that there were concerns "surrounding any perceived weaponization of space."[28] Notably, Trump's Undersecretary of Defense for Research and Engineering, Michael Griffin, played a key role in SDI's Delta 180 experiment and had served in SDIO.

In marked continuity with the 1980s, the Pentagon still has public diplomacy challenges concerning its space missions. After Trump established the Space Force, Netflix created a series called "Space Force" that caricatures the new military service.[29] Consequently, American defense leaders now must spend more time explaining how space technologies have become integral parts of the US warfighting infrastructure. Secrecy remains an even more basic problem that complicates the efforts of officials to communicate key aspects of American national security space activities, including how the US might defend its interests in space.[30] But despite these public relations difficulties, polls suggest that more than 60 percent of Americans approve of President Joe Biden's decision to keep the Space Force.[31] This situation is likely due in no small part to the perception that China is a "major threat" to the US's competitive edge in space, along with the growing number of public statements from senior US and foreign defense and intelligence officials about the critical role of space systems for the security of the US and its allies.[32]

A key discontinuity between the Cold War and today is that the community of spacefaring nations has grown considerably. There is now a vibrant and rapidly expanding commercial space sector as well. As the number of commercial space enterprises grows, linkages between the private and public space arenas are intensifying. Consequently, the lines between military and civilian space are blurrier than ever. The use of commercial space systems as part of the response to Russia's most recent invasion of Ukraine in 2022 is a case in point. Elon Musk's Starlink satellites have been described as a "lifeline [for Ukraine] in the war with Russia" by supporting everything from "artillery strikes to Zoom calls."[33] With these realities in mind, it is to be expected that commercial space systems will become targets for military strikes in future conflicts.

Since space systems are increasingly accessible to a larger number of state and non-state actors, more countries will seek out mechanisms for countering them. Consequently, the arena of ASAT-capable states is primed to expand. In response to Iran's growing interest in military satellites, Israeli officials have publicly identified the fact that their Arrow-3 missile defense system could be adapted for an ASAT role.[34] This Israeli example is an important reminder that the proliferation of advanced missile defense systems is simultaneously the proliferation of ASAT-capable systems. New Delhi underlined this reality when, in 2019, it used a modified missile defense capability to destroy one of its own satellites in low Earth orbit.[35] With this test, India became the fourth country, after the US, Russia, and China, to demonstrate an ASAT.

The politics of space in US relations with allies has also undergone significant changes in the post–Cold War era. In the late 1970s and 1980s, *overt* space militarization was highly controversial in NATO. Today, Britain, France, and Germany have dedicated military space commands.[36] In 2019, NATO declared that space is an "operational domain," but stopped short of calling it a warfighting domain.[37] In 2018, after a Russian satellite maneuvered close to a Franco-Italian communications satellite, Paris announced plans to deploy space laser weapons designed to protect French satellites.[38] Within NATO, there are, however, still more questions than answers regarding how the alliance is going to conduct multinational space operations. US allies in the Asia-Pacific region are investing more in space capabilities as well. Australia has created its own space command, and Japan now has space surveillance units.[39] These are yet more examples of the fact that space militarization has long expanded beyond the arena of superpowers.

In diplomacy, the use of precise language is essential; it can have direct bearing on peace and war. Troublingly, some policy experts today are making the same mistakes as their predecessors in using inaccurate terminology about the nature of space security. In the 1980s, many SDI critics alleged that the US was "militarizing outer space." We of course know that space was militarized from the beginning of the Space Age. Today, there is greater emphasis placed on the "weaponization of space."[40] Problematically, however, there is little agreement on what constitutes "space weaponization."[41] Even more importantly, Bleddyn Bowen shows that there is a tendency to "overstate the potential impact of *space-based* weapons on international stability and security" and to "overlook the potential impact of existing *Earth-based* space warfare technologies" (emphasis in original).[42] As Bowen notes, ground-based weapons have great potential with regard to space warfare. The notion that space-based weapons are a fundamental game changer in defense strategy and international security is highly misleading.

Just as in the late 1970s and 1980s, space arms control efforts at the present time face substantial obstacles. Since the number of spacefaring nations is growing, any new space treaty would likely need to be multilateral to be effective—an especially challenging prospect. In the post–Cold War era, space security involves a variety of military, economic, and environmental considerations. Today, there is growing awareness of the damaging effects of space debris that creates hazards for civil, military, and commercial space operations. Due to this reality, space sustainability is receiving more attention from both government officials and private interest groups.

With all these complicating factors in mind, what is to be done to promote stability in space? The greatest prospect for success in space arms control is limiting behaviors, such as debris-producing ASAT tests, rather than banning specific kinds of technologies that have multiple applications. There is still significant disagreement concerning what even constitutes a space weapon. Even more fundamentally, in light of the critical role of space technologies in modern conflict, spacefaring nations are unlikely to accept substantial limits on systems that can be used for space warfare. Spacefaring nations can, however, pledge not to carry out debris-producing ASAT tests without constraining their military capabilities. Importantly, debris is easily detectable, and any treaty banning such tests would therefore be verifiable.

Russia's heinous attack on Ukraine has underscored the urgency of space security. In the initial days of the invasion, there were media reports about Russia hacking Viasat, a satellite communications provider in Ukraine and other regions of the world.[43] Elon Musk has detailed attempts by Russian forces to jam Starlink.[44] There are reports of Russian interference with GPS signals in Ukraine as well.[45] Due to the importance of commercial satellite systems for Ukraine's defense, the Russian Ministry of Foreign Affairs warned that commercial satellites "may become a legitimate target for retaliation."[46] Kinetic space weapons, such as ASATs, will continue to receive the most attention, but we should expect a significant portion of space warfare actions to take place in the electromagnetic spectrum and cyber domain. The fundamental reality is that space warfare has already arrived, and future conflicts are almost certain to witness combat action extending into space.

Due to the growing number and sophistication of counter-space capabilities, the US government is prioritizing making its space systems more resilient. Currently, the Department of Defense is moving forward with an ambitious plan to deploy a National Space Defense Architecture (NSDA), which is "a tactical LEO [low Earth orbit satellite] network designed to communicate missile warnings; position, navigation, and timing data; and other vital information to wherever it's needed on the ground as quickly and securely as possible."[47] This network of satellites is intended to be able to persistently monitor increasingly advanced missile threats, including hypersonic weapons. US officials have stressed that NSDA will use a "large network of small satellites" rather than "the traditional handful of 'big, juicy targets."[48] In key respects, the NSDA infrastructure in space resembles the tracking satellites envisioned by SDIO in the late 1980s and early 1990s and the push for larger numbers of smaller satellites to increase the resiliency of the overall system.

There is clearly bipartisan support for expanding US space infrastructure to monitor and track threats on the ground, but space-based interceptors are the "third rail" that many US officials don't want to touch.[49] As the cost of deploying systems into space decreases and anxieties about advanced missile threats increase, it is very well possible that advocacy for boost phase missile defense using space-based interceptors will reemerge prominently. Promoters of space-based missile defense will likely point toward the lower cost of space launch as a critical step toward meeting the Nitze criterion of cost-effectiveness at the margin.[50] What can be stated with certainty is that if the US government decides at some future date to pursue space-based interceptors, it will once again become the subject of great domestic political and international controversy. Just like with SDI, technical considerations alone will not be the deciding factors. Both perception of threats and ideas about the proper role of space in US national strategy will be overwhelmingly powerful.

NOTES

THE "OTHER NIGHT SKY"

1. "Russian Direct-Ascent Anti-Satellite Missile Test Creates Significant, Long-Lasting Space Debris," US Space Command Public Affairs, November 15, 2021, available at https://www.spacecom.mil/Newsroom/News/Article-Display/Article/2842957/russian -direct-ascent-anti-satellite-missile-test-creates-significant-long-last/.

2. W. J. Hennigan, "Astronauts Take Shelter Aboard ISS after Russian Anti-Satellite Test, US Says," *Time*, November 15, 2021, available at https://time.com/6117840/astronauts -shelter-iss-russia-test/.

3. Brian Weeden and Victoria Samson, "India's ASAT Test Is Wake-Up Call for Norms of Behavior in Space," *SpaceNews*, April 8, 2019, available at https://spacenews.com /op-ed-indias-asat-test-is-wake-up-call-for-norms-of-behavior-in-space/; William J. Broad and David E. Sanger, "China Tests Anti-Satellite Weapon, Unnerving US," *New York Times*, January 18, 2007, available at https://www.nytimes.com/2007/01/18/world /asia/18cnd-china.html; Jim Wolf, "US Shot Raises Tensions and Worries over Satellites," *Reuters*, February 21, 2008, available at https://www.reuters.com/article/us -satellite-intercept-vulnerability/u-s-shot-raises-tensions-and-worries-over-satellites -idUSN2144210520080222.

4. David Hambling, "US and UK Accuse Russia of In-Orbit Test of Nesting Doll Anti-Satellite Weapon," *Forbes*, July 24, 2020, available at https://www.forbes.com/sites /davidhambling/2020/07/24/us-and-uk-accuse-russia-of-testing-in-orbit-anti-satellite -weaponry/?sh=37c6f8653f3e.

5. William Broad, "How Space Became the Next 'Great Power' Contest between the US and China," *New York Times*, January 24, 2021, available at https://www.nytimes.com /2021/01/24/us/politics/trump-biden-pentagon-space-missiles-satellite.html.

6. For an overview of the Outer Space Treaty, see Stephen Buono, "Merely a 'Scrap of Paper'? The Outer Space Treaty in a Historical Perspective," *Diplomacy and Statecraft* 31, no. 2 (2020): 350–372.

7. Kevin Chilton, "The Anti-Satellite Test Ban Must Not Undermine Deterrence," *Defense News*, April 29, 2022, available at https://www.defensenews.com/opinion/commentary/2022/04/29/the-anti-satellite-test-ban-must-not-undermine-deterrence/.

8. The two most comprehensive works to date on the origins of SDI are: Donald Baucom, *The Origins of SDI 1944–1983* (Lawrence: University of Kansas Press, 1992); Frances FitzGerald, *Way Out There in the Blue: Reagan, Star Wars, and the End of the Cold War* (New York: Simon and Schuster, 2001).

9. The first scholarly and comprehensive political history of the Space Age is Walter McDougall's *The Heavens and the Earth* (Baltimore: Johns Hopkins University Press, 1997). For a history of the militarization of space in the 1950s and 1960s, see Sean Kalic, *US Presidents and the Militarization of Space, 1946–1967* (College Station: Texas A&M Press, 2012). This general subject has received more scholarly attention from political scientists, see Bleddyn Bowen, *Original Sin: Power, Technology, and War in Outer Space* (London: Hurst, 2022); Bleddyn Bowen, *War in Space: Strategy, Spacepower, Geopolitics* (Edinburgh: Edinburgh University Press, 2020); Everett Dolman, *Astropolitik: Classical Geopolitics in the Space Age* (New York: Routledge, 2001); Joan Johnson-Freese, *Space Warfare in the 21st Century: Arming the Heavens* (New York: Routledge, 2016); Paikowsky Deganit, *The Power of the Space Club* (Cambridge: Cambridge University Press, 2017); Paul Stares, *The Militarization of Space, US Policy, 1945–1984* (Ithaca, NY: Cornell University Press, 1985); James C. Moltz, *The Politics of Space Security: Strategic Restraint and the Pursuit of National Interests, 3rd Edition* (Stanford, CA: Stanford University Press, 2019). In chapter 4 of his dissertation, William Schlickenmaier analyzes US ASAT development during the Cold War; see William Schlickenmaier, "Playing the General's Game: Superpowers, Self-Limiting, and Strategic Emerging Technologies" (Diss., Georgetown University, 2020).

10. Trevor Paglen, "The Other Night Sky," *Berkeley Art Museum and Pacific Film Archive*, exhibit dates: June 1–September 14, 2008, available at https://bampfa.org/program/trevor-paglen-other-night-sky-matrix-225.

11. For a survey of US spaceflight that includes military activities, see Asif Siddiqi, "American Space History: Legacies, Questions, and Opportunities for Research," in *Critical Issues in the History of Spaceflight*, ed. Steven J. Dick and Roger D. Launius (Washington, DC: National Aeronautics and Space Administration, 2006).

12. There have been multiple articles and books that have addressed aspects of US military and intelligence activities in space. For examples, see Aaron Bateman, "Mutually Assured Surveillance at Risk: Anti-Satellite Weapons and Cold War Arms Control," *Journal of Strategic Studies* 45, no. 1 (2022); Aaron Bateman, "Keeping the Technological Edge: The Space Arms Race and Anglo-American Relations in the 1980s," *Diplomacy and Statecraft* 33, no. 2 (2022): 355–378; *Eye in the Sky: The Story of the Corona Spy*

Satellites, ed. Dwayne Day, John Logsdon, Brian Latell (Washington, DC: Smithsonian Institution Press, 1997); Jeffrey Richelson, *America's Space Sentinels: DSP Satellites and National Security* (Lawrence: University of Kansas Press, 2001); James David, *Spies and Shuttles: NASA's Secret Relationships with the DoD and CIA* (Gainesville: University Press of Florida, 2015); and many other good technical histories of specific satellite systems written by Dwayne Day and published in *The Space Review*. For a history of space militarization and astroculture, see *Militarizing Outer Space*, ed. Alexander Geppert, Daniel Brandau, and Tilmann Siebeneichner (New York: Palgrave Macmillan, 2021).

13. Jeffrey Richelson, "Out of the Black: The Declassification of the NRO," National Security Archive Electronic Briefing Book No. 257, September 18, 2008, available at https://nsarchive2.gwu.edu/NSAEBB/NSAEBB257/index.htm.

14. For one example, see Michael J. Neufeld, "Cold War—But No War—In Space" in *Militarizing Outer Space*, ed. Alexander Geppert, Daniel Brandau, and Tilmann Siebeneichner (New York: Palgrave Macmillan, 2021), 46. For a history of the militarization of space in the 1950s and 1960s, see Kalic, *US Presidents and the Militarization of Space*. This general subject has received more scholarly attention from political scientists, two good examples are: Stares, *The Militarization of Space* and Moltz, *The Politics of Space Security*. In chapter 4 of his dissertation, William Schlickenmaier analyzes US ASAT development during the Cold War; see Schlickenmaier, "Playing the General's Game."

15. The Five Eyes include the US, the UK, Australia, Canada, and New Zealand. For an overview of the alliance, see Anthony Wells, *Between the Five Eyes: 50 Years of Intelligence Sharing* (Philadelphia: Casemate, 2020). A UK FCO study on ASATs described how all of NATO had become dependent on US space systems for command and control, intelligence, and navigation. See, FCO 66/1343, FCO briefing "Anti Satellite Weapons—The Current Situation," undated, TNA.

16. For an overview of SALT talks, see Matthew J. Ambrose, *The Control Agenda: A History of the Strategic Arms Limitation Talks* (Ithaca, NY: Cornell University Press, 2018); James Cameron, *The Double Game: The Demise of America's First Missile Defense System and the Rise of Strategic Arms Limitation* (Oxford: Oxford University Press, 2017).

17. Memorandum from David Elliot to Brent Scowcroft, "Final Report of the Ad Hoc NSC Space Panel—Part II: US Anti-Satellite Capabilities," November 3, 1976, Ford Library, 5 and 39.

18. William Broad, "Reagan's Legacy in Space: More Reach than Grasp," *New York Times*, May 8, 1988.

19. Andrew Butrica, *Single Stage to Orbit: Politics, Space Technology, and the Quest for Reusable Rocketry* (Baltimore: Johns Hopkins University Press, 2006).

20. Butrica details the connections between Mahan and the conservative space agenda; see part I in Butrica, *Single Stage to Orbit*.

21. Beth A. Fischer, "Nuclear Abolitionism, Strategic Defense Initiative, and the 1987 INF Treaty," in *The INF Treaty of 1987: A Reappraisal*, ed. Phillip Gassert, Tim Geiger,

and Hermann Wentker (Berlin: Vandenhoeck & Ruprecht, 2020), 48; FitzGerald, *Way Out There in the Blue*, 147–210; Paul Lettow, *Ronald Reagan and His Quest to Abolish Nuclear Weapons* (New York: Random House, 2005).

22. During his speech on September 20, 1963 to the UN General Assembly, Kennedy proposed that the US and the Soviet Union join together so that the first people to travel to the moon "would not be representatives of a single nation, but representatives of all our countries." John Logsdon observes that "with [Kennedy] died the possibility of US–Soviet cooperation in going to the moon." For more information on the details surrounding Kennedy's interest in an American-Soviet lunar expedition, see John M. Logsdon, *John F. Kennedy and the Race to the Moon* (New York: Palgrave, 2010), 182–193.

23. For a range of different views on the political significance of SDI, see Archie Brown, *The Human Factor: Gorbachev, Reagan, Thatcher and the End of the Cold War* (Oxford: Oxford University Press, 2020), 141; Henry Kissinger, *Diplomacy* (New York: Simon and Schuster, 1994), 782 and 787; Richard Rhodes, *Arsenals of Folly: The Making of the Nuclear Arms Race* (New York: Vintage, 2007), 179; John Lewis Gaddis, *The United States and the End of the Cold War* (New York: Oxford University Press, 1992), 43–44; Kenneth Adelman, *Reagan at Reykjavik: Forty-Eight Hours That Ended the Cold War* (New York: Broadside Books, 2014), 315. For more recent scholarship that uses Soviet sources to analyze the role of SDI in arms control negotiations and Soviet internal reforms, see Elizabeth Charles, "The Game Changer: Reassessing the Impact of SDI on Gorbachev's Foreign Policy, Arms Control, and US–Soviet Relations" (Diss., George Washington University, 2010); David Hoffman, *The Dead Hand: The Untold Story of the Cold War Arms Race and Its Dangerous Legacy* (New York: Anchor, 2010); Pavel Podvig, "Did Star Wars Help End the Cold War? Soviet Response to the SDI Program," *Science and Global Security* 25, no. 1 (2017): 3–27.

24. The Soviets often used the term "space-strike arms" in the context of SDI, although they did not always provide a precise definition. At a 1985 meeting with US counterparts, Eduard Shevardnadze placed ASATs "in the broad context of a ban on space-strike arms." See "Telegram from the US Embassy," November 6, 1985, *FRUS*, 1981–1988, Volume V, Soviet Union, March 1985–October 1986; Soviet General Nikolai Detinov described "space-strike arms" as including "space-based missile defense, the US ASAT, the Soviet ASAT, and 'space-based systems which can attack targets in space, in the air, on the earth or at sea.'" See "The Soviet Proposal at Geneva for a Blanket Moratorium," March 20, 1985, *FRUS*, 1981–1988, Volume V, Soviet Union, March 1985–October 1986. For additional background, see Peter Westwick, "'Space-Strike Weapons' Soviet Response to SDI and the Soviet Response to SDI," *Diplomatic History* 32, no. 5 (2008): 955–979.

25. Gabrielle Hecht observes that system design choices both reflect political values and shape political outcomes; see Gabrielle Hecht, *The Radiance of France: Nuclear Power and National Identity after World War II* (Cambridge, MA: MIT Press, 1998).

26. Richard Rhodes wrote that "SDI in Reagan's mind was never about any specific technology . . . strategic defense was a dream, a fantasy, an uninformed winner-take-all bet that American technology could make miracles happen"; see Rhodes, *Arsenals of Folly*, 179. Archie Brown observes that "years later it became even clearer than it was at the time that SDI was an expensive pipe dream"; see Brown, *The Human Factor*, 141. Frances FitzGerald argues that SDI was a product of "domestic politics, history, and mythology"; see FitzGerald, *Way Out There in the Blue*, 18. These perspectives gloss over the basis of technological decisions and overlook divergent ideas about what would indeed make a space-based missile defense system effective.

27. "Excerpts from Reagan's Speech at Jersey School," *New York Times*, June 20, 1986, available at https://www.nytimes.com/1986/06/20/world/excerpts-from-reagan-s-spe ech-at-jersey-school.html.

28. Rebecca Slayton provides an excellent explanation of the software and computing challenges associated with missile defense; see Rebecca Slayton, *Arguments that Count: Physics, Computing, and Missile Defense, 1949–2012* (Boston: MIT Press, 2013).

29. Janne Nolan details the consistent US pattern of situating new weapons programs into existing deterrence strategies; see Janne Nolan, *Guardians of the Arsenal: The Politics of Nuclear Strategy* (New York: Basic, 1989).

30. This observation is rooted in ideas about the social construction of technology. For an overview of important works on this subject, see Wiebe E. Bijker, Thomas P. Hughes, Trevor J. Pinch, ed., *The Social Construction of Technological Systems: New Directions in the Sociology and History of Technology* (Cambridge, MA: MIT Press, 1987); Merrit Roe Smith and Leo Marx, ed., *Does Technology Drive History? The Dilemma of Technological Determinism* (Cambridge, MA: MIT Press, 1995); Donald Mackenzie, *Inventing Accuracy: A Historical Sociology of Nuclear Missile Guidance* (Cambridge, MA: MIT Press, 1990); Wiebe Bijker, *Of Bicycles, Bakelites, and Bulbs: Toward a Theory of Sociotechnical Change* (Cambridge, MA: MIT Press, 1995).

31. Within the transatlantic alliance, aside from the US, I am devoting the majority of the attention to Britain, France, and West Germany, since they had the greatest impact on US military space policy and strategy. For background on Canada's position on SDI, see Joel J. Sokolsky and Joseph T. Jockel, "Canada and the Future of Strategic Defense," in *Perspectives on Strategic Defense*, ed. Steven W. Guerrier, Wayne C. Thompson, Zbigniew Brzezinski (New York: Routledge, 1987).

32. Asif Siddiqi writes about the need to move beyond purely nationalist narratives of spaceflight; see Asif Siddiqi, "Competing Technologies, National(ist) Narratives, and Universal Claims: Toward a Global History of Space Exploration," *Technology and Culture* 51, no. 2 (2010): 425–443.

33. PREM 19/1188, Heseltine and Howe Minute to MT, "Ballistic Missile Defence (BMD): UK Policy towards the US Strategic Defence Initiative," October 11, 1984, UK National Archives (TNA hereafter).

CHAPTER 1

1. "Agreement Concerning Cooperation in the Exploration and Use of Outer Space for Peaceful Purposes," May 24, 1972, available at https://history.nasa.gov/astp /documents/Agreement%20concerning%20coop%20(Nixon-Kosygin).pdf.

2. "Transcript of Nixon's Television Address to the Soviet People from the Great Kremlin Palace," *New York Times*, May 29, 1972, available at https://www.nytimes .com/1972/05/29/archives/transcript-of-nixons-television-address-to-the-soviet -people-from.html.

3. Philip J. Klass, "Keeping the Nuclear Peace," *New York Times*, September 3, 1972, available at https://www.nytimes.com/1972/09/03/archives/spies-in-the-sky-sky-spies -good-catch.html.

4. "Memo from David Elliot to Brent Scowcroft," Final Report of the Ad Hoc NSC Space Panel—Part II: US Anti-Satellite Capabilities, November 3, 1976, Ford Library, 5.

5. James Wieghart, "Russians Can Kill Us in Space, CIA Chief Says," *New York Daily News*, February 1, 1978, CREST, CIA-RDP99-00498R000100130069-5.

6. Tilmann Siebeneichner, "Spacelab: Peace, Progress and European Politics in Outer Space, 1973–85," in *Limiting Outer Space: Astroculture After Apollo*, ed. Alexander Geppert (New York: Palgrave Macmillan, 2018), 261.

7. McDougall, *The Heavens and the Earth*, 185.

8. Memorandum of Conversation, President Ford et al., April 13, 1976, Gerald Ford Presidential Library, available at https://www.fordlibrarymuseum.gov/library /document/0314/1553429.pdf.

9. Bernard Schriever, "The Battle for 'Space Superiority,'" *Air and Space Forces Magazine*, April 1, 1957, available at https://www.airandspaceforces.com/article/0457space /; Michael Neufeld, *Von Braun: Dreamer of Space, Engineer of War* (New York: Knopf, 2007), 260. Lyndon Johnson hyperbolically declared that "whoever gains the ultimate supremacy of space, gains control—total control over the earth for purposes of tyranny or for the service of freedom"; see Kalic, *US Presidents and the Militarization of Space*, 42.

10. Baucom, *The Origins of SDI*, 13.

11. Curtis Peebles, *High Frontier: The US Air Force and the Military Space Program* (Washington, DC: Air Force Historical Studies Office, 1997), 65.

12. Clayton Chun, "Shooting Down a 'Star': Program 437, the US Nuclear ASAT System and Present-Day Copycat Killers," Air University Cadre Paper no. 6, 4, available at https://media.defense.gov/2017/Nov/21/2001847039/-1/1/0/CP_0006_CHUN _SHOOTING_DOWN_STAR.PDF, 4.

13. Chun, "Shooting Down a 'Star.'"

14. Prior to the U-2 high-altitude reconnaissance aircraft in the mid-1950s, the US had very limited intelligence on the Soviet Union. This was because the Central

Intelligence Agency and its sister services were largely unsuccessful in penetrating the Iron Curtain. In 1969, the Defense Intelligence Agency estimated that "65% of what the US knows about the strength . . . and distribution of Soviet ground forces—90% for the Chinese—is attributable to satellite intelligence." See "The National Reconnaissance Program," June 4, 1969, NRO Digital Archive, available at https://www.nro.gov/Portals/65/documents/foia/declass/Archive/NARP/1969%20NARPs/SC-2018-00033_C05111857.pdf.

15. Robin Dickey, "The Rise and Fall of Space Sanctuary in US Policy," Aerospace Corporation, September 1, 2020.

16. Eisenhower established the Technological Capabilities Panel to examine potential solutions to the problem of strategic surprise. Among many recommendations, the panel strongly advocated the development of high-altitude aircraft and space reconnaissance systems. For an overview of this, see R. Cargill Hall, "Origins of US Space Policy: Eisenhower, Open Skies, and Freedom of Space," in *Exploring the Unknown: Selected Documents in the History of the US Civil Space Program*, John M. Logsdon, gen. ed., with Linda J. Lear, Jannelle Warren-Findley; and Ray A. Williamson, and Dwayne A. Day, *Exploring the Unknown: Selected Documents in the History of the US Civil Space Program, Volume I: Organizing for Exploration* (Washington, DC: NASA SP-4407, 1995), 213–229.

17. Hall, "Origins of US Space Policy," 56, and Wayne Austerman, *Program 437: The Air Force's First Anti-Satellite System* (Colorado Springs: Air Force Space Command History Office, 1991), 8.

18. McDougall, *The Heavens and the Earth*, 185; Neufeld, "Cold War—But No War—in Space," 54.

19. Clayton Laurie, "Leaders of the National Reconnaissance Office," May 1, 2002, available at https://www.nro.gov/Portals/65/documents/foia/docs/foia-leaders.pdf, 1.

20. The NRO was composed of different program elements. The Air Force managed Program A, the CIA oversaw Program B, and the Navy was in charge of Program C. Until the early 1970s, Program D included all covert aerial overflight efforts. The Air Force and the CIA were the two largest elements in NRO. The various program offices would play different roles in individual satellite projects. For example, the CIA could be tasked with payload development, whereas the Air Force might oversee systems integration, in addition to launch and operation. For an overview of the management and structural problems associated with the NRO and the NRP, see "Basic Authorities and Agreements which CIA Holds with USAF/NRO; Recommended Modifications and Suggestions regarding the NRO Organization," January 24, 1963, CREST, CIA-RDP86B00269R000800060006-1.

21. In March 1962, Deputy Secretary of Defense Roswell Gilpatric issued a Department of Defense directive concerning the "Security and Public Information Policy for Military Space Programs." Through this policy, the Kennedy administration obfuscated details about military and intelligence space capabilities, including the classification of all launch and recovery operations. The Pentagon and intelligence

community "sharply curtailed" the number of people who had a "blanket need-to-know" concerning military and intelligence space systems. For details, see Department of Defense Directive, "Security and Public Information Policy for Military Space Programs," March 23, 1962, National Security Archive, available at https://nsarchive2 .gwu.edu/NSAEBB/NSAEBB225/doc14.pdf. This subject will be discussed in greater detail in chapter 3.

22. For details concerning the secrecy of NRO systems, see Aaron Bateman, "Trust but Verify: Satellite Reconnaissance, Secrecy, and Cold War Arms Control," *Journal of Strategic Studies*, 2023; "Classification of Talent and Keyhole Information," January 16, 1964, CREST, CIA-RDP67R00587A000100140024-5.

23. Stephen Buono, "The Province of All Mankind: Outer Space and the Promise of Peace" (Diss., Indiana University, 2020), 231.

24. Baucom, *The Origins of SDI*, 17.

25. Alexander Flax, "Ballistic Missile Defense: Concepts and History," *Daedalus* 114, no. 2 (1985): 49.

26. Cameron, *The Double Game*, 40–45.

27. Chun, "Shooting Down a 'Star,'" 6.

28. The CIA remained convinced that "it would be technically feasible for the Soviets to launch [orbital bombardment] weapons of limited capability into orbit in the mid-1960s." Analysts nevertheless cautioned that the Kremlin would not have an "effective offensive [space] capability until the late 1960s [at the earliest]" and *lacked any* "evidence of Soviet plans or programs for the military use of space." See National Intelligence Estimate Number 11-8-62, CREST, CIA-DOC-0000267773; Chun, "Shooting Down a 'Star,'" 1.

29. Kalic, *US Presidents and the Militarization of Space*, 86.

30. Kalic, *US Presidents and the Militarization of Space*, 86.

31. Peebles, *High Frontier*, 62.

32. Buono, "The Province of all Mankind," 241.

33. Chun, "Shooting Down a 'Star,'" 14.

34. "Recommended Policy," undated, CREST, CIA-RDP66R00638R000100140017-0.

35. "Recommended Policy," 236.

36. "Recommended Policy," 236.

37. For an overview of the Outer Space Treaty, see Buono, "The Province of All Mankind."

38. Lyndon B. Johnson, "Remarks at the Signing of the Treaty on Outer Space," January 27, 1967, available at https://www.presidency.ucsb.edu/documents/remarks-the -signing-the-treaty-outer-space.

39. For a history of FOBS, see Asif Siddiqi, "The Soviet Fractional Orbiting Bombardment System (FOBS): A Short Technical History," *Quest: The History of Spaceflight Quarterly* 7, no. 4 (2000).

40. Buono, "The Province of All Mankind," 366.

41. Intelligence report form Thomas Hughes (INR) to the Secretary of State, August 1967, Wilson Center Digital Archive, available at https://digitalarchive.wilson center.org/document/134075.

42. Buono, "The Province of All Mankind," 366. Moltz argues that FOBS "would have violated the Outer Space Treaty"; see Moltz, *The Politics of Space Security*, 156.

43. Buono, "The Province of All Mankind," 366. Moltz argues that FOBS "would have violated the Outer Space Treaty"; see Moltz, *The Politics of Space Security*, 156.

44. Buono, "The Province of All Mankind," 366. Moltz argues that FOBS "would have violated the Outer Space Treaty"; see Moltz, *The Politics of Space Security*, 156.

45. Chun, "Shooting Down a 'Star,'" 26.

46. Chun, "Shooting Down a 'Star,'" 30.

47. Cameron, *The Double Game*, chapter 3.

48. Moltz, *The Politics of Space Security*, 152. Johnson was aware of the high-altitude nuclear tests that had unintentionally damaged and destroyed American satellites. The testing and use of nuclear-tipped ABMs could have placed critical US space reconnaissance systems in jeopardy. Johnson once let slip to a group of journalists that "I wouldn't want to be quoted on this but we've spent 35 or 40 billion dollars on the space program. And if nothing else had come out of it except the knowledge we've gained from space photography, it would be worth 10 times what the whole program has cost. Because tonight we know how many missiles the enemy has and, it turned out, our guesses were way off. We were doing things we didn't need to do. We were building things we didn't need to build. We were harboring fears we didn't need to harbor. Because of satellites, I know how many missiles the enemy has." Consequently, the president did not want to adopt any policy position that could have undermined space reconnaissance. Source for Johnson quote: Dwayne Day, John Logsdon, and Brian Latell, "Introduction," in *Eye in the Sky*, 1.

49. "Security and Space," May 9, 1967, CREST, CIA-RDP70B00501R000100160009-9.

50. Hal Brands, *What Good Is Grand Strategy? Power and Purpose in American Statecraft from Harry S. Truman to George W. Bush* (Ithaca, NY: Cornell University Press, 2014), 65.

51. Brands, *What Good Is Grand Strategy?*, 9.

52. Recent scholarship on Cold War arms control includes Cameron, *The Double Game*; James Cameron, "Soviet–American Strategic Arms Limitation and the Limits of Co-operative Competition," *Diplomacy and Statecraft* 33, no. 1 (2022); John Maurer, "The Purposes of Arms Control," *Texas National Security Review* 2, no. 1 (2018);

Michael Krepon, *Winning and Losing the Nuclear Peace: The Rise, Demise, and Revival of Arms Control* (Palo Alto, CA: Stanford University Press, 2022).

53. "Possible Disclosure of Satellite Reconnaissance," January 21, 1964, CREST, CIA-RDP67B00558R000100090003-8.

54. Bateman, "Trust but Verify."

55. John Logsdon, *After Apollo? Richard Nixon and the American Space Program* (New York: Palgrave Macmillan, 2015), 24.

56. For works on space cooperation and US foreign relations, see John Krige, *Sharing Knowledge, Shaping Europe: US Technological Collaboration and Nonproliferation* (Boston: MIT Press, 2016); John Krige, Angelina Long Callahan, and Ashok Maharaj, *NASA in the World* (New York: Palgrave, 2013).

57. Vance Mitchell, *Sharing Space—The Secret Interaction Between the National Aeronautics and Space Administration and the National Reconnaissance Office* (Chantilly, VA: National Reconnaissance Office, 2012), 64.

58. Logsdon, *After Apollo*, 165.

59. Mitchell, *Sharing Space*, 64. For an in-depth history of interactions between NASA and the Department of Defense, see David, *Spies and Shuttles*.

60. Mitchell, *Sharing Space*, 64; David, *Spies and Shuttles*.

61. Mitchell, *Sharing Space*, 64.

62. Memorandum from DCI Helms to President-Elect Nixon, January 6, 1969, *FRUS*, 1969–1976, Volume II, Organization and Management of US Foreign Policy, 1969–1972.

63. An intelligence study post invasion of Czechoslovakia affirmed "the need on an urgent basis of a near real time readout satellite reconnaissance system in support of the strategic warning process." See Memorandum for David Packard et al., September 23, 1969, NRO Electronic FOIA Room, available at https://www.nro.gov/Portals/65/documents/foia/declass/Archive/NARP/1969%20NARPs/SC-2018-00033_C05114965.pdf.

64. Dwayne Day, "Intersections in Real Time: The Decision to Build the KH-11 Kennen Reconnaissance Satellite (Part I)," *The Space Review*, September 9, 2019, available at https://www.thespacereview.com/article/3791/1.

65. Memorandum from Cline to Helms, July 20, 1971, NRO Electronic FOIA Room, available at https://www.nro.gov/Portals/65/documents/foia/declass/MAJOR%20NRO%20PROGRAMS%20&%20PROJECTS/NRO%20EOI/SC-2016-00001_C05096650.pdf.

66. These concepts included a digital near real-time satellite codenamed "Zaman," an interim near real-time capability called "Film Readout Gambit" (FROG), a Corona follow-on called "Hexagon," and a Manned Orbiting Laboratory (MOL). For overviews of these satellite systems, see Day, "Intersections in Real Time (Part I)"; Dwayne Day, "The Film Read Out Gambit Program," *The Space Review*, February 7, 2022, available

at https://www.thespacereview.com/article/4327/1; Frederick C. E. Oder, James C. Fitzpatrick, and Paul E. Worthman, *The Hexagon Story* (Chantilly, VA: National Reconnaissance Office, 1992); Dwayne Day, "Manned Orbiting Lab and the Search for a Military Role for Astronauts," *The Space Review*, June 17, 2019, available at https://www.thespacereview.com/article/3736/1. Helms advised approving Hexagon over MOL, at least in part, because it had greater utility for arms control verification, a compelling argument in the context of SALT; see Robert Perry, "A History of Satellite Reconnaissance," Volume IIIB—Hexagon, November 1973, Report produced by Headquarters Air Force, 74, available at https://www.nro.gov/Portals/65/documents/foia/docs/HOSR/SC-2017-00006e.pdf.

67. In 1971, Ray Cline, the head of intelligence at the State Department, expressed worry in a letter to Helms that "before [Kennen] is ready, we may well be in situations where the decision makers will urgently need more flexible satellite capabilities." He therefore pressed Helms to consider "a relatively inexpensive quick reaction system . . . Hopefully available within two years." See letter from Cline to Helms, July 20, 1971, NRO Electronic FOIA Room, available at https://www.nro.gov/Portals/65/documents/foia/declass/MAJOR%20NRO%20PROGRAMS%20&%20PROJECTS/NRO%20EOI/SC-2016-00001_C05096650.pdf. The White House would not support both FROG, the interim option, and Kennen, and ultimately chose the latter. For an overview of the Kennen decision process, see Day, "Intersections in Real Time."

68. National Intelligence Estimate 11-1-69, "The Soviet Space Program," National Security Archive, available at https://nsarchive2.gwu.edu/NSAEBB/NSAEBB501/docs/EBB-16a.pdf.

69. National Intelligence Estimate, "Soviet Strategic Defenses," February 25, 1971, *FRUS*, Volume XXIV, National Security Policy, 1969–1972, available at https://history.state.gov/historicaldocuments/frus1969-76v34/d178.

70. *FRUS* Editorial Note, undated, available at https://history.state.gov/historicaldocuments/frus1969-76v34/d197.

71. Memo from DCI Helms to Kissinger, "Limited Space Reconnaissance Assistance to PRC," October 24, 1971, CREST, CIA-LOC-HAK-450-5-11-4.

72. Memo from Kissinger on "Study of US Responses to Soviet Anti-Satellite Activities," November 26, 1971, CREST, CIA-LOC-HAK-537-6-7-2.

73. Robert Kilgo, "The History of the United States Anti-Satellite Program and the Evolution to Space Control and Offensive and Defensive Counterspace," *Quest* 11, no. 3 (2004): 32–33.

74. Kilgo notes that some senior Air Force leaders wondered if the Spike concept was "too sophisticated to attain without major technical advances"; see Kilgo, "The History of the United States Anti-Satellite Program," 33.

75. Amrom Katz, "Preliminary Thoughts on Crises: More Questions than Answers," March 1972, available at https://aerospace.csis.org/wp-content/uploads/2018/09/Amr

om-Katz-NRO-Paper-on-Satellite-Defenses-1972.pdf. Based on archival information from the former Soviet Union, Katz's observations were not unfounded. Asif Siddiqi wrote that the Soviets were investigating ASATs in the late 1950s, but that "the first major catalyst for the early Soviet ASAT program appears to have come from concurrent US plans to deploy an operational ASAT system." A 1962 article published in the then-classified Soviet MoD journal (*Voennaia mysl'* [*Military Thought*]) reveals that senior Soviet military officers believed that the US intended to deploy weapons in space that could be used to destroy targets on Earth and in space. Lack of accurate knowledge about strategic intent on both sides of the Iron Curtain was, therefore, fueling anxieties about space weapons. See Asif Siddiqi, "The Soviet Co-Orbital Anti-Satellite System: A Synopsis," June 1997, *British Interplanetary Society* 50, no. 6 (1997); and Major General P. Vysotskiy, "American Military Technical Means of Combat in Space," *Military Thought*, December 1, 1961, CIA-RDP33-02415A000500190010-4.

76. Katz, "Preliminary Thoughts on Crises"; Stares, *The Militarization of Space*, 164.

77. Katz, "Preliminary Thoughts on Crises," 164.

78. Katz, "Preliminary Thoughts on Crises," 164.

79. Katz, "Preliminary Thoughts on Crises," 164.

80. Siddiqi, "The Soviet Co-Orbital Anti-Satellite System."

81. "Treaty between the United States of America and the Union of Soviet Socialist Republics on the Limitation of Anti-Ballistic Missile Systems (ABM Treaty)," May 26, 1972.

82. Moltz, *The Politics of Space Security*, 170.

83. Moltz, *The Politics of Space Security*, 170.

84. During SALT-I era discussions on NTM, a Soviet representative "gestured overhead," and another mentioned the word "satellite." According to a US representative, "the Soviets behaved as though any explicit acknowledgment of this [satellite reconnaissance] activity—even in secret negotiations—would be injurious." For details, see "Views on Public Release of Information on US Satellite Reconnaissance," April 13, 1970, CREST, CIA-RDP79B01709A000200010007-3.

85. "Further Notes on NSAM 156 Arms Control Paper," September 1968, available at https://www.nro.gov/Portals/65/documents/foia/declass/Archive/NARP/1968%20 NARPs/SC-2018-00032_C05106064.pdf.

86. Klass, "Keeping the Nuclear Peace."

87. Memorandum from Rush to Clements, "US Responses to Soviet Anti-Satellite Activities," June 18, 1973, *FRUS*, 1969–1976, Volume E-3, Documents on Global Issues, 1973–1976.

88. See Thomas Ellis, "Reds in Space: American Perceptions of the Soviet Space Programme from Apollo to Mir 1967–1991" (Diss., University of Southampton, 2018).

89. "Telephone Conversation of President Gerald Ford with Apollo–Soyuz Test Project Crews," July 17, 1975, Ford Presidential Library, available at https://www .fordlibrarymuseum.gov/library/speeches/750412.htm.

90. John Noble, "Astronauts Say Their Good-Bys; Parting Is Today," *New York Times*, July 19, 1975, available at https://www.nytimes.com/1975/07/19/archives/astronauts -say-their-goodbys-parting-is-today-crews-of-apollo-and.html.

91. Siddiqi, "The Soviet Co-Orbital Anti-Satellite System," 233.

92. Siddiqi, "The Soviet Co-Orbital Anti-Satellite System," 233.

93. Sven Grahn, "Simulated War in Space—Soviet ASAT Tests," undated, available at http://www.svengrahn.pp.se/histind/ASAT/ASAT.htm#Mark.

94. "Interagency Intelligence Memorandum on Soviet Dependence on Space Systems," October 6, 1975, CREST, CIA-DOC_0000380221.

95. "Vast Soviet Naval Exercise Raises Urgent Questions for West," *New York Times*, April 28, 1975.

96. "Interagency Intelligence Memorandum on Soviet Dependence on Space Systems," October 6, 1975, CREST, CIA-DOC_0000380221.

97. "Interagency Intelligence Memorandum on Soviet Dependence on Space Systems."

98. Memorandum from Scowcroft to Ford, "Follow-Up on Satellite Vulnerability," March 16, 1976, *FRUS*, 1969–1976, Volume E-3, Documents on Global Issues, 1973–1976. Prior to this, Ford established a panel chaired by Charles Slitchter of the University of Chicago to examine the military use of space, focusing on satellite reconnaissance and tactical communications. For details, see Paul Stares, *Space Weapons and US Strategy* (New York: Routledge, 1985), chapter 8.

99. "Report of the NSC Ad Hoc Panel on Technological Evolution and Vulnerability of Space," October 1976, ii, US National Security Council Institutional Files 1974–1977, Gerald R. Ford Presidential Library.

100. In 1974, the NSC established a study group to examine using satellite reconnaissance (SIGINT and imagery) to support tactical needs; see "Ad Hoc Panel on Tactical Applications of National Reconnaissance Assets," July 11, 1974, CREST, CIA-RDP80M01082A000700050047-3. This ad hoc panel was part of the so-called Slichter Panel under the direction of Charles Slichter from the University of Illinois; for a description of this panel, see "Report on Activities of Technical Panels," November 8, 1974, CREST, LOC-HAK-54-5-13-2.

101. Memorandum from Scowcroft to Ford, "Soviet Anti-Satellite Capability," April 26, 1976, *FRUS*, 1969–1976, Volume E-3, Documents on Global Issues, 1973–1976. Satellite survivability was of even greater urgency because in 1976, the US launched its first near real-time satellite, KH-11. This was a much more sophisticated system that could provide timelier intelligence but was not easily replaceable. For details on KH-11, see

"Report of the NSC Ad Hoc Panel on Technological Evolution and Vulnerability of Space," October 1976, US National Security Council Institutional Files 1974–1977, Gerald R. Ford Presidential Library.

102. Memorandum of Conversation, Ford, Rumsfeld, Scowcroft, and Kissinger, April 13, 1976, Ford Library, available at https://www.fordlibrarymuseum.gov/library/document/0314/1553429.pdf.

103. For an overview of the "revolution in military affairs," see Andrew Krepinevich and Barry Watts, *The Last Warrior: Andrew Marshall and the Shaping of Modern Defense Strategy* (New York: Basic, 2015), 193–227.

104. Krepinevich and Watts, *The Last Warrior*, 193–227.

105. Krepinevich and Watts, *The Last Warrior*, 193–227.

106. Katz, "Preliminary Thoughts on Crises."

107. National Security Decision Memorandum 333 on "Enhanced Survivability of Critical US Military and Intelligence Space Systems," July 7, 1976, *FRUS*, 1969–1976, Volume E-3, Documents on Global Issues, 1973–1976.

108. "Report of the NSC Ad Hoc Panel on Technological Evolution and Vulnerability of Space," October 1976, 1, US National Security Council Institutional Files 1974–1977, Gerald R. Ford Presidential Library.

109. National Intelligence Estimate 11-1-73, "Soviet Space Programs," December 20, 1973, CREST, CIA-DOC_0000283822.

110. Memorandum from Smith to Scowcroft, "Final Report of the Ad Hoc NSC Space Panel—Part II: US Anti-Satellite Capabilities," November 3, 1976, *FRUS*, 1969–1976, Volume E-3, Documents on Global Issues, 1973–1976. It is not clear whether the intelligence community at this time fully understood the limitations of the Soviet radar satellite used to track US and allied naval vessels. An updated CIA report on this subject from 1983 said that Moscow's radar satellites could not "detect any ships in high seas or in rain" and that it could detect destroyer-sized ships "only under the best of conditions." For an overview of RORSAT's limitations, see "Key Conclusions about Present and Future Soviet Space Missions," a reference aid, June 1983, CREST, CIA DOC_0000498490, available at https://nsarchive2.gwu.edu/NSAEBB/NSAEBB501/docs/EBB-37.pdf.

111. Memorandum from Smith to Scowcroft.

112. Memorandum from Smith to Scowcroft.

113. Memorandum from Smith to Scowcroft, 39.

114. Memorandum from Smith to Scowcroft, 40.

115. Memorandum from Smith to Scowcroft.

116. A co-orbital ASAT was technologically "in hand" but could be defeated with countermeasures, for example greater maneuverability, required larger boosters to get

into space, and would require more significant upgrades to existing space tracking systems that would be used to target Soviet satellites. See Memorandum from Smith to Scowcroft, 27.

117. Memorandum from Smith to Scowcroft, 28.

118. Memorandum from Smith to Scowcroft, 31.

119. National Security Decision Memorandum 345, "US Anti-Satellite Capabilities," January 18, 1977, available at https://aerospace.org/sites/default/files/policy_archives /NSDM-345%20ASAT%20Jan77.pdf.

120. "Report of Secretary of Defense Donald H. Rumsfeld to the Congress on the FY 1978 Budget, FY 1979 Authorization Request and FY 1978–1982 Defense Programs," January 14, 1977, available at http://www.bits.de/NRANEU/others/strategy/1978 _DoD_Annual_Report-Sanitized.pdf.

121. Julian E. Zelizer, "Détente and Domestic Politics," *Diplomatic History* 33, no. 4 (2009): 658.

122. Anatoly Dobrynin, *In Confidence: Moscow's Ambassador to Six Cold War Presidents* (Seattle: University of Washington Press, 2001), 327–333.

123. Henry Kissinger, *Years of Renewal* (New York: Simon and Schuster, 1999), 299.

124. Kissinger, *Years of Renewal*, 299–300.

125. Kissinger, *Years of Renewal*.

126. Kissinger, *Years of Renewal*, 302.

127. Kissinger, *Years of Renewal*, 661.

128. For a history of the Euromissile Crisis, see Leopoldo Nuti, Frédéric Bozo, Marie-Pierre Rey, and Bernd Rother, ed., *The Euromissile Crisis and the End of the Cold War* (Washington, DC: Woodrow Wilson Center Press, 2015).

129. In a campaign speech delivered in Exeter, New Hampshire, on February 10, 1976, Reagan specifically said that "we compromised our clear technological lead in the anti-ballistic missile system, the ABM, for the sake of a deal" with the Soviet Union. See editorial note (69) in *FRUS*, 1969–1976, Volume XXXVIII, Part 1, Foundations of Foreign Policy, 1973–1976.

130. *FRUS*, 665.

131. "NSC Staff Request for Evaluations of the A Team-B Experiment," November 16, 1982, CREST, CIA-RDP85B00134R000200090002-8.

132. "NSC Staff Request for Evaluations of the A Team-B Experiment," 31.

133. Butrica, *Single Stage to Orbit*, 23.

134. Odd Arne Westad, "The Fall of Détente and the Turning Tides of History," in *The Fall of Détente: Soviet-American Relations during the Carter Years*, ed. Odd Arne Westad (Oslo: Scandinavian University Press, 1977), 9.

135. "Issues Paper Prepared by the PRM-23 Interagency Group," August 9, 1977, *FRUS*, 1977–1980, Volume XXVI, Arms Control and Nonproliferation.

136. Memorandum of Conversation, "Bilateral Matters Between Washington and Moscow," March 30, 1977, *FRUS*, Volume XXVI, Arms Control and Nonproliferation.

137. Memorandum of Conversation, "Bilateral Matters Between Washington and Moscow."

138. Letter from President Carter to Soviet General Secretary Brezhnev, March 4, 1977, *FRUS*, 1977–1980, Volume VI, Soviet Union.

139. Memorandum from Chairman of the Joint Chiefs of Staff Brown to Secretary of Defense, "Antisatellites," July 29, 1977, *FRUS*, Volume XXVI, Arms Control and Nonproliferation.

140. Interagency Intelligence Memorandum, "Soviet Dependence on Space Systems," November 1975, CREST, CIA-DOC_0000380221. There was, however, disagreement over this assessment. The chiefs' deterrent position was also out of line with the Ford-commissioned satellite vulnerability study that had concluded that a US ASAT would not enhance the survivability of American space systems through deterrence; see Memorandum from Scowcroft to President Ford, "Soviet Anti-Satellite Capability," April 26, 1976, *FRUS*, 1969–1976, Volume E-3, Documents on Global Issues, 1973–1976. The deterrence argument should not, however, be outrightly dismissed. The archival record reveals that there was disagreement about the extent of Soviet dependence on space and the potential for a US ASAT to serve as a deterrent. A memorandum on October 26, 1976, for the CIA member of the National Foreign Intelligence Board contained the position of the intelligence community on Soviet "attitudes and doctrine regarding interference with US space systems." It described the likelihood of Soviet interference with US satellites in peacetime as low and interference in the event of imminent conflict as high. There was, however, disagreement over the argument that the US was more dependent on space systems than the Soviet Union. According to the memorandum, the US Air Force would probably "contest any judgment that the US is or will be more dependent [on space] than the Soviets." Fundamentally, the belief that the Soviet Union might be (or might soon become) as dependent on space as the US could have served as justification for the argument that a new US ASAT might deter the Soviet Union from using its own ASAT *in certain situations*. See NFIB Review of Interagency Intelligence Memorandum, "Prospects for Soviet Interference with US Space Systems," October 26, 1967, CREST, CIA-RDP79M00467A002500010007-7.

141. Intelligence analysts pointed out in 1983 that radar satellites have "serious limitations . . . [they] are adversely affected by poor weather"; see NIE 11-1-83, "The Soviet Space Program," July 19, 1983, CIA-RDP00B00369R000100050007-1.

142. Memorandum from the Chairman of the Joint Chiefs of Staff to Secretary of Defense Brown, "Antisatellites," July 29, 1977, *FRUS*, Volume XXVI, Arms Control and Nonproliferation.

143. Stares, *The Militarization of Space*, 183–184.

144. "Summary of Significant Discussion and Conclusions of a Policy Review Committee Meeting, PRM/NSC-23, Coherent Space Policy," August 4, 1977, *FRUS*, Volume XXVI, Arms Control and Nonproliferation.

145. "Tactical Use of Reconnaissance Satellite Assets," June 27, 1977, CREST, CIA-RDP83M00171R001000190001-6.

146. "Tactical Use of Reconnaissance Satellite Assets."

147. "Intelligence as a Force Multiplier—Meeting with [Redacted] TENCAP Coordination Officer, April 24, 1986," April 25, 1986, CREST, CIA-RDP89B01330R000400750014-7.

148. "Superweapons: The Particle Beam and the Nuclear Arms Race," 1977, available at https://www.youtube.com/watch?v=Q_ZwqW2EFGo.

149. For example, see National Intelligence Estimate 11-3/8-77, "Soviet Capabilities for Strategic Nuclear Conflict Through the Late 1980s," CREST, CIA-DOC_0000268138.

150. Issues Paper Prepared by the PRM-23 Interagency Group, "Arms Control for AntiSatellite Systems Issues Paper," August 9, 1977, *FRUS*, 1977–1980, Volume XXVI, Arms Control and Nonproliferation.

151. "Arms Control for AntiSatellite Systems Issues Paper."

152. "Arms Control for AntiSatellite Systems Issues Paper."

153. Memorandum from Secretary of Defense Brown to President Carter, "Arms Control for Antisatellite Systems," August 19, 1977, *FRUS*, 1977–1980, Volume XXVI, Arms Control and Nonproliferation.

154. "Letter from the Special Advisor to the President for Science and Technology to Secretary of Defense," September 8, 1977, *FRUS*, 1977–1980, Volume XXVI, Arms Control and Nonproliferation.

155. The interagency study noted that "a relatively small level of successful cheating, e.g., a handful of ASATs and EW sites, could have a high payoff"; see "Arms Control for AntiSatellite Systems Issues Paper."

156. Memorandum from Brzezinski to Secretary of State et al., "Arms Control for ASAT Systems," September 23, 1977, *FRUS*, 1977–1980, Volume XXVI, Arms Control and Nonproliferation.

157. George S. Wilson, "Brown Says Some US Satellites Are Vulnerable to Soviet Hunters," *Washington Post*, October 5, 1977, available at https://www.washingtonpost.com /archive/politics/1977/10/05/brown-says-some-us-satellites-are-vulnerable-to-soviet -hunters/bac25e3d-4340-402d-8fb8-20e89f3e9f62/.

158. "The History of US Antisatellite Weapon Systems," undated, available at https:// fas.org/man/eprint/leitenberg/asat.pdf, 21.

159. Wilson, "Brown Says Some US Satellites Are Vulnerable to Soviet Hunters."

160. Memorandum of Conversation, "Arms Control Issues: SALT, ASATs, etc.," October 17, 1977, *FRUS*, 1977–1980, Volume XXVI, Arms Control and Nonproliferation.

161. Thomas O'Toole, "Space Wars," *Washington Post*, November 6, 1977, available at https://www.washingtonpost.com/archive/opinions/1977/11/06/space-wars/cb2edd42-cbb4-42ba-8514-202080dccafe/.

162. "Briefing on SALT Negotiations," Hearings before the Committee on Foreign Relations United States Senate, Ninety Fifth Congress, November 3 and 29, 1977, available at https://www.govinfo.gov/content/pkg/CHRG-95shrg24430/pdf/CHRG-95shrg24430.pdf.

163. National Intelligence Estimate 11-3/8-77, "Soviet Capabilities for Strategic Nuclear Conflict Through the Late 1980s," February 1978, CREST, CIA-DOC_0000268138.

164. Memorandum for Brzezinski, "Soviet and US High-Energy Laser Weapons Programs," November 28, 1977, *FRUS*, 1977–1980, Volume XXVI, Arms Control and Nonproliferation.

165. Richard Burt, "New Killer Satellites Make 'Sky-War' Possible," *New York Times*, June 11, 1978, available at https://www.nytimes.com/1978/06/11/archives/new-killer-satellites-make-skywar-possible.html.

166. Baucom, *The Origins of SDI*, 118.

167. Baucom, *The Origins of SDI*, 118.

168. Baucom, *The Origins of SDI*, 119.

169. Baucom, *The Origins of SDI*, 121 and 123.

170. Baucom, *The Origins of SDI*, 123 and 124.

171. Baucom, *The Origins of SDI*, 123 and 124.

172. FitzGerald, *Way Out There in the Blue*, 124.

173. Baucom, *The Origins of SDI*, 128.

174. Memorandum from the Director of the Arms Control and Disarmament Agency to Secretary of State Vance, "Anti-Satellite Limits," October 19, 1977, *FRUS*, 1977–1980, Volume XXVI, Arms Control and Nonproliferation.

175. Presidential Directive/NSC-33, March 10, 1978, *FRUS*, 1977–1980, Volume XXVI, Arms Control and Nonproliferation.

176. Memorandum to President Carter, "US Position on ASATs," February 24, 1979, *FRUS*, 1977–1980, Volume XXVI, Arms Control and Nonproliferation.

177. For a history of the INF dual-track decision, see Susan Colbourn, *Euromissiles: The Nuclear Weapons that Nearly Destroyed NATO* (Ithaca, NY: Cornell University Press, 2022).

178. Presidential Directive/NSC-37, May 11, 1978, *FRUS*, 1977–1980, Volume XXVI, Arms Control and Nonproliferation.

179. Maurer, "The Purposes of Arms Control," 9.

180. Maurer, "The Purposes of Arms Control," 19.

181. Cameron, *The Double Game*, 4–5.

182. "Telegram from US Embassy in Finland, ASAT First Round," June 20, 1978, *FRUS*, 1977–1980, Volume XXVI, Arms Control and Nonproliferation.

183. "Telegram from US Embassy in Finland, ASAT First Round."

184. "Telegram from US Embassy in Finland, ASAT First Round."

185. Alexander Geppert and Tilmann Siebeneichner, "Spacewar! The Dark Side of Astroculture," in *Militarizing Outer Space*, ed. Alexander Geppert, Daniel Brandau, and Tilmann Siebeneichner (New York: Palgrave Macmillan, 2021), 4.

186. Geppert and Siebeneichner, "Spacewar! The Dark Side of Astroculture," 15.

187. Siebeneichner, "Spacelab: Peace, Progress and European Politics in Outer Space," 261.

188. Pericles Gasparini Alves, "Prevention of an Arms Race in Outer Space: A Guide to the Discussions in the Conference on Disarmament," United Nations Institute for Disarmament Research, 1991, available at https://www.unidir.org/files/publications /pdfs/prevention-of-an-arms-race-in-outer-space-a-guide-to-the-discussions-in-the-cd -en-451.pdf, 4.

189. Memorandum for Deputy to the DCI for Resource Management, September 15, 1978, CREST, CIA-RDP85-00821R000100110001-3.

190. "Weekly Compilation of Presidential Documents," October 9, 1978, National Security Archives, available at https://nsarchive2.gwu.edu/NSAEBB/NSAEBB231/doc32 .pdf.

191. Summary of Conclusions of a Special Coordination Committee Meeting, "Anti-satellite Treaty," November 16, 1978, *FRUS*, 1977–1980, Volume XXVI, Arms Control and Nonproliferation.

192. Memorandum from Keeny to Aaron, "ASAT Instructions," January 18, 1979, *FRUS*, 1977–1980, Volume XXVI, Arms Control and Nonproliferation. Brzezinski did advise Carter that "the pursuit of limits on high energy laser ASAT applications should get high priority in follow-on negotiations." See Memorandum from Brzezinski to Carter, "Instructions to the Delegation to the 2nd Round of ASAT Talks," January 22, 1979, *FRUS*, 1977–1980, Volume XXVI, Arms Control and Nonproliferation. Carter's instructions for the next round of talks included: continuing to discuss ways to eliminate the existing Soviet ASAT system verifiably, suspend ASAT testing, and clarify "mutual understanding that might be codified in an initial ASAT agreement." See PD/NSC-45, "Instructions to the US Delegation to the ASAT Talks with the Soviets Commencing January 23 in Bern," *FRUS*, 1977–1980, Volume XXVI, Arms Control and Nonproliferation.

193. Memorandum from Slocombe to Brown, "PD-45/ASAT Negotiations—ACTION MEMORANDUM," February 1, 1979, *FRUS*, 1977–1980, Volume XXVI, Arms Control and Nonproliferation.

194. Memorandum from Vance to Carter, "Anti-Satellite (ASAT) Negotiations," undated, *FRUS*, 1977–1980, Volume XXVI, Arms Control and Nonproliferation.

195. FCO 66/1343, FCO briefing "Anti Satellite Weapons—The Current Situation," undated, TNA.

196. FCO 66/1137, "Minute from Killick to Mallaby, Anti-Satellite Weapons: US/ Soviet Exchanges," May 11, 1978, TNA.

197. FCO 66/1137, "Minute for Clay to O'Hara, Anti-Satellite Capabilities: US/Soviet Exchanges," June 5, 1978, TNA.

198. FCO 66/1343, FCO briefing "Anti Satellite Weapons—The Current Situation," undated, TNA.

199. John Fenske, "France and the Strategic Defence Initiative: Speeding up or Putting on the Brakes?" *International Affairs* 62, no. 2 (1986): 232.

200. On April 17, 1979, the US Mission to NATO shared Washington's ASAT arms control strategy with its transatlantic partners. According to American diplomats in Brussels, only the British Permanent Representative immediately replied and commented that the resumption of ASAT talks was "good news" and expressed worry about the Soviet ASAT system. See footnote 2 in Memorandum from Aaron to Mondale et al., "Instructions to the US Delegation to the 3rd Round of ASAT Talks with the Soviets Commencing April 23, 1979, in Vienna," April 20, 1979, *FRUS*, 1977–1980, Volume XXVI, Arms Control and Nonproliferation.

201. Memorandum for Brzezinski, "ASAT Negotiations," May 30, 1979, *FRUS*, 1977–1980, Volume XXVI, Arms Control and Nonproliferation.

202. Bart Hendricks and Dwayne Day, "Target Moscow: Soviet Suspicions about the American Space Shuttle (Part I)," *The Space Review*, January 27, 2020, available at https://www.thespacereview.com/article/3873/1.

203. Memorandum from Vance and Seignious to Carter, "Antisatellite Negotiations," June 6, 1979, *FRUS*, 1977–1980, Volume XXVI, Arms Control and Nonproliferation.

204. Memorandum from Vance and Seignious to Carter, "Antisatellite Negotiations."

205. Memorandum of Conversation, "Vance–Gromyko Discussion of Joint Communique," June 16, 1979, *FRUS*, 1977–1980, *FRUS*, 1977–1980, Volume XXVI, Arms Control and Nonproliferation. Although by May 1979 the two sides were working on a joint draft text, there were still important points of divergence. The Soviets wanted a provision stating that either party be permitted "to damage, destroy, or change the trajectory of space objects without violating the agreement simply by asserting that it [space object] was the target of what you call a 'hostile act.'" US delegates saw this provision as making an agreement "hollow." For details, see "Telegram from the Department

of State to the Embassy in the Soviet Union," May 18, 1979. For a detailed overview of unresolved ASAT issues, which primarily centered on the shuttle and a hostile acts agreement, see "Telegram from the Embassy in Austria to the Department of State," June 13, 1979, *FRUS*, 1977–1980, Volume XXVI, Arms Control and Nonproliferation. The Soviet side did not conceal its doubts about a comprehensive agreement. In telegram 4927 from Vienna on May 18, 1979, the Embassy reported that Khlestov said, "the very idea of a 'comprehensive' agreement has caused us to have serious doubts from the very beginning, and we have indicated this to the US Delegation many times"; see footnote 5 in "Telegram from the Embassy in Austria to the Department of State (4954)," May 21, 1979, *FRUS*, 1977–1980, Volume XXVI, Arms Control and Nonproliferation.

206. "Telegram from the Embassy in Austria to the Department of State," May 2, 1979, *FRUS*, 1977–1980, Volume XXVI, Arms Control and Nonproliferation.

207. Memorandum from Vance and Seignious to Carter, "Anti-Satellite Negotiations," May 19, 1979, *FRUS*, 1977–1980, Volume XXVI, Arms Control and Nonproliferation.

208. Memorandum for Secretary of State Vance, August 9, 1979, *FRUS*, 1977–1980, Volume XXVI, Arms Control and Nonproliferation.

209. CIA Report, "Soviet Interest in ASAT Talks," August 7, 1980, *FRUS*, 1977–1980, *FRUS*, 1977–1980, Volume XXVI, Arms Control and Nonproliferation.

210. For examples of arguments centered on the premise that reconnaissance satellites were fundamentally stabilizing, see John Lewis Gaddis, "Looking Back: The Long Peace," *The Wilson Quarterly* 13, no. 1 (1989): 54; John Lewis Gaddis, *The Long Peace: Inquiries into the History of the Cold War* (Oxford: Oxford University Press, 1989), 195–215; Neufeld, "Cold War—But No War—in Space," 62.

CHAPTER 2

1. Major Paul Viotti, ed., "Military Space Doctrine—The Great Frontier: Final Report for the United States Air Force Academy Military Space Doctrine Symposium," April 1–3, 1981, USAF Academy, 1981.

2. National Security Study Directive Number 13-82, "National Space Strategy," December 15, 1982, CREST, CIA-RDP85M00364R000400550064-1.

3. Brigadier General Peter Worden (USAF, ret.), interview with the author, June 10, 2019. Worden identified himself as a part of this informal group of "space zealots."

4. For different interpretations of SDI's origins, see Fischer, "Nuclear Abolitionism," 48; FitzGerald, *Way Out There in the Blue*, 147–210; Lettow, *Ronald Reagan and His Quest*; Edoardo Andreoni, "Ronald Reagan's Strategic Defense Initiative and Transatlantic Relations, 1983–1986" (Diss., Cambridge University, 2017), 27–28.

5. Fischer, "Nuclear Abolitionism," 48; Lettow, *Ronald Reagan and His Quest*, 21.

6. In his analysis of Reagan's space policy, John Logsdon does not cover SDI because he contends that it was not a space program; see Logsdon, *Ronald Reagan and the Space Frontier* (London: Palgrave Macmillan, 2019). In contrast, Butrica maintains that Reagan's military space, SDI in particular, and commercial space agendas were closely linked; see Butrica, *Single Stage to Orbit*, 47.

7. Richard Halloran, "US Plans Big Spending Increase for Military Operations in Space," *New York Times*, October 17, 1982, CREST, CIA-RDP92B00478R000800260003-3.

8. Butrica, *Single Stage to Orbit*, 47.

9. James Abrahamson, "SDI and the New Space Renaissance," *Space Policy* 1, no. 2 (1985), 119.

10. Fischer, "Nuclear Abolitionism," 50.

11. DeWitt Douglas Kilgore writes that astrofuturism stemmed from the "Euro-American preoccupation with imperial exploration and utopian speculation, which it recasts in the elsewhere and else *when* of outer space" (emphasis in original). He further writes that astrofuturism is built around the idea of the space frontier that is "an extension of the nation's expansion to continental and global power in the nineteenth and twentieth centuries." See DeWitt Douglas Kilgore, *Astrofuturism: Science, Race, and Visions of Utopia in Space* (Philadelphia: University of Pennsylvania Press, 2003), 1–2.

12. NSDD-42, "National Space Policy," July 4, 1982, CREST, CIA, RDP84B00148R 000100320009-2.

13. James G. Wilson, *The Triumph of Improvisation: Gorbachev's Adaptability, Reagan's Engagement, and the End of the Cold War* (Ithaca, NY: Cornell University Press, 2014).

14. Brands, *What Good Is Grand Strategy?*, 102.

15. Brands, *What Good Is Grand Strategy?*, 109.

16. Brands, *What Good Is Grand Strategy?*, 103.

17. Brands, *What Good Is Grand Strategy?*, 107.

18. For an overview of the Navy's investment in capabilities to exploit space systems for tactical operations, see Ivan Amato, *Taking Technology Higher: The Naval Center for Space Technology and the Making of the Space Age* (Washington, DC: NRL, 2022), 301–327. The military services increasingly viewed space systems as "force multipliers"; see "Intelligence as a Force Multiplier—Meeting with [Redacted] TENCAP Coordination Officer, April 24, 1986," April 25, 1986, CREST, CIA-RDP89B01330R000400 750014-7.

19. Viotti, ed., "Military Space Doctrine"; Thomas Karas, *The New High Ground, Strategies and Weapons of Space-Age War* (New York: Simon and Schuster, 1983), 9.

20. Viotti, ed., "Military Space Doctrine."

21. Viotti, ed., "Military Space Doctrine," 10.

22. Viotti, ed., "Military Space Doctrine," 13.

23. Sandra Erwin, "Trump Signs Defense Bill Establishing US Space Force: What Comes Next," *SpaceNews*, December 20, 2019, available at https://spacenews.com /trump-signs-defense-bill-establishing-u-s-space-force-what-comes-next/.

24. Peter L. Hays, "Struggling Towards Space Doctrine: US Military Space Plans, Programs, and Perspectives during the Cold War" (Diss., Fletcher School of Law and Diplomacy, Tufts University, 1994), 319.

25. Hays, "Struggling Towards Space Doctrine."

26. The president's July 1982 space directive called for the establishment of a new Senior Interagency Group (SIG) for space. Reagan's NSC used SIGs for interdepartmental coordination on national security matters. The assistant to the president for national security affairs was designated as the chair for SIG (Space). The following people were identified as its core members: the deputy or undersecretary of defense, the deputy or undersecretary of state, the director of the Arms Control and Disarmament Agency (ACDA), the deputy or undersecretary of commerce, the director of central intelligence, chairman of the joint chiefs, and the NASA administrator. Most significantly for the national security space system, the creation of this SIG elevated space (bureaucratically) to the same level as defense and intelligence policy. See NSDD-42, "National Space Policy," July 4, 1982, CREST, CIA-RDP84B00148R000100320009-2.

27. NSDD-42, "National Space Policy," 304.

28. NSDD-8, "Space Transportation System," November 13, 1981, CREST, CIA-RDP87B 01034R000600090002-9.

29. NASA's space shuttle was intended to be used for national security and civil purposes from the time of its approval in the early 1970s. The shuttle conducted multiple classified missions for the Department of Defense throughout its lifetime. SDI plans to use the shuttle to deliver weapons into space would have, however, made it an even greater potential military target for the Soviet Union. Jim David discusses NASA's relationship with the Department of Defense and intelligence community at length in his book *Spies and Shuttles*.

30. SIG Space Terms of Reference, July 28, 1982, CREST, CIA-RDP84B00049R0011 02740033-6.

31. General Lew Allen Biography, US Air Force, available at https://www.af.mil/About -Us/Biographies/Display/Article/107824/general-lew-allen-jr/.

32. NSDD-42, "National Space Policy," July 4, 1982, CREST, CIA-RDP88B00838R0 00300510026-0.

33. NSDD-42, "National Space Policy."

34. NSDD-42, "National Space Policy."

35. Richard Halloran, "Pentagon Draws Up First Strategy for Fighting a Long Nuclear War," *New York Times*, May 30, 1982.

36. "Air Force Space Command Facts," undated, available at https://www.afspc.af.mil /About-Us/.

37. "Air Force Space Command Facts." Paul Stares notes the emergence of a deterrence justification for the US ASAT but does not provide an explanation as to why this change occurred; see Stares, *The Militarization of Space*, 223.

38. Thomas Schelling, *Arms and Influence* (New Haven, CT: Yale University Press, 1966), 146.

39. Going back to the mid-1970s (at the latest), US national security officials were interested in whether there could be mutual American–Soviet vulnerability in space, and they wanted to know if such a situation would promote stability in space. DCI William Colby commissioned a study on this topic in 1975; see "DCI Proposal for Intelligence Assessment of Soviet Dependence on Spaceborne Systems," March 31, 1975, CREST, CIA-RDP80B01495R000600030024-0. A 1985 NIE reported that "we judge that, although the USSR is not at present overly dependent on space systems for the effective conduct of military operations, satellites become more important to the Soviets as the level of conflict increases"; see NIE, "Soviet Space Programs," December 1, 1985, CREST, CIA-RDP87T00495R000200130003-2. Donald Kerr maintained that "because of the heavy dependence of the superpowers on tactical satellites, the employment of ASAT against the tactical military space assets of the other side would be very tempting during a major conventional war. This fact has led to increasing attention to both satellite defense and to ASAT as either a deterrent to attacks on satellites or as an offensive tool"; see Donald Kerr, "Implications of Anti-Satellite Weapons for ABM Issues," George Keyworth Files, RAC Box 14, Ronald Reagan Presidential Library (RRPL hereafter).

40. Notably, every significant US space policy document during the Cold War officially adhered to the policy of using space for peaceful purposes. Despite the more outwardly militaristic character of Reagan's space policy, his administration still maintained that it was in line with the stated intent to use space for peaceful goals. The term "peaceful purposes" allowed significant room for interpretation as to what indeed constituted a peaceful space activity. Jimmy Carter's 1978 space policy specifically stated that "'peaceful purposes' allow[ed] for military and intelligence-related activities in pursuit of national security and other goals." See McDougall, *The Heavens and the Earth*, 185; PD/NSC-37, "National Space Policy," May 11, 1978.

41. Richard Halloran, "US Military Operations in Space to Be Expanded under Air Force," *New York Times*, June 22, 1982, available at https://www.nytimes.com/1982 /06/22/us/us-military-operations-in-space-to-be-expanded-under-air-force.html.

42. Testimony by General Lew Allen, Senate Foreign Relations Committee, July 11, 1979.

43. National Intelligence Estimate, "Soviet Military Capabilities and Intentions in Space," 1980, available at https://nsarchive2.gwu.edu/NSAEBB/NSAEBB501/docs/EBB -33.pdf, 51–52.

44. National Intelligence Estimate, "Soviet Military Capabilities and Intentions in Space." Early warning satellites, for example, were located in geosynchronous orbit, which is 22,236 miles away from the Earth's equator.

45. National Intelligence Estimate, "Soviet Military Capabilities and Intentions in Space."

46. DEFE 69/1204, "Defence Space Policy," August 23, 1982, TNA.

47. DEFE 69/1204, "Defence Space Policy."

48. Stares, *The Militarization of Space*, 221.

49. US Government Accounting Office, "US Antisatellite Program Needs a Fresh Look," GAO C-MASAD-83-5, (Washington, DC, 1983), available at https://www.gao.gov/products/C-MASAD-83-5.

50. US Government Accounting Office, "US Antisatellite Program Needs a Fresh Look."

51. George C. Wilson, "Soviets Reported Ready to Orbit Laser Weapons," *Washington Post*, March 1982, CREST, CIA-RDP84B00049R001503840047-5.

52. US Library Congressional Research Service, "Antisatellites (Killer Satellites)," Marcia S. Smith, IB81123 (1983), 5, available at https://archive.org/details/DTIC_ADA477965/page/n7/mode/2up?q=GAO+%22C-MASAD-82-10%22.

53. National Intelligence Estimate, "Soviet Space Program," July 19, 1983, CREST, CIA-RDP00B00369R000100050006-2.

54. Robert Newberry, "Space Doctrine for the 21st Century," Air Command and Staff College, 1997, 56, available at https://apps.dtic.mil/sti/pdfs/ADA398606.pdf.

55. Newberry, "Space Doctrine for the 21st Century," 57.

56. Hays, "Struggling Towards Space Doctrine," 310.

57. National Security Study Directive Number 13-82, "National Space Strategy," December 15, 1982, CREST, CIA-RDP85M00364R000400550064-1.

58. Gil Rye interview with the author, February 5, 2020.

59. National Security Study Directive Number 13-82, "National Space Strategy," December 15, 1982, CREST, CIA-RDP85M00364R000400550064-1.

60. National Security Study Directive Number 13-82, "National Space Strategy."

61. Stares, *The Militarization of Space*, 223. The archival record shows that Ford believed that restraint in space was hampering US national security interests. Even Carter, who wanted an ASAT ban, affirmed the right of "self-defense in space" and precluded non-kinetic (e.g., jamming) capabilities from inclusion in ASAT talks. See chapter 1 and PD/NSC-37, "National Space Policy," May 11, 1978, available at https://fas.org/spp/military/docops/national/nsc-37.htm.

62. See Lettow, *Ronald Reagan and His Quest*; Brands, *What Good Is Grand Strategy?*, 109; Andreoni, "Ronald Reagan's Strategic Defense Initiative," 39.

63. Lettow, *Ronald Reagan and His Quest*, 39; FitzGerald, *Way Out There in the Blue*, 121.

64. Andreoni, "Ronald Reagan's Strategic Defense Initiative," 39.

65. "The Real War in Space: We Are Engineering the Post-Nuclear Era," *BBC Panorama*, 1978, available at https://www.bbc.co.uk/programmes/p02864t8.

66. Donald Baucom, *Origins of the Strategic Defense Initiative Organization* (Washington, DC: Pentagon, 1989), 251.

67. Daniel Graham, *Confessions of a Cold Warrior* (Fairfax, VA: Preview Press, 1995).

68. Baucom, *Origins of The Strategic Defense Initiative Organization*, 255.

69. Baucom, *Origins of The Strategic Defense Initiative Organization*, 255.

70. Martin Anderson, *Revolution* (San Diego, CA: Harcourt, 1988), 85–88.

71. Sanford Lakoff and Herbert F. York, *A Shield in Space? Technology, Politics, and the Strategic Defense Initiative* (Berkeley: University of California Press, 1989), 9.

72. Guy Cook interview with the author, August 27, 2020.

73. Lakoff and York, *A Shield in Space?*, 9.

74. Ellis, "Reds in Space," 141.

75. Baucom, *Origins of the Strategic Defense Initiative Organization*, 253.

76. Daniel Graham, "Toward a New US Strategy: Bold Strokes Rather Than Increments," *Strategic Review*, Spring 1981, 9–16.

77. Steven Rearden, "Special Report for Ballistic Missile Defense Organization, Congress and SDIO, 1983–1989," Rearden (1997), 37, available at https://apps.dtic.mil/dtic/tr/fulltext/u2/a338619.pdf.

78. Steven Rearden, "Special Report for Ballistic Missile Defense Organization, Congress and SDIO, 1983–1989," 38.

79. Rearden, "Special Report for Ballistic Missile Defense Organization, Congress and SDIO, 1983–1989," 37.

80. Rearden, "Special Report for Ballistic Missile Defense Organization, Congress and SDIO, 1983–1989," 229.

81. Rearden, "Special Report for Ballistic Missile Defense Organization, Congress and SDIO, 1983–1989," 229.

82. Rearden, "Special Report for Ballistic Missile Defense Organization, Congress and SDIO, 1983–1989," 229.

83. Michael Michaud, *Reaching For the High Frontier: The American Pro-Space Movement, 1972–84* (Westport, CT: Praeger, 1986), 228. While kinetic interceptors could be built from existing capabilities, Graham generally overlooked the software and information technology requirements for putting together a functioning missile defense *system*. Most of the missile defense advocates at this stage did not have a background

in technical program management for the development of large systems. What did unite them all was a belief in the limitless potential of technology and the need to push into space to achieve a decisive strategic advantage over the Soviet Union.

84. Gregg Herken, "The Earthly Origins of Star Wars," *Bulletin of the Atomic Scientists*, October 1987, 22. The author could not locate any transcripts or memoranda concerning this meeting at the Reagan Presidential Library.

85. Guy Cook interview with the author, August 27, 2020.

86. Guy Cook interview with the author, August 27, 2020.

87. Guy Cook interview with the author, August 27, 2020.

88. "Space Based Lasers—Can They Be Used as a National Defense?" National Defense Policy Task Force, October 2, 1981, George Keyworth Files, Box 11, RRPL.

89. "Space Based Lasers—Can They Be Used as a National Defense?"

90. Rearden, "Special Report for Ballistic Missile Defense Organization, Congress and SDIO, 1983–1989," 44–45.

91. Angelo Codevilla, *While Others Build: The Commonsense Approach to the Strategic Defense Initiative* (New York: Free Press, 1988), 87.

92. US Government Accountability Office, "DoD's Space-Based Laser Program: Potential Progress and Problems," GAO C-MASAD-82-10, Washington, DC, 1982.

93. DeLauer testimony, September 20, 1982, US Congress, Senate, Committee on Foreign Relations, Subcommittee on Arms Control, Oceans, International Operations and Environment, Hearings: Arms Control and the Militarization of Space, Government Printing Office, Washington, DC, 1982.

94. The lack of high-level support did not deter Wallop, Graham, and their fellow missile defense boosters. Networks of space advocates inside and outside of government institutions have long played an important role in promoting space projects in different national contexts. Siddiqi has shown how an informal network of Soviet space enthusiasts kept the dream of spaceflight alive and ultimately catalyzed the USSR's space program. Similarly, missile defense and space enthusiasts validated Reagan's deep-rooted belief in the possibilities of technological progress. The enthusiasm for using advanced technologies to shift the strategic balance in the US favor was accompanied by a fear of Soviet technological surprise. The latter would become an especially important rhetorical tool for promoting SDI at home and abroad. See Asif Siddiqi, *The Red Rockets' Glare: Spaceflight and the Russian Imagination, 1857-1957* (Cambridge: Cambridge University Press, 2014), 10.

95. FitzGerald, *Way Out There in the Blue*, 18.

96. John Prados, "The Strategic Defense Initiative: Between Strategy, Diplomacy and United States Intelligence Estimates," in *Crisis of Detente in Europe: from Helsinki to Gorbachev, 1975–1985*, ed. Leopoldo Nuti (London: Routledge, 2009), 90–95.

97. FitzGerald, *Way Out There in the Blue*, 157. In a May 1981 memorandum to the president, Casey wrote that "[CIA] analysis has been academic, soft, not sufficiently relevant and realistic" and that he "frequently found that [he received] better intelligence judgments from the streetwise, on the ground Operations staff than [he got] from the more academic Analytical staff in Washington." Many CIA analysts were academics at heart. At one time, there were enough PhD historians and political scientists at the CIA to staff a small liberal arts college. The operations officers—the people who actually recruited spies in the back alleys of Moscow and Vienna—were not, however, trained to analyze large-scale geopolitical trends. They were (and remain today) in the people business; their task during the Cold War was to find people who were willing to betray their country and to provide secrets to the US. While operations officers most certainly developed expertise in the countries in which they lived and worked under cover, they did not have the broader perspective of the analysts. Additionally, the CIA analytical staff used intelligence from other sources (e.g., imagery satellites and electronic intercepts) to put human intelligence into context and to determine its credibility. See Memorandum for Reagan from William Casey, "Progress at the CIA," May 6, 1981. Reagan initialed the document, available at http://web.archive .org/web/20130413105819/foia.cia.gov/sites/default/files/document_conversions/17 /19810506.pdf.

98. A 1980 NIE stated that the Soviets were "working on a new sensor for their current nonnuclear orbital interceptor. We expect the Soviets to continue design and engineering of a space-based laser system that would have significant advantages over their orbital interceptor for the antisatellite function. They conceivably could have a prototype spaceborne laser weapon for antisatellite testing by the mid-to-late1980s." See NIE 11-1-80, "Soviet Military Capabilities and Intentions in Space," August 6, 1980, available at https://www.cia.gov/readingroom/docs/DOC_0000284010.pdf.

99. Memorandum for the NIO/SP, "Regarding DeLauer's Comments on Soviet Space Lasers," March 3, 1982, CREST, CIA-RDP86R00893R000100090043-5.

100. Intelligence Estimate, "Outlook for Rapid Expansion of Soviet Space Programs through 1986," October 1982, available at http://nsarchive.gwu.edu/NSAEBB/NSAE BB501/docs/EBB-35.pdf.

101. Ellis, "Reds in Space," 167.

102. Memorandum for Casey, "The Soviet ABM Program," June 10, 1981, available at http://web.archive.org/web/20130413105925/foia.cia.gov/sites/default/files/document _conversions/17/19810717.pdf.

103. Email from Alan Thomson (retired CIA analyst) to author, February 21, 2021.

104. National Intelligence Estimate, "Soviet Ballistic Missile Defense," October 13, 1982, CREST, CIA-RDP00B00369R000100040001-8. The estimative language used in the report was not definitive. Terms such as "unlikely" or "could be able" created significant ambiguity. Richards J. Heuer is highly critical of such terms and calls them

"empty shells" because the policymaker "fills them with meaning though the context in which they are used and what is already in the [reader's] mind." See Richards J. Heuer Jr., *Psychology of Intelligence Analysis* (Washington, DC: Center for the Study of Intelligence, Central Intelligence Agency, 1999), 87.

105. National Intelligence Estimate, "Soviet Ballistic Missile Defense," October 13, 1982, CREST, CIA-RDP00B00369R000100040001-8.

106. National Intelligence Estimate, "Soviet Ballistic Missile Defense."

107. Melanie Brand, "Intelligence, Warning, and Policy: The Johnson Administration and the 1968 Soviet Invasion of Czechoslovakia," *Cold War History*, May 6, 2020, 12; Cynthia M. Grabo, *Anticipating Surprise: Analysis for Strategic Warning* (Washington, DC: Joint Military Intelligence College's Centre for Strategic Intelligence Research, 2002), 4.

108. Interagency Intelligence Memorandum, "Soviet Ballistic Missile Defense," July 17, 1981, CREST, CIA-RDP83b00140r000100060013-5.

109. Aleksandar Matovski, "Strategic Intelligence and International Crisis Behavior," *Security Studies* 29, no. 5 (2020): 964–990.

110. Brands, *What Good Is Grand Strategy?*, 107.

111. Joyce Wadler and Merrill Brown, "New York Rally Draws Half-Million," *Washington Post*, June 13, 1982, available at https://www.washingtonpost.com/archive /politics/1982/06/13/new-york-rally-draws-half-million/2a1258be-a17f-4e41-bf12 -9368766b32cf/.

112. Ronald Reagan, "Address to the Nation on Strategic Arms Reduction and Nuclear Deterrence," RRPL, November 22, 1982, available at https://www.reaganlibrary.gov /archives/speech/address-nation-strategic-arms-reduction-and-nuclear-deterrence.

113. Lettow, *Ronald Reagan and His Quest*, 85; and Reagan, "Address to the Nation on Strategic Arms Reduction and Nuclear Deterrence."

114. FitzGerald, *Way Out There in the Blue*, 191.

115. FitzGerald, *Way Out There in the Blue*, 191.

116. Lettow, *Ronald Reagan and His Quest*, 86

117. Mira Duric, *The Strategic Defense Initiative: US Policy and the Soviet Union* (Abingdon, UK: Routledge, 2003), 7.

118. Lettow, *Ronald Reagan and His Quest*, 82.

119. Lettow, *Ronald Reagan and His Quest*, 82.

120. Lettow, *Ronald Reagan and His Quest*, 82, 93.

121. Edmund Morris, *Dutch: A Memoir of Ronald Reagan* (New York: Modern Library, 1999), 799.

122. Lettow, *Ronald Reagan and His Quest*, 100.

123. Lettow, *Ronald Reagan and His Quest*, 100.

124. Lettow, *Ronald Reagan and His Quest*, 102

125. Edward Reiss, *The Strategic Defense Initiative* (Cambridge: Cambridge University Press, 1992), 106.

126. Fischer notes that except for Weinberger, most of Reagan's advisors opposed SDI; see Beth Fischer, "Nuclear Abolitionism," 49.

127. George Shultz, *Turmoil and Triumph: Diplomacy, Power, and the Victory of the American Deal* (New York: Scribner, 1993), 249–250.

128. Two days before Reagan's address to the nation, Shultz received a copy of it and was shocked. He was able to convince Reagan to insert language that research would be consistent with treaty obligations because Shultz was especially worried about the implications of the president's proposal for the ABM Treaty. See *FRUS*, 1981–1988, Volume IV, Soviet Union, January 1983–March 1985, 84.

129. Shultz, *Turmoil and Triumph*, 204.

130. Ronald Reagan, "Address to the Nation on Defense and National Security," RRPL, March 23, 1983, available at https://www.reaganlibrary.gov/archives/speech/address -nation-defense-and-national-security.

131. Lettow, *Ronald Reagan and His Quest*, 100.

132. Telegram from the US Embassy in Moscow to State Department HQ, "Soviet Reaction to President's Speech," March 25, 1983; see *FRUS*, 1981–1988, Volume IV, Soviet Union, January 1983–March 1985; Rhodes, *Arsenals of Folly*, 158.

133. NSDD-85, "Eliminating the Threat from Ballistic Missiles," March 25, 1983, available at https://aerospace.csis.org/wp-content/uploads/2019/02/NSDD-85-Eliminating -the-Threat-From-Ballisitc-Missiles.pdf.

134. Rearden, "Special Report for Ballistic Missile Defense Organization, Congress and SDIO, 1983–1989," 52–53.

135. US Government Accounting Office, *Strategic Defense Initiative Program: Better Management Direction and Controls Needed*, GAO NSIAD-88-26, Washington, DC, 1987, available at https://www.gao.gov/assets/150/145900.pdf.

136. "Hearing on H.R. 3073 People Protection Act," Research and Development Subcommittee and Investigations Subcommittee of the Committee on Armed Services House of Representatives, First Session November 10, 1983, Government Printing Office, Washington, DC, 1984, 19–20.

137. "Hearing on H.R. 3073 People Protection Act," 19.

138. Donald Hafner, "Assessing the President's Vision: The Fletcher, Miller, and Hoffman Panels," *Daedalus* 114, no. 2 (1985): 94. The content of the Fletcher study remains classified.

139. "Hearing on H.R. 3073 People Protection Act," 22.

140. Hafner, "Assessing the President's Vision."

141. Krepinevich and Watts, *The Last Warrior*, 150.

142. "Ballistic Missile Defenses and US National Security," Summary Report, Fred Hoffman Study Director, CREST, CIA-RDP90R00961R000100090004-7. In addition to the Fletcher and Hoffman panels, the White House commissioned an interagency study under Franklin Miller, director of Strategic Forces Policy in the Pentagon; it has not been declassified. For details concerning the Miller study, see Hafner, "Assessing the President's Vision," 96.

143. Douglas Waller, James Bruce, and Douglas Cook, "SDI: Progress and Challenges," Report submitted to Senator William Proxmire, Senator J. Bennett Johnston, and Senator Lawton Chiles, March 17, 1986, 3, available at https://apps.dtic.mil/dtic/tr/fulltext/u2/a351714.pdf.

144. NSDD-119, "Strategic Defense Initiative," January 6, 1984, available at https://aerospace.csis.org/wp-content/uploads/2019/02/NSDD-119-Strategic-Defense-Initiative.pdf.

145. Rearden, "Special Report for Ballistic Missile Defense Organization, Congress and SDIO, 1983–1989," 70.

146. "Talking Points for Meeting with Cap Weinberger, meeting scheduled for February 10, 1984," George Keyworth Files, Box 11, RRPL.

147. Bateman, "Keeping the Technological Edge," 355–378.

148. "Strategic and Theater Nuclear Forces Part 7," Hearings Before the Committee on Armed Services United States Senate, Ninety Ninth Congress, 78, 1985.

149. Lou Cannon, "President Seeks Future Defense against Missiles," *Washington Post*, March 24, 1983.

150. Ronald Reagan, "Address before a Joint Session of the Congress on the State of the Union," January 25, 1984, available at https://www.presidency.ucsb.edu/documents/address-before-joint-session-the-congress-the-state-the-union-4.

151. Butrica, *Single Stage to Orbit*, 23.

152. Logsdon, *Ronald Reagan and the Space Frontier*, 121–122.

153. Telegram from the US Embassy in Moscow to State Department HQ, "Andropov on President's Defense Speech," March 28, 1983, *FRUS*, 1981–1988, Volume IV, Soviet Union, January 1983–March 1985.

154. Westwick, "'Space-Strike Weapons,'" 955.

155. Westwick, "'Space-Strike Weapons,'" 955.

156. Christopher Andrew and Vasili Mitrokhin, *The Sword and the Shield: The Mitrokhin Archive and the Secret History of the KGB* (New York: Basic Books, 1999), 214.

157. "Spravka," undated, Strategic Arms Reduction Talks Miscellaneous Documents, Box 5, Kataev Archive, Hoover Institute.

158. Westwick, "'Space-Strike Weapons,'" 955.

159. Telegram from the US Embassy in Moscow to State Department HQ, "Pell Discussion with Andropov," August 18, 1983, *FRUS*, 1981–1988, Volume IV, Soviet Union, January 1983–March 1985.

160. US Embassy in Moscow to State Department HQ, "Pell Discussion with Andropov."

161. Intelligence Memorandum, "Soviet Interest in Arms Control Negotiations in 1984," March 23, 1984, *FRUS*, 1981–1988, Volume IV, Soviet Union, January 1983–March 1985.

162. Siddiqi, "The Soviet Co-Orbital Anti-Satellite System," 234.

163. "Letter to President from 106 Congressmen Regarding ASAT," August 8, 1983, CREST, CIA-RDP87M00539R001001390023-0.

164. "Letter to President from 106 Congressmen Regarding ASAT."

165. Report to the Congress, "US Policy on ASAT Arms Control," March 31, 1984, available at http://insidethecoldwar.org/sites/default/files/documents/report%20to%20the%20Congress%20US%20policy%20on%20ASAT%20Arms%20Control%20March%2031%2C%201984.pdf.

166. Memorandum for Reagan, "Secretary Weinberger's Views on an ASAT Initiative," June 18, 1984, *FRUS*, 1981–1988, Volume IV, Soviet Union, January 1983–March 1985.

167. "Soviet Interest in Arms Control Negotiations in 1984," March 23, 1984, *FRUS*, 1981–1988, Volume IV, Soviet Union, January 1983–March 1985.

168. Donald Kerr, "Implications of Anti-Satellite Weapons for ABM Issues," George Keyworth Files, RAC Box 14, RRPL.

169. Memorandum from Iklé to Weinberger, "Arms Control Diplomacy," July 3, 1984, *FRUS*, Volume IV, Soviet Union, January 1983–March 1985.

170. Ronald Reagan, "Remarks and a Question-and-Answer Session with Area High School Seniors in Jacksonville, Florida," December 1, 1987, available at https://www.reaganlibrary.gov/archives/speech/remarks-and-question-and-answer-session-area-high-school-seniors-jacksonville.

171. Minutes of a National Security Planning Group Meeting, "US–Soviet Arms Control Objectives," December 5, 1984, *FRUS*, 1981–1988, Volume IV, Soviet Union, January 1983–March 1985.

172. Memorandum, "Comments on Draft National Space Strategy," February 1, 1984, available at https://fas.org/irp/offdocs/nsdd/23-2278a.gif.

173. Memorandum, "Comments on Draft National Space Strategy."

174. Donald Kerr, "Implications of Anti-Satellite Weapons for ABM Issues," George Keyworth Files, RAC Box 14, RRPL.

175. GAO, "Ballistic Missile Defense: Records Indicate Deception Program Did Not Affect 1984 Test Results," June 1984, GAO/NSIAD-94-219 Homing Overlay Experiment Deception.

176. GAO, "Ballistic Missile Defense: Records Indicate Deception Program Did Not Affect 1984 Test Results."

177. An interagency report on space arms control observed that "modest ABM interceptors . . . such as demonstrated by HOE . . . could be quite effective against undefended satellites at low altitude." For details, see "Alternative Near-Term ASAT Arms Control Approaches," August 1, 1984, Executive Secretariat NSC Box 3, RRPL.

178. Memorandum for Director of Central Intelligence and Deputy Director of Central Intelligence, "DS&T Weekly Activities for 17 August 1984," August 17, 1985, CREST, CIA-RDP86M00886R002700020002-2.

179. "Arms Control Chronology: Last November's Walkout to This Month's 'Let's Talk,'" *Christian Science Monitor*, November 27, 1984.

180. "The ASAT Question," *Washington Post*, June 12, 1984, available at https://www .washingtonpost.com/archive/politics/1984/06/12/the-asat-question/16194bb8-0832 -4888-8c16-ff9e38961118/.

181. Memorandum from Lehman to McFarlane, "ASAT," June 13, 1984, Executive Secretariat NSC Box 3, RRPL.

182. Memorandum from Cooper for National Security Advisor, "Possible ASAT Arms Control Initiatives," June 16, 1984, Executive Secretariat NSC Box 3, RRPL.

183. Memorandum from Lehman to McFarlane, "ASAT," June 13, 1984, Executive Secretariat NSC Box 3, RRPL.

184. Caspar Weinberger made this point very clearly during a September 1984 NSPG meeting. For details, see Minutes of an NSPG Meeting, "Next Steps in the Vienna Process," September 18, 1984, *FRUS* 1981–1988, Volume IV, Soviet Union, January 1983–March 1985.

185. Simon Miles, *Engaging the Evil Empire: Washington, Moscow, and the Beginning of the End of the Cold War* (Ithaca, NY: Cornell University Press, 2020), 101.

186. Miles, *Engaging the Evil Empire*, 101.

187. Miles, *Engaging the Evil Empire*, 101.

188. Miles, *Engaging the Evil Empire*, 101.

189. See chapter 6 for the details concerning the nuclear and space talks.

190. Memorandum for McFarlane, "Unclassified Version of the National Space Strategy," August 16, 1984, available at https://aerospace.csis.org/wp-content/uploads /2019/02/NSDD-144-National-Space-Strategy.pdf.

191. This will be discussed at length in chapter 5.

192. Intelligence Report, "Allied Attitudes towards the Strategic Defense Initiative and US Development of Anti-Satellite Weapons," June 20, 1984, CREST, CIA-RDP 85T00287R001100280001-1.

193. Intelligence Report, "Allied Attitudes towards the Strategic Defense Initiative and US Development of Anti-Satellite Weapons."

194. Martin Schram, "Mondale Ads Take Aim at 'Star Wars' Plan," *Washington Post*, October 20, 1984, available at https://www.washingtonpost.com/archive/politics /1984/10/20/mondale-ads-take-aim-at-star-wars-plan/b619768d-d09b-4dd3-944e -960a4fdb182a/.

195. Martin Schram, "Mondale Ads Take Aim at 'Star Wars' Plan."

196. "Transcript of the Reagan–Mondale Debate on Foreign Policy," *New York Times*, Soviet Union, January 1983–March 1985, available at https://www.nytimes.com/1984 /10/22/us/transcript-of-the-reagan-mondale-debate-on-foreign-policy.html.

197. "Transcript of the Reagan–Mondale Debate on Foreign Policy."

198. Minutes of a National Security Planning Group Meeting, "Next Steps in the Vienna Process," September 18, 1984, *FRUS*, 1981–1988, Volume IV, Soviet Union, January 1983–March 1985.

199. Thomas W. Graham and Bernard M. Kramer, "The Polls: ABM and Star Wars: Attitudes Toward Nuclear Defense, 1945–1985," *The Public Opinion Quarterly* 50, no. 1 (1986): 125–134.

200. Ellis, "Reds in Space," 169.

201. Minutes of a National Security Planning Group Meeting, "Discussion of Substantive Issues for Geneva," December 17, 1984, *FRUS*, 1981–1988, Volume IV, Soviet Union, January 1983–March 1985.

202. Logsdon, *Ronald Reagan and the Space Frontier*, 97. The space station was ultimately approved, but it cost Rye his career. Neither the Air Force nor senior advisors to Reagan were overwhelmingly supportive of this proposal.

203. Logsdon, *Ronald Reagan and the Space Frontier*, 97.

204. Minutes of a National Security Planning Group Meeting, "Discussion of Substantive Issues for Geneva," December 17, 1984, *FRUS*, 1981–1988, Volume IV, Soviet Union, January 1983–March 1985.

205. PREM 19/1444, "Record of Discussion between Abrahamson and European Allies at the Pentagon," July 17, 1984, TNA.

206. Bateman, "Intelligence and Alliance Politics," 4–5.

207. Bateman, "Intelligence and Alliance Politics," 4–5.

208. National Security Study Directive Number 13-82, "National Space Strategy," December 15, 1982, CREST, CIA-RDP85M00364R000400550064-1.

CHAPTER 3

1. "Pope Calls for Accords on Peaceful Space Use," *New York Times*, June 21, 1986, available at https://www.nytimes.com/1986/06/21/world/pope-calls-for-accords-on -peaceful-space-use.html.

2. George Weigel, *Witness to Hope: The Biography of John Paul II* (New York: Harper Collins, 1990), 500; Marie Gayte, "The Vatican and the Reagan Administration: A Cold War Alliance?" *The Catholic Historical Review* 97, no. 4 (2011): 726.

3. Weigel, *Witness to Hope*, and Wilson, *The Triumph of Improvisation*, 7.

4. Gayte, "The Vatican and the Reagan Administration," 727.

5. For two examples, see Slayton, *Arguments that Count*, and Benjamin T. Wilson, "Insiders and Outsiders: Nuclear Arms Control Experts in Cold War America" (Diss., MIT, 2014).

6. John F. Kennedy, "Excerpt from the Special Message to the Congress on Urgent National Needs," May 25, 1961, available at https://www.nasa.gov/vision/space /features/jfk_speech_text.html.

7. John F. Kennedy, "Rice Stadium Moon Speech," September 12, 1962, available at https://er.jsc.nasa.gov/seh/ricetalk.htm.

8. Logsdon, *John F. Kennedy and the Race to the Moon*, 157.

9. Roger Launius, *Apollo's Legacy: Perspectives on the Moon Landings* (Washington, DC: Smithsonian Books, 2019), 21.

10. "NSA in Space" (Fort Meade: National Security Agency, 1975), 58, available at https://www.governmentattic.org/26docs/NSAinSpaceViaNRO_1975.pdf.

11. Ken Renshaw, "From the Front Office: The Retirement of BYEMAN," Security Newsletter, National Reconnaissance Office, August/September 2004, issue 4, available at https://nsarchive2.gwu.edu/NSAEBB/NSAEBB225/doc06.pdf.

12. Mary Battiata, "Aviation Week's Soaring Success," *Washington Post*, March 10, 1986, available at https://www.washingtonpost.com/archive/lifestyle/1986/03/10 /aviation-weeks-soaring-success/991801df-4bb5-4091-a0d9-04864be8d5b0/.

13. "Implications of the Soviet Satellite Reconnaissance Program," June 9, 1964, CREST, CIA-RDP79R01095A000800030008-3.

14. Memorandum from Peterson AFB to HQ USAF, "Knowledge of the DSP System," National Security Archive, available at https://nsarchive2.gwu.edu/NSAEBB/NSAEBB 235/14.pdf.

15. Bateman, "Trust but Verify."

16. "Report of 156 Committee Regarding SALT," November 12, 1971, CREST, CIA-RDP82M00531R000800050004-2.

17. See the section "Satellites Policing the Superpowers" in chapter 1.

18. Donald Morris, "Spies in the Sky Keep Two Big Powers in Check," *Houston Post*, October 8, 1972, CREST, CIA, CIA-RDP80-01601R000100120001-0.

19. "How the US Keeps Tabs on the Red World," *US News and World Report*, October 8, 1972, CREST, CIA, CIA-RDP80-01601R000100120001-0.

20. For details regarding the plaque on the moon, see "We Came in Peace for All Mankind," July 20, 2020, available at https://www.nasa.gov/image-feature/we-came-in-peace-for-all-mankind.

21. Richard D. Lyons, "US officials Fear Soviet's Lead in Hunter-Killer and Spy Satellites," *New York Times*, June 30, 1978, available at https://www.nytimes.com/1978/01/30/archives/us-officials-fear-soviets-lead-in-hunterkiller-and-spy-satellites.html.

22. Memorandum from Scowcroft to Ford, "US Anti-Satellite Capabilities," December 16, 1976, *FRUS*, 1969–1976, Volume XXXV, National Security Policy, 1973–1976.

23. The History of US Antisatellite Weapon Systems, undated, available at https://fas.org/man/eprint/leitenberg/asat.pdf, 21.

24. Jimmy Carter, "Remarks at the Congressional Space Medal of Honor Awards Ceremony," October 1, 1978, National Security Archive, available at https://nsarchive2.gwu.edu/NSAEBB/NSAEBB231/doc32.pdf.

25. "Prevention of an Arms Race in Outer Space," Federation of American Scientists, available at https://programs.fas.org/ssp/nukes/ArmsControl_NEW/nonproliferation/NFZ/NP-NFZ-PAROS.html.

26. Paul Meyer, "The CD and PAROS: A Short History," April 2011, available at https://www.unidir.org/files/publications/pdfs/the-conference-on-disarmament-and-the-prevention-of-an-arms-race-in-outer-space-370.pdf.

27. John Noble Wilford, "Military's Future in Space: A Matter of War or Peace," *New York Times*, October 19, 1982, CREST, CIA-RDP92B00478R000800260001-5.

28. Wilford, "Military's Future in Space: A Matter of War or Peace."

29. Gray, *American Military Space Policy*, 24–25.

30. James Lardner, "The Call of the Hawk's Hawk," *Washington Post*, May 14, 1982, available at https://www.washingtonpost.com/archive/lifestyle/1982/05/14/the-call-of-the-hawks-hawk/d64f98df-5b56-4b58-a56d-990ce8b22919/; General Advisory Committee on Arms Control and Disarmament Members, undated, CREST, CIA-RDP86M00886R000800130024-7.

31. Richard Garwin, "Reagan's Riskiness," *New York Times*, March 30, 1983, available at https://www.nytimes.com/1983/03/30/opinion/reagan-s-riskiness.html.

32. Gray, *American Military Space Policy*, 17.

33. Gray, *American Military Space Policy*, 18.

34. Talking points for a meeting between Keyworth and Weinberger detail how the Air Force "has taken [the] stand [sic] that we should go ahead and push for [an] ASAT Treaty . . . AF claims that [MHV ASAT] 'cost effectiveness' too low to warrant not negotiating [an] ASAT Treaty." The talking points furthermore stated that "the very organization deemed most suitable to take on this space mission . . . wants no part of it." See "Talking Points for Meeting with Cap Weinberger," meeting scheduled for February 10, 1984, George Keyworth Files, RAC Box 14, Box 11, RRPL.

35. Paul Stares, "Déjà Vu: The ASAT Debate in Historical Context," *Arms Control Today* 13, no. 11 (1983).

36. Gray, *American Military Space Policy*, 52–53.

37. John Pike, "Anti-Satellite Weapons and Arms Control," *Arms Control Today* 13, no. 11 (1983): 6.

38. Pike, "Anti-Satellite Weapons and Arms Control."

39. Gray, *American Military Space Policy*, 75–76.

40. Donald L. Hafner, "Outer Space Arms Control: Unverified Practices, Unnatural Acts?" *Survival* 25, no. 6 (1983): 242.

41. Gray, *American Military Space Policy*, 12.

42. Anette Stimmer, "Star Wars or Strategic Defense: What's in a Name?" *Journal of Global Security Studies* 4, no. 4 (2019).

43. Linenthal, *Symbolic Defense*, 15.

44. Linenthal, *Symbolic Defense*.

45. Alexander Geppert, "European Astrofuturism, Cosmic Provincialism: Historicizing the Space Age," in *Imagining Outer Space: European Astroculture in the Twentieth Century*, ed. Alexander Geppert (New York: Palgrave, 2012), 8.

46. John Moakley, "Space Weapons and Congress," *Arms Control Today* 13, no. 11 (1983): 9.

47. Rip Bulkeley and Graham Spinardi, *Space Weapons: Deterrence or Delusion?* (Oxford: TJ Press, 1986), 3.

48. Gray, *American Military Space Policy*, 49.

49. Brad Knickerbocker, "Military's Role in Space," *Christian Science Monitor*, January 15, 1985, CREST, CIA-RDP90-00965R000403530003-7.

50. Joe Moakley, "An Arms Lesson From History," *New York Times*, August 29, 1985, available at https://www.nytimes.com/1985/08/29/opinion/an-arms-lesson-from-history.html.

51. Arms Control in Outer Space, "Hearings before the Subcommittee on International Security and Scientific Affairs of the Committee on Foreign Affairs, 98th Congress," 1983.

52. "Hearings before the Subcommittee on International Security and Scientific Affairs of the Committee on Foreign Affairs, 98th Congress."

53. Memorandum for the record, "Briefing on the Soviet Anti-Satellite System," September 11, 1985, CREST, CIA-RDP87M01152R000400530006-9. In 1989, John Kerry introduced legislation called the Satellite Security Act that included long-term restrictions on testing ASATs in space. Kerry said that ASATs threatened "the single most important means we have of monitoring the Soviet Union." For details, see the CIA memorandum that includes attachment with Congressional testimony, July 20, 1989, CREST, CIA-RDP92M00732R001000080023-4.

54. Zbigniew Brzezinski, Robert Jastrow, and Max Kampelman, "Defense in Space is not 'Star Wars," *New York Times*, January 27, 1985, available at https://www.nytimes.com/1985/01/27/magazine/defense-in-space-is-not-star-wars.html.

55. Hugh Thomas, "Mr. Gorbachev's Own Star Wars," Centre for Policy Studies, March 1986, available at https://www.cps.org.uk/files/reports/original/111028090946-MrGorbachevsownStarWars1986.pdf.

56. Carlos Chagas and Vittorio Canuto, ed. "The Impact of Space Exploration on Mankind," Vatican City, 1986, available at https://www.pas.va/content/dam/casinapioiv/pas/pdf-volumi/scripta-varia/sv58pas.pdf, conference hosted by the Pontifical Academy of Sciences, Rome, Italy, October 1–5, 1984.

57. Chagas and Canuto, "The Impact of Space Exploration on Mankind."

58. Chagas and Canuto, "The Impact of Space Exploration on Mankind."

59. "Transcript of the Reagan–Mondale Debate on Foreign Policy," October 22, 1984, available at https://www.nytimes.com/1984/10/22/us/the-candidates-debate-transcript-of-the-reagan-mondale-debate-on-foreign-policy.html.

60. Ellis, "Reds in Space," 170.

61. Ronald Reagan, NSDD-172, "Presenting the Strategic Defense Initiative," May 30, 1985, available at https://irp.fas.org/offdocs/nsdd/nsdd-172.pdf.

62. Minutes of a National Security Planning Group Meeting, "Discussion of Substantive Issues for Geneva," December 17, 1984, *FRUS*, 1981–1988, Volume IV, Soviet Union, January 1983–March 1985.

63. Abrahamson, "SDI and the New Space Renaissance."

64. Abrahamson, "SDI and the New Space Renaissance," 120.

65. Peter J. Westwick, "From the Club of Rome to Star Wars: The Era of Limits, Space Colonization and the Origins of SDI," in *Limiting Outer Space: Astroculture After Apollo*, ed. Alexander Geppert (New York: Palgrave Macmillan, 2018), 283.

66. Neil M. Maher, *Apollo in the Age of Aquarius* (Cambridge: Harvard University Press, 2017), 65.

67. Columba Peoples, *Justifying Ballistic Missile Defence: Technology, Security, Culture* (Cambridge: Cambridge University Press, 2010), 142.

68. Ashton Carter, "Directed Energy Missile Defense in Space," background paper, April 4, 1984, available at http://www.princeton.edu/~ota/disk3/1984/8410/8410.PDF.

69. Carter, "Directed Energy Missile Defense in Space," 46.

70. Hans Bethe, Richard Garwin, Kurt Gottfried, Henry Kendall, "Space-Based Missile Defense," A Report by the Union of Concerned Scientists (Cambridge, MA: UCS, 1984).

71. Bethe et al., "Space-Based Missile Defense," 48.

72. Bethe et al., "Space-Based Missile Defense," 85.

73. Memorandum for the Record, "The SDIO: A Security Threat to NRO Technology," August 15, 1985, CREST, CIA-RDP97M00248R000500220031-8.

74. Tom Diaz, "SDI Top Soviet Propaganda Target, Ex-KGB Disinformation Expert Says," *Washington Times*, December 17, 1985, CREST, CIA-RDP90-00965R000201560009.

75. Erik K. Pratt, *Selling Strategic Defense: Interests, Ideologies, and the Arms Race* (Princeton: Princeton University Press, 1990), 97.

76. Ellis, "Reds in Space," 146.

77. PREM 19/1188, joint MoD–FCO report "Ballistic Missile Defence: UK Policy Towards the US Strategic Defence Initiative," October 1984, A-3, TNA.

78. "Ballistic Missile Defence: UK Policy Towards the US Strategic Defence Initiative."

79. "Ballistic Missile Defence: UK Policy Towards the US Strategic Defence Initiative," A-5.

80. Barry Goldwater, "Ten Ways to Catch Up in the Space Race," *Los Angeles Times*, January 15, 1963.

81. Ellis, "Reds in Space," 149.

82. Ellis, "Reds in Space," 154.

83. Gennadii Gerasimov, *Keep Space Weapons-Free* (Moscow: Novosti Press, 1983), 23, Woodrow Wilson Center Digital Archive, available at https://digitalarchive.wilsoncenter.org/document/110904.

84. Thomas C. Brandt, "The Military Uses of Space" in *American Plans for Space* (Washington, DC: National Defense University Press, 1986), 81–91.

85. Gerasimov, *Keep Space Weapons Free*, 49.

86. Members of the Reagan administration were sincerely worried about Soviet disinformation efforts targeting SDI, but they could not be certain about the effectiveness of Moscow's anti-SDI campaigns. A 1986 DIA assessment concluded that "Soviet active

measures against SDI have not achieved any significant successes. They have not kept US allies such as West Germany, Great Britain, Italy, and probably Japan from joining SDI." See Sven Kraemer files, Box 19, "The Soviet Active Measures Campaign Against the Strategic Defense Initiative," DIA intelligence assessment, October 1986, RRPL.

87. Intelligence report, "Soviet Responses to the US Strategic Defense Initiative," undated, CIA-RDP86B00420R000901700012-2.

88. "Soviet Responses to the US Strategic Defense Initiative."

89. KGB PGU upravlenie "RT" January 15, 1986, No. 167/466, O dopolnitel'nix zadachax v informatsionnoi rabote po voenno-strategicheskoi problematike, file: delo operativnoi perepiski No. 2113, Po voenno-stragicheskoi problematike, Lithuanian Special Archives.

90. Intelligence Memorandum, "Star Wars Study by Soviet Scientists Reported in the 7 January 1985 Issue of the Washington Post," January 8, 1985, CREST, CIA-RDP88R 01225R000100090001-9.

91. R. Jeffrey Smith, "Physicists Fault SDI Timetable," Washington Post, April 23, 1987, available at https://www.washingtonpost.com/archive/politics/1987/04/23/physicists -fault-sdi-timetable/42067485-7fae-40b9-937e-0339f0362514/; Hans Bethe et al., "Space-Based Missile Defense."

92. Westwick, "Space-Strike Weapons," 955–959.

93. "Letter from Charles Wick to Reagan," January 4, 1982, CREST, CIA-RDP83M 00914R002100120055-0.

94. Peter Grier, "'Soviet Military Power': Basis for Pentagon Charges," Christian Science Monitor, April 3, 1985, CIA-RDP90-00965R000302550016.

95. Memorandum for Stephen Bosworth, "The Strategic Weapons Spiral: Soviet Reactions to US Initiatives?" August 25, 1983, CREST, CIA-RDP00B00369R000200210006-3.

96. Memorandum for the Record, "DCI/DDCI Meeting with Assistant to the President for National Security Affairs," January 3, 1985, CREST, CIA-RDP87M00539R000800 990046-3.

97. "DCI/DDCI Meeting with Assistant to the President for National Security Affairs."

98. Memorandum from Larry Gershwin et al. to DCI/DDCI, "Another UNCLASSIFIED SDI Paper," June 10, 1986, CREST, CIA-RDP97M00248R000500220015-6.

99. Memorandum for the Record, "Meeting with Deputy Secretary of Defense Taft," November 14, 1984, CREST, CIA-RDP86M00886R000700050004-9.

100. "Letter from Teller to Casey," February 7, 1986, CREST, CIA-RDP88G01116 R000800900008-7.

101. Ted Agres, "Win Race in Space or Face Blackmail, Weinberger Warns," Washington Times, May 13, 1985, CIA-RDP90-00965R000100060015-2.

102. "Moscow TV on 'Soviet Military Power' Pamphlet," April 4, 1985, CREST, CIA-RDP09-00997R000100300001-5.

103. Membership would include Robert Linhard, Gil Rye, John Douglass, Ronald Sable, Thomas Donnelly, John Grimes, and Steven Steiner, in addition to working level representatives from ACDA, OSDI, State/PM, State/EUR, Abe's Office, and OSTP. See Robert Linhard Files, Executive Secretariat NSC Box 107, Memorandum for McFarlane from Linhard, "SDI Bible," December 14, 1984, RRPL.

104. Robert Linhard Files, Executive Secretariat NSC Box 107, Memorandum for McFarlane from Linhard, "SDI Bible," December 14, 1984, RRPL.

105. Robert Linhard Files, Executive Secretariat NSC Box 107, Working Papers, "Strategic Defense Initiative Bible (Policy Points)," undated, RRPL.

106. Ronald Reagan, NSDD 172, "Presenting the Strategic Defense Initiative," May 30, 1985, available at https://irp.fas.org/offdocs/nsdd/nsdd-172.pdf.

107. Robert Linhard Files, Executive Secretariat NSC Box 107, "What Is the President's Strategic Defense Initiative?," public diplomacy pamphlet part of "SDI Bible," RRPL.

108. Robert Linhard Files, Executive Secretariat NSC Box 107, Memorandum for McFarlane from Linhard and Steve Steiner, "Public Affairs Plan for SDI," December 10, 1984, RRPL.

109. Letter from Jim Courter to Robert Gates, March 27, 1987, CREST, CIA-RDP90 M00004R000300090020-4.

110. Robert Linhard Files, Executive Secretariat NSC Box 107, Memorandum for McFarlane from Linhard and Steve Steiner, "Public Affairs Plan for SDI," December 10, 1984, RRPL.

111. Michael Krepon, "Nitze's Strategic Concept," July 28, 2010, available at https://www.armscontrolwonk.com/archive/402822/nitzes-strategic-concept/.

112. PREM 19/1404, "Ministry of Defence Record of Conversation (Heseltine–Weinberger)," February 28, 1984, TNA.

113. Andrew and Mitrokhin, *The Sword and the Shield*, 491–496.

114. Memorandum for the Record, "Report on European SDI Briefing January 20 to 24, 1986," February 3, 1986, CREST, CIA-RDP87T01145R000100130001-2.

115. Gayte, "The Vatican and the Reagan Administration," 726.

116. Gayte, "The Vatican and the Reagan Administration," 727.

117. Weigel, *Witness to Hope*, 500.

118. Gayte, "The Vatican and the Reagan Administration," 726.

119. Gayte pushes back on the assertion that there was a "holy alliance" between the US and the Holy See during the presidency of Ronald Reagan; see Gayte, "The Vatican and the Reagan Administration," 720.

120. Gayte, "The Vatican and the Reagan Administration."

CHAPTER 4

1. Susan Colbourn and Matthias Haeussler, "Once More, with Feeling: Transatlantic Relations in the Reagan Years," in *The Reagan Moment: America and the World in the 1980s*, ed. Jonathan R. Hunt and Simon Miles (Ithaca, NY: Cornell University Press, 2021), 123–143.

2. Intelligence Report, "Allied Attitudes towards the Strategic Defense Initiative," February 8, 1985, CREST, CIA-RDP85T01058R000202390001-0.

3. Richard Bernstein, "Mitterrand Seeks the Future in Silicon Valley," *New York Times*, March 26, 1984, available at https://www.nytimes.com/1984/03/26/world/mitterrand -seeks-the-future-in-silicon-valley.html.

4. Bateman, "Keeping the Technological Edge"; Anthony Eames, "A 'Corruption of British Science?': The Strategic Defense Initiative and British Technology Policy," *Technology and Culture* 62, no. 3 (2021); Peter Westwick, "The International History of the Strategic Defense Initiative: American Influence and Economic Competition in the Late Cold War," *Centaurus* 52 (2010): 338–351.

5. Erich Reimann, "Mitterrand Criticizes West German Stance on Manned Space Flight," UPI, October 21, 1987, available at https://apnews.com/article/26c44688dce3 d01059704d14a6a92a5c.

6. For a history of ESA's early years, see John Krige, Arturo Russo, and Sebasta, Laurenza, *A History of the European Space Agency, Vol. 1* (Noordwijk: European Space Agency, 2000).

7. John M. Logsdon, *Together in Orbit: The Origins of International Participation in the Space Station* (Washington, DC: NASA History Division, 2005), 33.

8. Bateman, "Keeping the Technological Edge." A 1969 UK space policy review observed that the UK "has confined its interests in the technical possibilities of space technology to the use of space vehicles for the gathering and transmission of information . . . they are not an end in themselves either political, economic, or technological." See, FCO 55/348, "European Space Cooperation," February 19, 1969, TNA.

9. Steven Breyman, "SDI, The Federal Republic of Germany, and NATO: Political, Economic, and Strategic Implications," University of California, San Diego: Institute on Global Conflict and Cooperation, 1987, available at https://escholarship.org/uc/item /8w4236cs.

10. PREM 19/1443, Minute from Wright to Howe, "Strategic Defence Initiative," January 29, 1985, TNA.

11. Krige et al., *NASA in the World*, 83.

12. Krige et al., *NASA in the World*, 96.

13. Krige et al., *NASA in the World*, 108–109, 120–121.

14. Krige et al., *NASA in the World*, 118.

15. Krige et al., *A History the European Space Agency* (Vol. 1), 274.

16. Krige et al., *A History the European Space Agency* (Vol. 1), 120.

17. "OMB Study on International Competitiveness in Launch Services," February 12, 1985, CREST, CIA-RDP92B00181R001701630001-2.

18. Butrica, *Single Stage to Orbit*, 40.

19. For details about the US perception of Soviet commercial launch capabilities, see "Prospects for Soviet Commercial Exploitation of Space Systems and Related Services," December 1983, CREST, CIA-RDP84T00926R000100170006-0.

20. "Prime Minister Mauroy letter to Thatcher, Skynet Satellites and Ariane Launcher," November 16, 1983, Thatcher Archive, available at https://www.margaretthatcher.org /document/132066.

21. Bateman, "Keeping the Technological Edge."

22. For a history of SAMRO, see Sébastien Matte la Faveur, "The Interest and Opposition of the French Military in Satellite Reconnaissance for France: A Talk with a General Officer of the French Forces," *Space Chronicle* 59, suppl. 1 (2006): 1–9.

23. For a brief history of SPOT, see John Krige, Arturo Russo, and Laurenza Sebesta, *A History of the European Space Agency 1958–1987, Vol. II* (Noordwijk: European Space Agency, 2000), 56; and "History of the French Reconnaissance System," November 6, 2016, available at https://satelliteobservation.net/2016/11/06/history-of-the-french -reconnaissance-system/.

24. Niklas Reinke, *The History of German Space Policy: Ideas, Influence, and Interdependence 1932–2002* (Paris: Beauchesne, 2007), 305.

25. A. Langemeyer and F. Pene, "The Symphonie System and its Utilization," The Space Congress Proceedings, 1976, available at https://commons.erau.edu/cgi/viewcontent .cgi?article=2805&context=space-congress-proceedings.

26. Reinke, *The History of German Space Policy*, 306.

27. Reinke, *The History of German Space Policy*, 306–307.

28. Frederico Bozo, "The Sanctuary and the Glacis: France, the Federal Republic of Germany, and Nuclear Weapons in the 1980s (Part 1)," *Journal of Cold War Studies* 22, no. 3 (2020): 166.

29. Krige, *Sharing Knowledge, Shaping Europe*, 80.

30. Krige, *Sharing Knowledge, Shaping Europe*.

31. Fenske, "France and the Strategic Defence Initiative," 232.

32. Martin Bailey, "Plan for a UN Spy in Space to Police World," *London Observer*, September 13, 1981, CREST, CIA-RDP90-00806R000201110090-7; A. Walter Dorn, "Peacekeeping Satellites: The Case for International Surveillance Verification," 1987, available at https://www.walterdorn.net/pub/62-peacekeeping-satellites-chapter-iv-the -proposals.

33. Memorandum from the Acting Director of the Defense Communications Agency (Layman) to Secretary of Defense Weinberger, "US Communications Satellite Support to UK Naval Forces," April 9, 1982, *FRUS*, 1981–1988, Volume XIII, Conflict in the South Atlantic, 1981–1984.

34. Bateman, "Keeping the Technological Edge," 375.

35. Ilaria Parisi, "France's Reaction Towards the Strategic Defence Initiative (1983–1986): Transforming a Strategic Threat into a Technological Opportunity," in *NATO and the Strategic Defence Initiative*, ed. Luc-André Brunet (New York: Routledge, 2023), 121–124.

36. PREM 19/2067, Minute from Howe to Thatcher, "Launcher for Skynet 4," December 12, 1983, TNA.

37. After the 1986 *Challenger* disaster, the UK turned to Ariane to launch Skynet, underscoring the value of having an alternative to US launch services.

38. Fenske, "France and the Strategic Defence Initiative," 231.

39. For in-depth analysis of the initial reaction of Western European officials to Reagan's March 1983 SDI speech, see Andreoni, "Ronald Reagan's Strategic Defense Initiative," 48–72.

40. For an overview of the European reactions to SDI, see Sean N. Kalic, "Reagan's SDI Announcement and the European Reaction: Diplomacy in the Last Decade of the Cold War," in *Crisis of Detente in Europe: from Helsinki to Gorbachev, 1975–1985*, ed. Leopoldo Nuti (London: Routledge, 2009); Westwick, "The International History of the Strategic Defense Initiative," 341.

41. Memorandum from Lehman to National Security Council Staff, "Shultz, Nitze, McFarlane, Lehman Conversation," December 13, 1985, *FRUS*, 1981–1988, Volume IV, Soviet Union, 1983–1985.

42. For an overview of the Trident negotiation, see Suzanne Doyle, "A Foregone Conclusion? The United States, Britain and the Trident D5 Agreement," *Journal of Strategic Studies* 40, no. 6 (2017); and Suzanne Doyle, "The United States Sale of Trident to Britain, 1977–1982: Deal Making in the Anglo-American Nuclear Relationship," *Diplomacy and Statecraft* 28, no. 3 (2017). For a more in-depth overview of the Anglo-American Trident negotiations, see Kristan Stoddart, *Facing Down the Soviet Union: Britain, the USA, NATO and Nuclear Weapons, 1976–1983* (Basingstoke, UK: Palgrave Macmillan, 2014).

43. CAB 164/1664, Note from Mottram to Thatcher, March 29, 1983, TNA.

44. Andreoni, "Ronald Reagan's Strategic Defense Initiative," 52. For a history of the political tensions surrounding the deployment of Pershing IIs and cruise missiles in Europe, see Colbourn, *Euromissiles*.

45. The UK MoD originally believed that Reagan's SDI ambitions might prove to be "more rhetorical than substantial"; see PREM 19/1188, MoD Letter to No. 10, "President's Reagan's Speech on Defensive Technology," March 29, 1983, TNA.

46. PREM 19/1188, MOD Letter to No. 10, "President's Reagan's Speech on Defensive Technology."

47. Jeremy Stocker, *Britain and Ballistic Missile Defence, 1942–2002* (London, 2004), 149.

48. Andreoni, "Ronald Reagan's Strategic Defense Initiative," 76.

49. Andreoni, "Ronald Reagan's Strategic Defense Initiative," 69

50. Andreoni, "Ronald Reagan's Strategic Defense Initiative," 69.

51. PREM 19/1188, "Howe letter to Heseltine on Military Developments in Space," June 14, 1984, TNA.

52. Parisi, "France's Reaction Towards the Strategic Defence Initiative," 118–119.

53. "Letter to Jacques Attali from Bud McFarlane," June 14, 1984, Sven Kraemer Box 1, RRPL.

54. Thatcher speech to the European Atlantic Group, July 11, 1984, Margaret Thatcher Foundation, available at https://www.margaretthatcher.org/document/105722.

55. Intelligence Memorandum, "Allied Attitudes Towards the Strategic Defense Initiative and US Development of Anti-Satellite Weapons," June 20, 1984, CREST, CIA-RDP85T00287R001100280001-1.pdf.

56. PREM 19/1188, "No. 10 Letter to FCO on ASATs," February 24, 1984, TNA.

57. PREM 19/1188, "Cradock Letter to Powell on ASATs and Arms Control," July 2, 1984, TNA.

58. PREM 19/1188, "BMD: UK Policy Towards SDI," October 11, 1984, TNA.

59. PREM 19/1188, "No. 10 Record of Conversation," July 16, 1984, TNA. Oddly, the meeting notes incorrectly stated how Thatcher had suggested during her speech to the European Atlantic Group that negotiating restrictions on high-altitude ASATs should be the focus of space arms control efforts; in reality, she made no such comments.

60. Eames, "A 'Corruption of British Science?'" 814.

61. Eames, "A 'Corruption of British Science?'"

62. PREM 19/1188, "No. 10 Record of Conversation," July 16, 1984, TNA.

63. PREM 19/1188, "No. 10 Record of Conversation," July 16, 1984, TNA.

64. Margaret Thatcher, *The Downing Street Years* (New York: Harper Collins, 1993), 466.

65. PREM 19/1188, Powell Minute to Thatcher, "Ballistic Missile Defence (BMD): UK Policy towards the US Strategic Initiative," October 11, 1984, TNA. Powell sent a memorandum to the MoD saying that Thatcher's general view was "that the case against the strategic defence initiative is not so open and shut as suggested in the conclusions . . . of the paper." Due to Soviet advances in missile defense research, the "Americans have little option but to push ahead at least to the point where they can be confident that they are matching the Soviet Union." See PREM 19/1188,

No. 10 Letter to MoD, "Ballistic Missile Defence (BMD): UK Policy towards the US Strategic Initiative," October 15, 1984, TNA. Thatcher became increasingly convinced that SDI could serve as useful leverage with the Soviet Union. In a briefing on SDI provided by Private Secretary for Overseas Affairs Bryan Cartledge, Thatcher underlined the observation that "there is ample evidence that the Soviet leadership is more worried about United States intentions in the BMD field than any other aspect of United States military activity." Thatcher would repeat this theme in future policy discussions. See PREM 19/1188, Cartledge Briefing for Thatcher, "Ballistic Missile Defence (BMD): UK Policy towards the US Strategic Initiative," November 2, 1984, TNA.

66. PREM 19/1188, Heseltine and Howe Minute to MT, "Ballistic Missile Defence (BMD): UK Policy Towards the US Strategic Defence Initiative," October 11, 1984, TNA.

67. "Declaration by the WEU Foreign and Defence Ministers," October 27, 1984, available at https://www.cvce.eu/en/obj/declaration_by_the_weu_foreign_and_defence _ministers_rome_27_october_1984-en-c44c134c-aca3-45d1-9e0b-04d4d9974ddf.html.

68. DEFE 13/2066, "The Military Use of Space," WEU Report, May 15, 1984, TNA.

69. DEFE 13/2066, Memorandum for DUS, "Anglo-French Cooperation in the SATCOM Field," November 5, 1984, TNA.

70. DEFE 13/2066, "WEU Report: Military Use of Space," September 10, 1984, TNA.

71. Colin Gray, "Strategic Defense, Deterrence, and the Prospects for Peace," *Ethics* 95, no. 3 (1985): 659.

72. Brown, *The Human Factor*, 127.

73. Camp David Declaration (draft), NSC: Records Thatcher Visit—December 84 Box 90902, RRPL.

74. PREM 19/1444, "Record of Discussion at the Pentagon," July 17, 1984, TNA.

75. A 1985 CIA report noted that the Europeans were seeking "greater exchange of technology than they received in past joint efforts—most notably Spacelab." For details, see "Summit Issues: Space Station Cooperation," April 23, 1985, CREST, CIA-RDP85T01058R000202790001.

76. PREM 19/1188, UK Embassy in Washington to FCO, "Consultations with the Americans on Space," July 18, 1984, TNA.

77. PREM 19/1444, Record of Discussion at the Pentagon, July 17, 1984, TNA.

78. For the evolution of the SDI systems concept, see Donald Baucom, "The Rise and Fall of Brilliant Pebbles," *The Journal of Social, Political, and Economic Studies* 29, no. 2 (2004): 149.

79. Jon Agar, *Science Policy under Thatcher* (London: UCL Press, 2020), 205.

80. British officials observed that "we have received a number of US briefings on the SDI program, but most of these have concentrated on strategic and political issues

and aims, rather than technical achievements"; see PREM 19/1444, "US and USSR BMD Programmes," attached to briefing for the prime minister on BMD/SDI, January 7, 1985, TNA.

81. PREM 19/1444. A description of this meeting is contained within US and USSR BMD Programmes, attached to briefing for the prime minister on BMD/SDI, January 7, 1985, TNA.

82. "Warning and Forecast Report for Europe," January 29, 1985, CREST, CIA-RDP87R00529R000300250048-3.

83. "Warning and Forecast Report for Europe."

84. Memorandum by Ministry of Foreign Affairs, "Partecipazione europea alla Iniziativa di Difesa Strategica. Implicazioni politiche," January 1, 1985, History and Public Policy Program Digital Archive, Instituto Luigi Sturzo, Archivio Giulio Andreotti, NATO Series, Box 182, Subseries 5-1, Folder 001, available at https://digitalarchive.wilsoncenter.org/document/155254.

85. Memorandum from Adelman to Shultz, Weinberger, and Casey, "Cooperating with the Allies in SDI Research," CREST, CIA-RDP87M00220r000100030016-1.

86. Memorandum from Adelman to Shultz, Weinberger, and Casey, "Cooperating with the Allies in SDI Research," CREST, CIA-RDP87M00220r000100030016-1.

87. Abrahamson, "SDI and the New Space Renaissance."

88. John Krige, "Hybrid Knowledge: The Transnational Co-Production of the Gas Centrifuge for Uranium Enrichment in the 1960s," *The British Journal for the History of Science* 45, no. 3 (2012).

89. PREM 19/1444, "Letter from Weinberger to Heseltine," March 26, 1985, TNA.

90. PREM 19/1444, "Letter from Weinberger to Heseltine."

91. Luc-André Brunet, "The Strategic Defence Initiative and the Atlantic Alliance in the 1980s," in *NATO and the Strategic Defence Initiative*, ed. Luc-André Brunet (New York: Routledge, 2023), 4.

92. Proposed talking points for DCI meeting with national security principals, undated, CREST, CIA-RDP87M00539R0008009900260-5.

93. Will Winkelstein, "SDI Poses Problems for Japan," *Chicago Tribune*, April 12, 1986, available at https://www.chicagotribune.com/news/ct-xpm-1986-04-12-8601270004-story.html.

94. PREM 19/1468, "Thatcher Meeting with Nakasone," May 4, 1985, TNA.

95. "National Security Policy Group Meeting," October 11, 1985, CREST, CIA-RDP88B00443R000200910003-0.

96. "National Security Policy Group Meeting."

97. Breyman, "SDI, The Federal Republic of Germany, and NATO."

98. For a history of COCOM, see John H. Henshaw, "The Origins of COCOM: Lessons for Contemporary Proliferation control Regimes," May 1993, available at https://www.stimson.org/wp-content/files/file-attachments/Report7_1.pdf.

99. Breyman, "SDI, The Federal Republic of Germany, and NATO."

100. NIE, Major NATO Allies: Perspectives on the Soviet Union, September 1, 1985, CREST, CIA-RDP87T00495R001001110008-9; Memorandum from McFarlane to Reagan, "US–USSR Economic Working Group of Experts Meetings in Moscow," January 4, 1985, *FRUS, 1981–1988*, Volume IV, Soviet Union, 1983–1985, available at https://static.history.state.gov/frus/frus1981-88v04/pdf/frus1981-88v04.pdf, 1274.

101. Timothy Sayle, *Enduring Alliance: A History of NATO and the Postwar Global Order* (Ithaca, NY: Cornell University Press, 2019), 115–116.

102. Ralph Dietl, *The Strategic Defense Initiative: Ronald Reagan, NATO Europe, and the Nuclear and Space Talks, 1981–1988* (London: Lexington Books, 2018), 53.

103. Colbourn, *Euromissiles*.

104. PREM 19/1444, "Message to FCO on NPG," March 26, 1985, TNA.

105. PREM 19/1444, "Heseltine to Howe, SDI Participation," March 27, 1985, TNA.

106. PREM 19/1444, "Heseltine to Howe, SDI Participation."

107. PREM 19/1444, "Powell to PM," March 27, 1985, TNA.

108. PREM 19/1444, undated note (likely March 27, 1985) from Powell to Thatcher marked strictly personal, TNA.

109. PREM 19/1444, "Memo from Powell to Mottram," March 28, 1985, TNA.

110. Charles Powell interview with the author via telephone, February 4, 2020; Agar, *Science Policy under Thatcher*, 189. Although she was circumspect in talks with European politicians about involvement in SDI, Thatcher recognized the benefits of an exchange of information with European governments to ascertain their intentions regarding SDI participation; see PREM 19/1444, "Memo from Powell to Mottram," March 28, 1985, TNA.

111. "France Tries to Close the Technology Gap," July 8, 1985, CREST, CIA-RDP85T01058R000303110001-8.

112. PREM 19/1467, "Nicholson Briefing for Thatcher on EUREKA," April 28, 1985, TNA.

113. "The Earth Imperilled," 1986, History and Public Policy Program Digital Archive, Vsevolod S. Avduyevsky and Anatoli I. Rudev, Novosti Press Agency Publishing House, 1986, available at http://digitalarchive.wilsoncenter.org/document/110196, 23.

114. Robert Gilpin, *France in the Age of the Scientific State* (Princeton: Princeton University Press, 1968), 3.

115. Parisi, "France's Reaction Towards the Strategic Defence Initiative," 121–124.

116. PREM 19/1467, "Nicholson Briefing for Thatcher on EUREKA," April 28, 1985, TNA.

117. PREM 19/1444, UK Embassy in Bonn to FCO, "German Views on SDI," February 12, 1985, TNA.

118. PREM 19/1444, UK Embassy in Bonn to FCO, "German Views on SDI."

119. Andreoni, "Ronald Reagan's Strategic Defense Initiative," 103.

120. Reinke, *The History of German Space Policy*, 290–291; Christoph Bluth, "SDI: The Challenge to West Germany," *International Affairs* 62, no. 2 (1986): 250.

121. Horst Teltschik interview with the author via telephone, October 14, 2020.

122. Horst Teltschik interview with the author via telephone; Krige et al., *NASA in the World*, 85.

123. Dietl, *The Strategic Defense Initiative*, 67.

124. National Security Policy Group Meeting, October 11, 1985, CREST, CIA-RDP88B 00443R000200910003-0.

125. Lakoff and York, *A Shield in Space?*, 221.

126. Reinke, *The History of German Space Policy*, 296.

127. "Report by Ambassador Petrignani to the Minister of Foreign Affairs Andreotti," May 24, 1985, History and Public Policy Program Digital Archive, Istituto Luigi Sturzo, Archivio Giulio Andreotti, NATO Series, Box 182, Subseries 5-1, Folder 001, available at https://digitalarchive.wilsoncenter.org/document/155251.

128. "Report by Ambassador Petrignani to the Minister of Foreign Affairs Andreotti," History and Public Policy Program Digital Archive, 297.

129. David Dickson, "A European Defense Initiative: The Idea that European Nations Band Together for a Strictly European Version of SDI Is Gaining Support," *Science* 229, no. 4719 (1985): 1243.

130. Dickson, "A European Defense Initiative."

131. Jerome Paolini, "Politique Spatiale Militaire Française et Coopération Européenne," *Politique Étrangère* 52, no. 2 (1987): 435; Wayne Sandholtz, *High-Tech Europe: The Politics of International Cooperation* (Berkley: University of California Press, 1992), 270.

132. Dickson, "A European Defense Initiative," 1243.

133. David Dickson, "Europe Tries a Strategic Technology Initiative," *Science* 229, no. 4709 (1985): 143.

134. "Mitterrand Suggests Space Station," *UPI*, February 7, 1984, available at https://www.upi.com/Archives/1984/02/07/Mitterrand-suggests-space-station /7817444978000/.

135. David Yost, "Les Inquiétudes Européennes Face aux Systèmes de Défense Anti-Missiles: un Point de Vue Américain," *Politique étrangère* 49, no. 2 (1984): 396.

136. Krige et al., *NASA in the World*, 254.

137. Krige et al., *NASA in the World*.

138. Intelligence Report, "Summit Issue: Space Station Cooperation," April 23, 1985, CREST, CIA-RDP85T01058R000202790001-6.

139. Intelligence Report, "Summit Issue: Space Station Cooperation."

140. Siebeneichner, "Spacelab: Peace, Progress and European Politics in Outer Space," 273.

141. PREM 19/1468, "Economic Summit: Supplementaries on Foreign Policy Issue," May 1985, TNA.

142. Horst Teltschik interview with the author via telephone, October 14, 2020.

143. Reinke, *The History of German Space Policy*, 291.

144. William Drozdiak, "Moscow Warns Bonn on SDI," *New York Times*, November 14, 1985, available at https://www.washingtonpost.com/archive/politics/1985/11/14/moscow-warns-bonn-on-sdi/0b9a8a86-4f52-4ff2-ac28-c68807d5ea36/.

145. Dietl, *The Strategic Defense Initiative*, 69.

146. Horst Teltschik interview with the author via telephone, October 14, 2020; Elizabeth Pond, "W. Germany Looks to Commercial Uses for 'Star Wars,'" *Christian Science Monitor*, September 5, 1985, available at https://www.csmonitor.com/1985/0905/orib.html.

147. PREM 19/1445, "Heseltine Minute to Thatcher on SDI Participation," May 28, 1985, TNA.

148. PREM 19/1445, "Heseltine Minute to Thatcher on SDI Participation."

149. Agar, *Science Policy under Thatcher*, 204.

150. PREM 19/1445, "Minutes from OD Meeting on SDI Participation," July 10, 1985, TNA.

151. PREM 19/1445, "Minutes from OD Meeting on SDI Participation."

152. PREM 19/1445, "Minutes from OD Meeting on SDI Participation."

153. PREM 19/1445, "Powell Minute to Thatcher, SDI Participation," July 11, 1985, TNA.

154. PREM 19/1445, "Powell Minute to Thatcher, SDI Participation," 1985.

155. PREM 19/1445, "Ministry of Defence Record of Conversation (Heseltine–Weinberger)," February 28, 1984, TNA.

156. PREM 19/1445, "British Participation in SDI Research," July 23, 1985, TNA.

157. PREM 19/1445, "Record of Meeting between the Defence Secretary and Mr. Caspar Weinberger," July 22, 1985, TNA.

158. PREM 19/1445, "Minute from Powell to Mottram, UK Participation in SDI Research," July 22, 1985, TNA.

159. PREM 19/1445, "Heseltine Minute to Thatcher, British Participation in SDI Research," July 23, 1985, TNA.

160. Agar, *Science Policy under Thatcher*, 206.

161. PREM 19/1661, "PM Meeting with the United States Secretary of Defense at the Pentagon," July 26, 1985, TNA.

162. PREM 19/1661, "PM Meeting with Vice President Bush," July 26, 1985, TNA.

163. PREM 19/1445, "Letter from Heseltine to Weinberger," August 1, 1985, TNA.

164. PREM 19/ 1445, "Mottram to Powell, Annex A of Areas Identified for UK SDI Research," September 25, 1985, TNA.

165. PREM 19/1445, "Mottram to Powell, SDI Research: UK Participation," September 25, 1985, TNA.

166. PREM 19/1445, "JB Unwin to Mallaby, SDI Research: UK Participation," September 26, 1985, TNA.

167. Michael Griffin interview with the author via telephone, October 30, 2020.

168. PREM 19/1445, "Unsigned Letter from Weinberger to Heseltine," undated, TNA.

169. PREM 19/1445, "Discussion on UK Participation in SDI Research," undated, TNA.

170. DEFE 72/457, "Anglo-French Defence Staff Talks," September 26 and 17, 1985, TNA.

171. PREM 19/1445, Mallaby Minute to Thatcher, Strategic Defence Initiative Research: United Kingdom Participation," October 25, 1985, TNA.

172. PREM 19/1445, "Owen to Thatcher, SDI: UK Participation," October 25, 1985, TNA.

173. PREM 19/1445, "UK Embassy in Washington Telegram to FCO," October 26, 1985, TNA.

174. The distinction between research and development remained a difficult subject for British officials. The ABM Treaty forbade the transfer of missile defense technologies to "third parties." Even though the US maintained that no ABM components would be shared with allies, Geoffrey Howe was concerned about even the appearance that there might be a breach of the ABM Treaty and advised that the UK should "continue to discuss with the Americans the legal implications." Howe and Heseltine feared that any movement beyond SDI research could jeopardize the ABM Treaty. Identifying a boundary between SDI research and development would become an especially contentious issue at the Reykjavik Summit in 1986. See PREM 19/1445, "Minute from Howe to Heseltine, SDI Participation: Legal and European Angles," October 29, 1985, TNA.

175. PREM 19/1445, "Minute from Heseltine to Thatcher, Strategic Defence Initiative Research: United Kingdom Participation," October 31, 1985, TNA.

176. PREM 19/1445, "Strategic Defence Initiative Research: United Kingdom Participation."

177. CAB 130/1303, "Report of the US–UK Joint Working Group Concerning Cooperative Research for the Strategic Defense Initiative," October 15, 1985, TNA.

178. PREM 19/1445, "Powell Minute to MT," November 13, 1985, TNA.

179. PREM 19/1445, "Minute from Powell to Mottram, Strategic Defence Initiative Research: United Kingdom Participation," November 1, 1985, TNA.

180. CAB 130/1303, "US–UK Memorandum of Understanding on Conducting Research for the Strategic Defence Initiative," December 6, 1985, TNA.

181. PREM 19/2068, "Transcript from Hearing in Parliament on SDI," December 9, 1985, TNA.

182. PREM 19/2068, "Cable from UK Embassy in Moscow to FCO, Soviet Demarche on UK/US Memorandum of Understanding," December 24, 1985, TNA.

183. Andreoni, "Ronald Reagan's Strategic Defense Initiative," 138.

184. Gorbachev wrote a letter to Kohl warning against participation in SDI on the grounds that it could harm relations between Bonn and Moscow as the American–Soviet arms control negotiations approached. See Memorandum for McFarlane, "Soviet Activities Leading to the Summit," November 15, 1985, CREST, CIA-RDP87M00539R001301650002-1.

185. Memorandum for McFarlane, "Soviet Activities Leading to the Summit," 106–107.

186. Alexander Higgins, "West Germany Gives 'Star Wars' Its Qualified Backing," *Associated Press*, March 27, 1986, available at https://apnews.com/article/2790f0cb5a787f2c072b4f09a53401da.

187. Breyman, "SDI, The Federal Republic of Germany, and NATO."

188. Andreoni, "Ronald Reagan's Strategic Defense Initiative," 139.

189. Andreoni, "Ronald Reagan's Strategic Defense Initiative," 138.

190. PREM 19/1764, "MoD Record of Conversation (Younger and German Defence Minister)," September 22, 1986, TNA.

191. Developments in Europe, May 1986: Hearing and Markup before the House Committee on Foreign Affairs, 99th Congress, 1986.

192. CAB 130/1322, "Official Group on the Strategic Defence Initiative," July 23, 1986, TNA.

193. CAB 130/1303, "Memorandum of Understanding between the Government of the United States of America and the Government of the United Kingdom of Great

Britain and Norther Ireland Relating to Cooperative Research for the Strategic Defense Initiative," December 6, 1985, TNA.

194. CAB 130/1322, "Strategic Defence Initiative Research: United States/United Kingdom Administrative Arrangements," March 20, 1986, TNA. In January 1986, Britain won its first SDI research contracts (worth a little more than one million dollars), which were awarded to Heriot-Watt University in Edinburgh and Ferranti, a British defense company. Both organizations would use the American funds to do research into optical computers, which use photons for computation. National laboratories, such as Rutherford Appleton, had secured contracts to examine issues related to lethality and hardening (i.e., survivability) for SDI components. See CAB 130/1322, "Updated Transcript of Conversation," 1986, TNA.

195. Robert McCartney, "Terrorist Group Kills Executive Near Munich," *Washington Post*, July 10, 1986, available at https://www.washingtonpost.com/archive/politics /1986/07/10/terrorist-group-kills-executive-near-munich/da623290-a2db-448f-ac1c -f18bda0da97f/.

196. Richard Hofer, "Terrorist Bomb Kills West German Industrialist," *Associated Press*, July 9, 1986, available at https://apnews.com/article/318bdddd11180d71395de36b69 b3f14a.

197. FCO 46/552, "Meeting of the Special Working Group of the WEU," February 21, 1986, TNA.

198. Vie-publique.fr, "Discours de Jacques Chirac, Président du RPR, sur l'Attitude de l'Europe et de la France Face à l'Initiative de Défense Stratégique," Paris, October 18, 1985; Richard Bernstein, "'Star Wars' Splits Two French Leaders," *New York Times*, May 28, 1986, available at https://www.nytimes.com/1986/05/28/world/star-wars -splits-2-french-leaders.html.

199. Andreoni, "Ronald Reagan's Strategic Defense Initiative," 127.

200. "First Such Switch: Britain to Use Ariane Instead of US Shuttle," *Los Angeles Times*, May 29, 1986, available at https://www.latimes.com/archives/la-xpm-1986-05 -29-mn-7891-story.html.

201. Strategic Defense Initiative Organization, "1990 Report to Congress on the Strategic Defense Initiative," May 1990, B-2-B-4, available at https://apps.dtic.mil/dtic/tr /fulltext/u2/a224950.pdf.

202. Strategic Defense Initiative Organization, "1990 Report to Congress on the Strategic Defense Initiative," 1990.

203. Strategic Defense Initiative Organization, "1990 Report to Congress on the Strategic Defense Initiative."

204. Westwick, "The International History of the Strategic Defense Initiative," 346.

205. Horst Teltschik interview with the author via telephone, October 14, 2020.

206. Table contained within "Strategic Defense Initiative Organization," 1990 Report to Congress on the Strategic Defense Initiative, May 1990, B-2, available at https://apps.dtic.mil/dtic/tr/fulltext/u2/a224950.pdf.

207. For details concerning Israeli involvement in SDI, see Or Rabinowitz, "'Arrow' Mythology Revisited: The Curious Case of the Reagan Administration, Israel and SDI Cooperation," *The International History Review* 43, no. 6 (2021).

208. Siebeneichner, "Spacelab: Peace, Progress and European Politics in Outer Space," 274.

CHAPTER 5

1. Paul Nitze, *From Hiroshima to Glasnost: At the Center of Decision* (New York: Grove Atlantic, 1989), 435.

2. Jack Matlock, *Reagan and Gorbachev: How the Cold War Ended* (New York: Random House, 2005), 229.

3. Minutes of a National Security Planning Group, "Arms Control Follow-up to Reykjavik," October 27, 1986, *FRUS*, 1981–1988, Volume VI, Soviet Union, October 1986–January 1989.

4. PREM 19/1444, "Record of Discussion at the Pentagon on SDI: July 17, 1984," attached to a briefing for the PM on BMD/SDI, January 4, 1985, TNA.

5. PREM 19/1444, "Record of Discussion at the Pentagon on SDI." A primary problem that remained to be addressed was the vulnerability of SDI space-based systems to attack by Soviet ASATs. Because of the technological linkage between SDI and ASATs, the US found itself in an undesirable predicament: curbing ASATs could harm SDI, but the Soviet Union developing more sophisticated ASATs in response to SDI would only "complicate the environment threat for the SDI": that is, make it more vulnerable. For more details, see PREM 19/1444, "Record of Discussion at the Pentagon on SDI," July 17, 1984, attached to a briefing for the PM on BMD/SDI, January 4, 1985, TNA.

6. Robert Linhard Collection, undated concept paper on SDI, SDI-Mafia file, January 6, 1986, RAC 8, Box 3, RRPL.

7. Executive Secretariat NSC Box 107, "Strategic Defense Initiative," undated paper on the SDI program, RRPL.

8. Executive Secretariat NSC Box 107, "Strategic Defense Initiative," undated paper on the SDI program.

9. Minutes of a National Security Planning Group Meeting, "Discussion of Substantive Issues for Geneva," December 17, 1984, *FRUS*, 1981–1988, Volume IV, Soviet Union, January 1983–March 1985.

10. Memorandum from Iklé to Weinberger, "Arms Control Diplomacy," July 3, 1984, document 235, *FRUS*, Volume IV, Soviet Union, January 1983–March 1985.

11. Memorandum from Weinberger to McFarlane, June 28, 1984, Executive Secretariat NSC Box 106, RRPL.

12. Minutes of a National Security Planning Group Meeting, Soviet Defense and Arms Control Objectives, *FRUS*, 1981–1988, Volume IV, Soviet Union, January 1983–March 1985.

13. Intelligence Memorandum, "Allied Attitudes Towards the Strategic Defense Initiative and US Development of Anti-Satellite Weapons," June 20, 1984, CREST, CIA-RDP85T00287R001100280001-1.pdf.

14. Minutes of a National Security Planning Group Meeting, "Discussion of Substantive Issues for Geneva," December 17, 1984, *FRUS*, 1981–1988, Volume IV, Soviet Union, January 1983–March 1985.

15. Consultations with the British, French, Germans, and Italians on ASAT/SDI, July 17–18, 1985, Executive Secretariat NSC Box 106, RRPL.

16. Memorandum from Weinberger to McFarlane, June 28, 1984, Executive Secretariat NSC Box 106, RRPL. Notably, Weinberger's concern about Soviet high-altitude ASATs was, in the mid-1980s, hypothetical, since US intelligence analysts did not yet have any indications that the Soviets were developing such a capability; see NIE 11-1-83, "The Soviet Space Program," July 19, 1983, CREST, CIA-RDP00B00369R000100050006-2.

17. Anatoly Zak, "IS Anti-Satellite System," *RussianSpaceWeb*, available at https://www.russianspaceweb.com/is.html.

18. For indications of a Soviet high-altitude intelligence satellite system, see *Russian SpaceWeb*.

19. Memorandum from Under Secretary of Defense for Policy (Iklé) to Secretary of Defense Weinberger, "Arms Control Diplomacy," July 3, 1984, *FRUS*, 1981–1988, Volume IV, Soviet Union, January 1983–March 1985.

20. Ronald Reagan, White House Diaries, June 26, 1984, available at https://www.reaganfoundation.org/ronald-reagan/white-house-diaries/diary-entry-06261984/.

21. Letter from Secretary of Defense Weinberger to the President's Assistant for National Security Affairs (McFarlane), June 29, 1984, *FRUS*, 1981–1988, Volume IV, Soviet Union, January 1983–March 1985.

22. Minutes of a National Security Planning Group Meeting, "Next Steps in the Vienna Process," September 18, 1984, *FRUS*, 1981–1988, Volume IV, Soviet Union, January 1983–March 1985.

23. Minutes of a National Security Planning Group Meeting, "US–Soviet Arms Control Objectives," December 5, 1984, *FRUS*, 1981–1988, Volume IV, Soviet Union, January 1983–March 1985.

24. Minutes of a National Security Planning Group Meeting, "US–Soviet Arms Control Objectives,"

25. British diplomats noted in a summary of a July 1984 meeting with US and European officials at the Pentagon that "there is some evidence of doubts in USAF and Navy circles about the US F15 low level ASAT program, which could lead them to favor a moratorium on all further ASAT testing." For details, see PREM 19/1188, cable from UK Embassy Washington to FCO, "Consultations with the Americans on Space," July 18, 1984, TNA. For the US discussion of ASATs, see Minutes of a National Security Planning Group Meeting, "US–Soviet Arms Control Objectives," December 5, 1984, *FRUS*, 1981–1988, Volume IV, Soviet Union, January 1983–March 1985.

26. Minutes of a National Security Planning Group Meeting, "Discussion of Geneva Format and SDI," December 10, 1984, *FRUS*, 1981–1988, Volume IV, Soviet Union, January 1983–March 1985.

27. Minutes of a National Security Planning Group Meeting, "US–Soviet Arms Control Objectives," December 5, 1984, *FRUS*, 1981–1988, Volume IV, Soviet Union, January 1983–March 1985.

28. Caspar Weinberger, *In the Arena: A Memoir of the 20th Century* (Washington, DC: Regnery, 2001), 284.

29. NSDD 165, "Instructions for the First Round of US/Soviet Negotiations in Geneva," March 8, 1985, *FRUS*, 1981–1988, Volume IV, Soviet Union, January 1983–March 1985.

30. NSDD 165, "Instructions for the First Round of US/Soviet Negotiations in Geneva," 1985.

31. Memorandum of Conversation, "Shultz and Gromyko," January 7, 1985, *FRUS*, 1981–1988, Volume IV, Soviet Union, January 1983–March 1985.

32. *FRUS*, 1981–1988, Volume IV, Soviet Union, 1985.

33. Memorandum for Robert McFarlane from Jack Matlock and Ron Lehman, "Revised Statement on Vienna Talks," August 1, 1984, Jack F. Matlock Files, Box 5, RRPL.

34. Thomas C. Schelling and Morton H. Halperin, *Strategy and Arms Control* (New York: Twentieth Century Fund, 1961), 50.

35. On January 18, 1985, Shultz held a press conference at the White House to announce the US delegation to the Nuclear and Space Talks set to begin in Geneva on March 12. He said that Ambassador Maynard W. Glitman, a minister-counselor of the Foreign Service of the US, will be nominated as the US negotiator on intermediate-range nuclear arms. Ambassador Max M. Kampelman will be nominated as US negotiator on space and defensive arms. Ambassador Kampelman will also serve as head of the US delegation. Ambassador Paul H. Nitze and Ambassador Edward L. Rowny will serve as special advisors to the president and to the secretary of state on arms reduction negotiations. Notably, Weinberger wanted Edward Teller appointed as the lead for the space and defensive arms forum because "no one else could be trusted to be totally committed to SDI." See editorial note in *FRUS*, 1981–1988, Volume IV, Soviet Union, January 1983–March 1985, 365.

36. Memorandum of Conversation, "Shultz Phone Call with Reagan," January 8, 1985, *FRUS*, 1981–1988, Volume IV, Soviet Union, January 1983–March 1985.

37. Don Oberdorfer, "US, Soviets to Resume Arms Talks," *Washington Post*, January 9, 1985, available at https://www.washingtonpost.com/archive/politics/1985/01/09/us -soviets-to-resume-arms-talks/ea807b41-6748-46c2-9813-3ecc89574c34/.

38. Don Oberdorfer, "US, Soviets to Resume Arms Talks."

39. Memorandum from Lenczowski to McFarlane, March 12, 1985, *FRUS*, 1981–1988, Volume V, Soviet Union, March 1985–October 1986.

40. CIA Report, "Gorbachev the New Brook," June 1985, CREST, CIA-RDP88b 00443R001704320004-4. Miles highlights the significance of this report; see Miles, *Engaging the Evil Empire*, 121.

41. Telegram from US Embassy in Moscow to Department of State, Memorandum of Conversation after Chernenko's Funeral, March 14, 1985, *FRUS*, 1981–1988, Volume V, Soviet Union, March 1985–October 1986.

42. Page 2 of KGB PGU upravlenie "RT" January 15, 1986, No. 167/466, O dopolnitel'nix zadachax v informatsionnoi rabote po voenno-strategicheskoi problematike, file: delo operativnoi perepiski No. 2113, Po voenno-stragicheskoi problematike, Lithuanian National Archives. On May 21, 1985, Yuli Kvitsinskiy, the chief Soviet negotiator for SDI (according to CIA document in *FRUS*), said to a group of KGB officers that "SDI remained the central problem in the [arms control] negotiations." A GRU (military intelligence) officer then gave a more technical presentation on SDI and said that "although the SDI system was very complicated, it was not unrealistic to think that the US would be able to implement it, sooner or later." He further stated that "it was possible that SDI would eventually be able to intercept 90 percent of the Soviet Union's strategic missiles." For details, see discussion by Soviet officials of the SDI and other arms control issues, September 25, 1985, *FRUS*, 1981–1988, Volume V, Soviet Union, March 1985–October 1986. Soviet defector Oleg Gordievsky (former senior KGB officer) informed the British that "the principal Soviet concern over SDI is not so much that they [Soviets] consider it a threat as that they feel that it forces them to accelerate their own program in a way that they cannot afford." The latter is McFarlane's description of Gordievsky's position in a Memorandum to Reagan rather than direct quotes from the former KGB officer. See Memorandum for the President, "Gordiyevsky's [McFarlane's spelling] Suggestions," October 30, 1985, available at https://www.thereaganfiles.com/19851030 -gordiyevsky.pdf.

43. Kvitsinskiy, "SDI remained the central problem in the [arms control] negotiations."

44. Much of the analysis concerning SDI clearly came from open-source US materials. For details, review the files associated with delo operativnoi perepiski No. 2113, Po voenno-stragicheskoi problematike, Lithuanian National Archives.

45. Peter Schweizer, *Reagan's War: The Epic Story of His Forty-Year Struggle and Final Triumph Over Communism* (New York: Anchor, 2003), 152. Going back to the early 1960s, Soviet military leaders were concerned that the US would develop space weapons for destroying other satellites, striking ground and naval targets, and defending against ballistic missiles. A 1962 article in the Soviet Ministry of Defense's classified in-house journal *Voennaia mysl'* (*Military Thought*) predicted that the US would "create space weapons for antimissile and antispace defense, which can be employed as a sort of shield preventing access to space and the carrying out by the enemy of his military space programs." The author of the article also argued that the view in the US was that "future space weapons systems will strengthen offensive, and not defensive capabilities." See Lt Gen N. Korenevskiy, "The Role of Space Weapons in a Future War" (CIA translation), September 7, 1962, CREST, CIA-RDP33-02415A000500190011-3.

46. V. V. Afinov, "Razvitie v SShA vysokotochnogo oruzhiya i perspektivy sozdaniya razvedyvatel'no-udarnykh kompleksov," *Voennaia mysl'* 4 (1983): 63–71.

47. Michael J. Sterling, *Soviet Reactions to NATO's Emerging Technologies for Deep Attack* (Santa Monica: RAND, 1985), 3–4.

48. Arnold Buchholz, "The Scientific-Technological Revolution (STR) and Soviet Ideology," *Studies in Soviet Thought* 30, no. 4 (1985): 342.

49. Leslie Gelb, "Who Won the Cold War?" *The New York Times*, August 20, 1992, available at https://www.nytimes.com/1992/08/20/opinion/foreign-affairs-who-won -the-cold-war.html.

50. Charles, "The Game Changer," 70, and Vladislav Zubok, *A Failed Empire: The Soviet Union in the Cold War from Stalin to Gorbachev* (Chapel Hill: University of North Carolina Press, 2009), 273.

51. Miles, *Engaging the Evil Empire*, 118.

52. Podvig, "Did Star Wars Help End the Cold War?," 11.

53. Jonathan Haslam, *Russia's Cold War: From the October Revolution to the Fall of the Wall* (New Haven, CT: Yale University Press, 2013), 355.

54. Memorandum for Brzezinski, "Soviet and US High-Energy Laser Weapons Programs."

55. NIE 11-1-83, "Soviet Space Program," July 19, 1983, CREST, CIA-RDP00B0036 9R000100050007-1.

56. Memorandum of Conversation, "Meeting with Vladimir Shcherbitsky," March 7, 1985, *FRUS*, 1981–1988, Volume IV, Soviet Union, January 1983–March 1985.

57. Ronald Reagan, NSDD 172, "Presenting the Strategic Defense Initiative," May 30, 1985, available at https://www.reaganlibrary.gov/public/archives/reference/scanned -nsdds/nsdd172.pdf.

58. PREM 19/1444, "Record of Discussion at the Pentagon on SDI," July 17, 1984, attached to a briefing for the PM on BMD/SDI, January 4, 1985, TNA.

59. PREM 19/1444, "Record of Discussion at the Pentagon on SDI."

60. Alan Sherr, "Sound Legal Reasoning or Policy Expedient? The 'New Interpretation' of the ABM Treaty," *International Security* 11, no. 3 (1986–1987): 73.

61. US SALT Delegation, Memcon "SALT," January 26, 1972, National Security Archive Electronic Briefing Book No. 60, available at https://nsarchive2.gwu.edu/NSAEBB /NSAEBB60/.

62. Robert Linhard Collection, RAC Box 8, Box 3, "SDI Issues," SDI-Mafia File, January 6, 1986, RRPL.

63. Robert Linhard Collection, RAC Box 8, Box 3, "SDI Issues."

64. Robert Linhard Collection, RAC Box 8, Box 3, "SDI Issues."

65. PREM 19/1444, "FCO Letter to No. 10, SDI Testing-ABM Treaty," February 28, 1985, TNA.

66. PREM 19/1445, "FCO Minute to No. 10, SDI and ABM Treaty," May 13, 1985, TNA.

67. PREM 19/1444, "FCO Letter to No. 10, SDI Testing-ABM Treaty," February 28, 1985, TNA.

68. See editorial note 69 in *FRUS*, 1969–1976, Volume XXXVIII, Part 1, Foundations of Foreign Policy, 1973–1976.

69. PREM 19/1444, "FCO Letter to No. 10, SDI Testing-ABM Treaty," February 28, 1985, TNA.

70. John Darnton, "Weinberger Says ABM Pact May Ultimately Need Amending," *New York Times*, March 25, 1983, available at https://www.nytimes.com/1983/03/25 /world/weinberger-says-abm-pact-may-ultimately-need-amending.html.

71. NIE 11-3/8-83, "Soviet Capabilities for Strategic Nuclear Conflict," 1983–1989, available at http://insidethecoldwar.org/sites/default/files/documents/NIE%2011-3% 208-81%20Soviet%20Capabilities%20for%20Strategic%20Nuclear%20Conflict%20 1983-1993%20March%206%2C%201984_0.pdf.

72. Michael Gordon, "CIA Is Skeptical that New Soviet Radar Is Part of an ABM Defense System," *National Journal*, March 9, 1985, CREST, CIA-RDP94b00280r000700 020008-2.

73. Response to Senator Helms' Soviet ABM Compliance Questions, August 14, 1987, CREST, CIA-RDP90G00152r000600770012-7.

74. PREM 19/1444, "Heseltine Minute to Thatcher, Strategic Defence Initiative," March 27, 1985, TNA.

75. PREM 19/1444, "Strategic Defence Initiative."

76. PREM 19/1444, "Strategic Defence Initiative."

77. Final Communique, NATO Nuclear Planning Group, March 26 and 27 1985, available at https://www.nato.int/docu/comm/49-95/c850327a.htm.

78. "Soviets Admit Radar Station Violated Pact," *Los Angeles Times*, October 23, 1989, available at https://www.latimes.com/archives/la-xpm-1989-10-23-mn-543-story.html.

79. Boyce Rensberger, "Space Shuttle To Be Used in 'Star Wars' Laser Test," *Washington Post*, May 24, 1985, available at https://www.washingtonpost.com/archive/polit ics/1985/05/24/space-shuttle-to-be-used-in-star-wars-laser-test/b60ddbb6-a55a-4929 -9a4d-39085577d7a5/.

80. William J. Broad, "Laser Test Fails to Strike Mirror in Space Shuttle," *New York Times*, June 20, 1985, available at https://www.nytimes.com/1985/06/20/us/laser-test -fails-to-strike-mirror-in-space-shuttle.html.

81. Broad, "Laser Test Fails to Strike Mirror in Space Shuttle," 1985.

82. Thomas O'Toole, "Space Shuttle Succeeds in Laser Test," *Washington Post*, June 22, 1985, available at https://www.washingtonpost.com/archive/politics/1985/06/22/space -shuttle-succeeds-in-laser-test/a85ddc9a-05ff-4036-b435-8354e467af33/.

83. Don Oberdorfer, "White House Revises Interpretation of ABM Treaty," *Washington Post*, October 9, 1985, available at https://www.washingtonpost.com/archive/politics /1985/10/09/white-house-revises-interpretation-of-abm-treaty/20619e6d-17cd-4329 -b559-a061a71f2b6b/.

84. Sherr, "Sound Legal Reasoning or Policy Expedient?"

85. Ronald Reagan, NSDD 192, "The ABM Treaty and the SDI Program," October 11, 1985, available at https://fas.org/irp/offdocs/nsdd/nsdd-192.pdf.

86. Miles, *Engaging the Evil Empire*, 122.

87. Miles, *Engaging the Evil Empire*.

88. Podvig, "Did Star Wars Help End the Cold War?," 8–9.

89. Podvig, "Did Star Wars Help End the Cold War?"

90. Podvig, "Did Star Wars Help End the Cold War?"

91. Memorandum of Conversation, "Verification, Krasnoyarsk, NST Instructions, Defense and Space," December 6, 1987, *FRUS*, 1981–1988, Volume XI, START I.

92. Tim Ahern, "Senate Approves Antisatellite Tests," *Washington Post*, May 25, 1985, CREST, CIA-RDP90-00965R000100070005-2.

93. Ahern, "Senate Approves Antisatellite Tests," 1985.

94. Moltz, *The Politics of Space Security*, 202.

95. "Arms Controllers Win a Year-Long Ban on Anti-Satellite (ASAT) Missile Tests," in *CQ Almanac 1985*, 41st ed., 162-64, Washington, DC: Congressional Quarterly, 1986, available at http://library.cqpress.com/cqalmanac/cqal85-11475.

96. "Soviet Proposes UN Space Talks," *New York Times*, August 17, 1985, available at https://www.nytimes.com/1985/08/17/world/soviet-proposes-un-space-talks.html.

97. CIA Cable, "Druzhinin Commentary on US Decision to Begin Combat Test of Anti-Satellite Weapons in Space," August 21, 1985, CREST, CIA-RDP88-00733R000 200230041-3.

98. Memorandum of Conversation, "Shultz, Shevardnadze et al.," September 25, 1985, *FRUS*, 1981–1988, Volume V, Soviet Union, March 1985–October 1986.

99. Siddiqi, "The Soviet Co-Orbital Anti-Satellite System," 230–234.

100. Siddiqi, "The Soviet Co-Orbital Anti-Satellite System," 235.

101. Bateman, "Mutually Assured Surveillance at Risk," 12.

102. Richard Matlock email to the author, October 11, 2020.

103. Reagan–Gorbachev meeting in Geneva, second plenary meeting, November 19, 1985, *FRUS*, 1981–1988, Volume V, Soviet Union, March 1985–October 1986.

104. "Reagan–Gorbachev meeting in Geneva."

105. CIA Report, "USSR: Cost of the Space Program," April 1, 1985, CREST, CIA-RDP86T00591R000200200004-7.

106. CIA Report, "USSR: Cost of the Space Program."

107. CIA Report, "USSR: Cost of the Space Program."

108. Peter Westwick, "'Space-Strike Weapons,'" 955.

109. Gray, *American Military Space Policy*, 35.

110. Reagan–Gorbachev Meeting in Geneva, Third Plenary meeting, November 20, 1985, *FRUS*, 1981–1988, Volume V, Soviet Union, March 1985–October 1986.

111. Lettow, *Ronald Reagan and His Quest*, 118.

112. FitzGerald, *Way Out There in the Blue*, 305

113. Adelman, *Reagan at Reykjavik*, 94.

114. Points on Sharing SDI Technology with the Soviets, undated, *FRUS*, 1981–1988, Volume V, Soviet Union, March 1985–October 1986.

115. Letter from Keyworth to Reagan, December 17, 1985, CREST, CIA-RDP88G0 1117R000903020003-4; Formal Study on Sharing SDI: Memorandum for Casey on Sharing SDI, May 9, 1986, CREST, CIA-RDP97M00248R000500220019-2.

116. Letter from Casey to Weinberger on Sharing SDI, November 5, 1985, CREST, Doc_0005433431.

117. See "Inviting the Allies to Participate" in chapter 4 for details about Western European views on American COCOM policies.

118. Reagan–Gorbachev Meeting in Geneva, Third Plenary Meeting, November 20, 1985, *FRUS*, 1981–1988, Volume V, Soviet Union, March 1985–October 1986.

119. Miles, *Engaging the Evil Empire*, 126.

120. Memorandum from Nitze to Shultz, "NST Issues," November 26, 1985, *FRUS*, 1981–1988, Volume V, Soviet Union, March 1985–October 1986.

121. Letter from Reagan to Gorbachev, November 28, 1985, *FRUS*, 1981–1988, Volume V, Soviet Union, March 1985–October 1986.

122. Letter from Gorbachev to Reagan, December 24, 1985, *FRUS*, 1981–1988, Volume V, Soviet Union, March 1985–October 1986.

123. PREM 19/20168, "Meeting with Abrahamson and Thatcher," 14 July 1986, TNA.

124. Fred Hiatt, "Space Launch Needs of SDI Are Estimated," *Washington Post*, June 27, 1986, available at https://www.washingtonpost.com/archive/politics/1986/06/27/spa ce-launch-needs-of-sdi-are-estimated/a9b71961-0f28-466f-a058-bf606d53c5a6/.

125. Hiatt, "Space Launch Needs of SDI Are Estimated," 1986.

126. National Security Council Meeting, "Program Briefing on SDI," July 1, 1986, available at https://www.thereaganfiles.com/19860701-nsc-132-sdi.pdf.

127. Missile Defense Agency, "Historical Funding for MDA FY85-17," available at https://www.mda.mil/global/documents/pdf/FY17_histfunds.pdf.

128. Michael Griffin and Michael Rendine, "Delta 180/Vector Sum: The First Powered Space Intercept," AIAA 26th Aerospace Sciences Meeting, January 11–14, 1988.

129. John Dassoulas and Michael Griffin, "The Creation of the Delta 180 Program and Its Follow-Ons," *Johns Hopkins APL Technical Digest*, Volume 11, Numbers 1 and 2 (1990), 88, available at https://www.jhuapl.edu/Content/techdigest/pdf/V11-N1-2/11 -01-Dassoulas.pdf.

130. Michael Griffin interview with the author, October 30, 2020. In late September 1986, the US and the Soviet Union were still solidifying details for a superpower summit, including settling on the location. On September 30, the US and the USSR announced that a summit would take place October 10–12 in Reykjavik. Consequently, Abrahamson could not have been certain prior to the Delta 180 test that the summit would indeed take place in the next few weeks. Nevertheless, even if the quote from Abrahamson about "gun camera film on the table at Reykjavik" is apocryphal, it certainly was in line with the general's belief that the US needed to demonstrate SDI technologies to maintain resolve in the face of Soviet advances to kill SDI in the arms control negotiations. The test also gave Reagan even more confidence that the program was progressing. For details on American–Soviet discussions on the location for the summit, see Memorandum of Conversation, "Shultz and Shevardnadze," September 20, 1986, *FRUS*, 1981–1988, Volume V, Soviet Union, March 1985–October 1986.

131. SNIE 11-9-86, "Gorbachev's Policy Toward the United States, 1986–1988," September 12, 1986, *FRUS*, 1981–1988, Volume V, Soviet Union, March 1985–October 1986.

132. Memorandum of Conversation, "Shultz–Shevardnadze Plenary on Arms Control," September 19, 1986, *FRUS*, 1981–1988, Volume V, Soviet Union, March 1985–October 1986.

133. Memorandum from Poindexter to Reagan, "NST Experts Meetings," September 5–6, September 9, 1986, *FRUS*, 1981–1988, Volume V, Soviet Union, March 1985–October 1986.

134. Memorandum from Poindexter to Reagan, "NST Experts Meetings."

135. Memorandum of Conversation (10:40am–12:30pm), "Reagan, Gorbachev et al.," October 11, 1986, *FRUS*, 1981–1988, Volume V, Soviet Union, March 1985–October 1986.

136. Shultz, *Turmoil and Triumph*, 760.

137. Shultz, *Turmoil and Triumph*.

138. Memorandum of Conversation (3:30pm–5:40pm), "Reagan, Gorbachev et al.," October 11, 1986, *FRUS*, 1981–1988, Volume V, Soviet Union, March 1985–October 1986.

139. Memorandum of Conversation (3:30pm–5:40pm), "Reagan, Gorbachev et al."

140. Memorandum of Conversation (8:00pm–4:00am), "Ridgway, Bessmertnykh et al.," October 11–12, 1986, *FRUS*, 1981–1988, Volume V, Soviet Union, March 1985–October 1986.

141. Memorandum of Conversation (3:25pm–4:30pm and 5:30pm–6:50pm), "Reagan, Gorbachev et al.," October 12, 1986, *FRUS*, 1981–1988, Volume V, Soviet Union, March 1985–October 1986.

142. Charles, "The Game Changer," 222.

143. Adam Clymer, "Summit Aftermath: What The Public Thinks; First Reaction: Poll Shows Arms-Control Optimism and Support for Reagan," *New York Times*, October 16, 1986, available at https://www.nytimes.com/1986/10/16/world/summit-aftermath-what-public-thinks-first-reaction-poll-shows-arms-control.html.

144. Thatcher–Reagan (record of conversation), Executive Secretariat, NSC: System File, 8607413, October 86, RRPL.

145. Poindexter Memo for President Reagan, October 17, 1986, RAC Box 3 (Alton Keel Files), RRPL.

146. Rhodes, *Arsenals of Folly*, 272.

147. Rhodes, *Arsenals of Folly*.

148. Record of Conversation, "Akhromeyev, Brown, Vance, Kissinger, and Jones," February 4, 1987, Wilson Center Digital Archive, available at https://www.wilsoncenter.org/sites/default/files/media/uploads/documents/Euromissiles_Reader_PartIV.pdf.

149. Charles, "The Game Changer," 240.

150. Krepinevich and Watts, *The Last Warrior*, 193–227.

151. Anatoly Chernyaev, "Notes from Politburo Session," October 30, 1986, in The Reykjavik File.

152. Chernyaev, "Notes from Politburo Session."

153. Baucom, "The Rise and Fall of Brilliant Pebbles."

154. Baucom, "The Rise and Fall of Brilliant Pebbles," 146.

155. Strategic Defense Initiative Organization, "Report to Congress on The Strategic Defense System Architecture," December 17, 1987, 18, US National Archives and Records Administration, available at https://www.archives.gov/files/declassification/iscap /pdf/2009-033-doc1.pdf. Because SDI would have required the placement of satellites into geosynchronous orbit to track ballistic missiles, the program could have incentivized the USSR to place greater emphasis on the development of ASATs that could reach higher orbits. Soviet investment in high-altitude ASATs could have threatened US early warning satellites that were used to detect ICBMs, which could have led to more rapid crisis escalation. An attack on an early warning satellite could have been interpreted by US officials as a pretext for a nuclear first strike.

156. Kenneth Adelman email to the author, March 8, 2020.

157. From Kataev: Vipiska iz protokola No. 66 zasedaniya Poliburo TsK KPSS ot 19 Maya 1987 goda, Box 5, Kataev Archive.

158. SNIE, "Soviet Actions to Counter the Strategic Defense Initiative," February 1, 1986, CREST, CIA-RDP09T00367R000300070001-8.

159. Brendan M. Greeley, "SDI Finds Staged Deployment Would Inhibit Soviet Attack," *Aviation Week and Space Technology*, April 27, 1987.

160. PREM 19/2068, Prime Minister's Meeting with General Abrahamson, March 14, 1987, TNA.

161. PREM 19/2068, "Prime Minister's Meeting with General Abrahamson."

162. According to Charles Powell, Thatcher was "entranced by the [Abrahamson] briefings," and as a result, "she grew steadily more favorable to [SDI]." Consequently, Abrahamson "pretty well had a regular pass to Number 10 Downing Street for several years, to come and tell us about every latest development." For details, see Michael D. Kandiah and Gillian Staerck, ed., "The British Response to SDI," 2005, 32, available at https://www.kcl.ac.uk/sspp/assets/icbh-witness/sdi.pdf.

163. Baucom, "The Rise and Fall of Brilliant Pebbles," 148.

164. R. Jeffrey Smith, "SDI Decision 'May Be Nearing,'" *Washington Post*, January 23, 1987, available at https://www.washingtonpost.com/archive/politics/1987/01/23/sdi -decision-may-be-nearing/086ebe04-5cb5-4779-9d03-26fd0074cad7/.

165. Smith, "SDI Decision 'May Be Nearing.'"

166. PREM 19/3056, "No. 10 Conversation Record (MT-Italian PM Craxi)," February 11, 1987, TNA.

167. PREM 20/1968, "Armstrong Minute to Powell, The Strategic Defence Initiative and the ABM Treaty," April 2, 1987, TNA.

168. PREM 20/1968, "The Strategic Defence Initiative and the ABM Treaty."

169. Podvig, "Did Star Wars Help End the Cold War?," 13.

170. For more information on Skif-DM, see Pavel Podvig, "Protivoraketnaya oborona kak faktor strategicheskik vzaimootnoshenii SSSR/Rossii i SShA v 1945–2003 gg" (Diss., Moskva, 2004), 126–127.

171. Podvig, "Protivoraketnaya oborona kak faktor strategicheskik vzaimootnoshenii SSSR/Rossii i SShA v 1945–2003 gg."

172. Ronald Reagan, NSDD-258, "Anti-Satellite (ASAT Program)," February 6, 1987, available at https://aerospace.csis.org/wp-content/uploads/2019/02/NSDD-258-Anti -Satellite-Program.pdf.

173. Ronald Reagan, NSDD-258, "Anti-Satellite (ASAT Program)."

174. Aleksandr Yakovlev, Memorandum for Gorbachev, "Toward an Analysis of the Fact of the Visit of Prominent American Political Leaders to the USSR (Kissinger, Vance, Kirkpatrick, Brown, and others)," February 25, 1987, in the INF Treaty, 1987–2019, National Security Archive.

175. Charles, "The Game Changer," 345.

176. Aleksandr Savel'yev and Nikolay Detinov, *The Big Five: Arms Control Decision-Making in the Soviet Union* (New York: Praeger, 1995), 86.

177. Memorandum of Conversation, "Shultz, Shevardnadze et al.," April 15, 1987, *FRUS*, 1981–1988, Volume XI, START I.

178. Memorandum for Carlucci, undated, document 202, *FRUS*, 1981–1988, Volume XI, START I.

179. Ellis, "Reds in Space," 146.

180. Ronald Reagan, "The US Anti-Satellite (ASAT) Program: A Key Element in the National Strategy of Deterrence," May 11, 1987, available at https://fas.org/spp/mil itary/program/asat/reag87.html.

181. David C. Morrison, "Year of Decision for ASAT Program," *Science*, June 19, 1987, 236, no. 4808, available at https://science.sciencemag.org/content/236/4808 /1512.

182. Brian Weeden, "Through A Glass, Darkly: Chinese, American, and Russian Anti-Satellite Testing in Space," *The Space Review*, March 17, 2014, available at https://www .thespacereview.com/article/2473/1. The MHV was also not an especially useful system from the standpoint of military utility. It could only reach 30 percent of satellites on the joint chiefs of staff's target list; see Nancy Gallagher, "Towards a Reconsideration of the Rules for Space Security," undated paper, 12, available at https://cissm .umd.edu/sites/default/files/2019-08/space_rules2006.pdf.

183. Memorandum from Nitze and Timbie to Shultz, "START," July 7, 1987, *FRUS*, 1981–1988, Volume XI, START I.

184. Minutes of NSPG Principals Meeting, "Arms Control—Shultz Meeting in Moscow," April 3, 1987, *FRUS*, 1981–1988, Volume XI, START I.

185. "Paper Prepared by the Arms Control Support Group," November 19, 1987, *FRUS*, 1981–1988, Volume XI, START I. Advisors to the president discussed creating a "test range in space," in which a "limited number of test vehicles could function within that specifically delineated orbit and they could be observed by both parties." For details, see Memorandum from Kampelman to Shultz, October 28, 1987, *FRUS*, 1981–1988, Volume XI, START I.

CHAPTER 6

1. Ricky B. Kelly, "Centralized Control of Space: The Use of Space Forces by a Joint Force Commander" (MA thesis, School of Advanced Air Power Studies, 1993), 1, available at https://media.defense.gov/2017/Dec/29/2001861991/-1/-1/0/T_KELLY_CENTRALIZED_CONTOL_OF_SPACE.PDF.

2. Gordon R. Mitchell, "Placebo Defense: Operation Desert Mirage? The Rhetoric of Patriot Missile Accuracy in the 1991 Persian Gulf War," *Quarterly Journal of Speech* 86, no. 2 (2000), available at http://www.pitt.edu/~gordonm/JPubs/PatriotQJS.pdf.

3. George H. W. Bush, "Address Before a Joint Session of the Congress on the Persian Gulf Crisis and the Federal Budget Deficit," September 11, 1990, available at https://bush41library.tamu.edu/archives/public-papers/2217.

4. "National Military Strategy of the United States," January 1, 1992, available at https://history.defense.gov/Portals/70/Documents/nms/nms1992.pdf?ver=AsfWYUHa-HtcvnGGAuWXAg%3d%3d; "National Security Strategy of the United States," August 1991, available at https://history.defense.gov/Portals/70/Documents/nss/nss1991.pdf?ver=3sIpLiQwmknO-RplyPeAHw%3d%3d.

5. Peter Westwick, *Into the Black: JPL and the American Space Program 1976–2004* (New Haven, CT: Yale University Press, 2007), 179.

6. Memorandum on Rationale for Space Policy Revision, June 19, 1987, CREST, CIA-RDP92B00181R001701640053-4.

7. Memorandum from Lawrence Gershwin to Deputy Director of Central Intelligence, "NSC Space Policy Review," CREST, CIA-RDP92B00181R001701640052-5.

8. Intelligence Report, "Probable SL-16 Space Launch Vehicle Program at Plesetsk Missile and Space Test Center, USSR," September 1987, CREST, CIA-RDP91T01115R000400590001-7.

9. NIE, "Soviet Forces Capabilities for Strategic Nuclear Conflict Through the Late 1990s," July 1, 1987, CREST, CIA-RDP09T00367R000200280001-6.

10. Martin Sieff, "Russia's New Rocket Could Tilt Space Power Balance to Soviets," *Washington Times*, May 18, 1987, CREST, CIA-RDP90-00965R000706130001-7.

11. Memorandum on Rationale for Space Policy Revision, June 19, 1987, CREST, CIA-RDP92B00181R001701640053-4.

12. John Rhea, "The Scaled Down Look of Star Wars," *Air Force Magazine*, October 1989, available at https://www.airforcemag.com/PDF/MagazineArchive/Documents/1989 /October%201989/1089StarWars.pdf.

13. Frank Carlucci official biography, OSD History, available at https://history.defense .gov/Multimedia/Biographies/Article-View/Article/571285/frank-c-carlucci/. In the bill that Reagan vetoed, Congress directed that no more than $85 million be spent on the development of space-based interceptors; see Colin Norman, "Congress Reins in, Redirects SDI," *Science* 241, no. 4862 (1988).

14. Ronald Reagan, "Remarks on the Veto of the National Defense Authorization Act, Fiscal Year 1989," August 3, 1988, available at https://www.reaganlibrary.gov /archives/speech/remarks-veto-national-defense-authorization-act-fiscal-year-1989 -and-question-and.

15. William J. Broad, "US Promoting Offensive Role for 'Star Wars,'" *New York Times*, November 27, 1988, available at https://www.nytimes.com/1988/11/27/world/us -promoting-offensive-role-for-star-wars.html.

16. Broad, "US Promoting Offensive Role for 'Star Wars,'" 1988.

17. FCO 46/6939, UK Embassy to FCO, "US Anti-Satellite System," January 18, 1989, TNA.

18. FCO/46/6939, UK Embassy DC to FCO, "US Anti-Satellite Systems," January 9, 1988, TNA.

19. Presidential Directive on National Space Policy, February 11, 1988, available at https://aerospace.org/sites/default/files/policy_archives/National%20Space%20 Policy%20Feb88.pdf.

20. FCO/46/6939, UK Embassy DC to FCO, "US Anti-Satellite Systems."

21. FCO 46/6939, UK Embassy to FCO, "US Anti-Satellite System."

22. Stockholm International Peace Research Institute, *SIPRI Yearbook 1991* (Oxford: Oxford University Press, 1991), 59, available at https://www.sipri.org/sites/default /files/SIPRI%20Yearbook%201991.pdf.

23. Charles Monfort, "ASATs: Star Wars on the Cheap," *Bulletin of the Atomic Scientist* 45 (1989).

24. George H. W. Bush, "Address Accepting the Presidential Nomination at the Repub-lican National Convention in New Orleans," August 18, 1988, available at https:// www.presidency.ucsb.edu/documents/address-accepting-the-presidential-nomination -the-republican-national-convention-new.

25. Gerald M. Boyd, "Bush Is Cautious about Deploying Missile Defense," August 26, 1988, *New York Times*, available at https://www.nytimes.com/1988/08/26/us/bush-is -cautious-about-deploying-a-missile-defense.html.

26. William F. Buckley Jr., "Bush and SDI What's Going On?" August 30, 1988, *Washington Post*, available at https://www.washingtonpost.com/archive/opinions/1988/08/30/bush-and-sdi-whats-going-on/de047e59-69eb-411f-abe9-e178a0660e54/.

27. Bill Peterson, "Bush Acts to Reassure Conservatives on SDI," August 31, 1988, *Washington Post*, available at https://www.washingtonpost.com/archive/politics/1988/08/31/bush-acts-to-reassure-conservatives-on-sdi/09e9b039-6239-4881-b0f5-b15b58c55be2/.

28. Peterson, "Bush Acts to Reassure Conservatives on SDI."

29. PREM 19/2614, "Implications for the United Kingdom of Strategic Defences," November 11, 1988, TNA.

30. PREM 19/2614, "Minute from Powell to Thatcher, Strategic Defence Initiative," December 23, 1988, TNA. Powell further said in this minute that "at some stage [SDIO must] move from theory and the laboratory to practical application," suggesting that modifications of to the ABM Treaty would be required and not necessarily undesirable.

31. PREM 19/2614, "Strategic Defence Initiative," January 16, 1989, TNA.

32. PREM 19/2890, "No. 10 Conversation Record, PM–Bush," June 1, 1989, TNA.

33. George H. W. Bush, "Address on Administration Goals Before a Joint Session of Congress," February 9, 1989, available at https://www.presidency.ucsb.edu/documents/address-administration-goals-before-joint-session-congress.

34. Dan Quayle, "Major Policy Address on SDI," June 29, 1989, available at https://www.c-span.org/video/?8199-1/major-policy-address-sdi.

35. Baucom, "The Rise and Fall of Brilliant Pebbles," 148.

36. Taylor Dinerman, "Missile Defense, RLVs, and the Future of American Spacepower," *The Space Review*, December 1, 2003, available at https://www.thespacereview.com/article/66/1.

37. Baucom, "The Rise and Fall of Brilliant Pebbles," 146.

38. PREM 19/2614, Annex A of "Briefing for Prime Minister's Meeting with Lt Gen Abrahamson," February 4, 1988, TNA.

39. PREM 19/2614, "Briefing for Prime Minister's Meeting with Lt Gen Abrahamson."

40. PREM 19/2614, "Implications for the United Kingdom of Strategic Defences," November 11, 1988, TNA.

41. PREM 19/2614, Annex A of "Briefing for Prime Minister's Meeting with Lt Gen Abrahamson."

42. Harry Anderson and John Barry, "A Start on Star Wars," *Newsweek*, February 8, 1988.

43. PREM 19/2614, Annex A of "Briefing for Prime Minister's Meeting with Lt Gen Abrahamson."

44. "Report of the Defense Science Board Task Force Subgroup on Strategic Air Defense," May 19, 1988, 3, available at https://apps.dtic.mil/dtic/tr/fulltext/u2/a200164.pdf.

45. "Report of the Defense Science Board Task Force Subgroup on Strategic Air Defense."

46. Slayton, *Arguments that Count*, 195.

47. Slayton, *Arguments that Count*.

48. Office of Technology Assessment, "SDI: Technology, Survivability, and Software," US Government Printing Office, Washington, DC, May 1988, 3, available at https://ota.fas.org/reports/8837.pdf.

49. Office of Technology Assessment, "SDI: Technology, Survivability, and Software," 3.

50. Office of Technology Assessment, "SDI: Technology, Survivability, and Software," 4–5. For more details, see Slayton, *Arguments that Count*, 196.

51. PREM 19/2614, "Thatcher Meeting with Abrahamson," January 12, 1989, TNA.

52. Charles Powell interview with the author via telephone, February 4, 2020.

53. Baucom, "The Rise and Fall of Brilliant Pebbles," 153.

54. Robert Mackay, "$20 Million More for Anti-Satellite Weapon," April 27, 1989, *UPI*, available at https://www.upi.com/Archives/1989/04/27/20-Million-More-for-Anti-Satelite-Weapon/2091609652800/.

55. Congressional Record—Senate, July 17, 1989, available at https://www.govinfo.gov/content/pkg/GPO-CRECB-1989-pt11/pdf/GPO-CRECB-1989-pt11-2-1.pdf.

56. Congressional Record—Senate.

57. Roger C. Hunter, "A US ASAT Policy for a Multipolar World" (MA thesis, School of Advanced Air Power Studies, 1992), available at https://apps.dtic.mil/sti/pdfs/ADA425653.pdf.

58. Hunter, "A US ASAT Policy for a Multipolar World," 36.

59. Stockholm International Peace Research Institute, *SIPRI Yearbook 1990* (Oxford: Oxford University Press, 1990), 68, available at https://www.sipri.org/sites/default/files/SIPRI%20Yearbook%201990.pdf.

60. See "Détente Unravels and a New (Limited) Space Competition Begins" in chapter 1.

61. FCO 46/6939 UK Embassy to FCO, "ASATs," November 22, 1989, TNA.

62. "Letter from Ronald Peterson to Senator John Chafee," July 28, 1989, CREST, CIA-RDP92M00732R001000080006-3.

63. "Letter from Ronald Peterson to Senator John Chafee."

64. Joseph S. Nye and James A. Schear, ed., *Seeking Stability in Space: Anti-Satellite Weapons the Evolving Space Regime* (Lanham, MD: University Press of America, 1987).

65. Daniel Quayle, "Remarks of the Vice President to the Navy League of the United States Sea-Air-Space Exposition," Sheraton Hotel, Washington, DC, March 23, 1989, private papers of Joseph DeSutter.

66. R. Jeffrey Smith, "Year of Lobbying Turned 'Brilliant Pebbles' into Top SDI Plan," *Washington Post*, April 26, 1989, available at https://www.washingtonpost.com/archive /politics/1989/04/26/year-of-lobbying-turned-brilliant-pebbles-into-top-sdi-plan /2f70b1e9-0b5a-408d-a5d0-19167462228a/.

67. Michael R. Gordon, "Nunn Seeks Shield for Missiles Fired in Error," *New York Times*, January 20, 1988, available at https://www.nytimes.com/1988/01/20/world /nunn-seeks-shield-for-missiles-fired-in-error.html.

68. George C. Wilson, "SDI was 'Oversold,' Cheney Says," *Washington Post*, March 29, 1989, available at https://www.washingtonpost.com/archive/politics/1989/03/29/sdi -was-oversold-cheney-says/920e44ca-fa73-4b5d-8082-2db2447bdb67/.

69. James Gerstenzang, "'Star Wars' Cuts Confirmed by White House," *LA Times*, September 8, 1989, available at https://www.latimes.com/archives/la-xpm-1989-09 -08-mn-1839-story.html.

70. Gerstenzang, "Star Wars' Cuts Confirmed by White House."

71. Gerstenzang, "Star Wars' Cuts Confirmed by White House."

72. Anderson and Barry, "A Start on Star Wars."

73. NSD-14 remains classified and fully redacted. Baucom, in his official capacity as SDIO historian, was able to see the document and quote from sections of it. See Baucom, "The Rise and Fall of Brilliant Pebbles," 154–155.

74. Baucom, "The Rise and Fall of Brilliant Pebbles," 164.

75. Baucom, "The Rise and Fall of Brilliant Pebbles," 156.

76. Baucom, "The Rise and Fall of Brilliant Pebbles."

77. Baucom, "The Rise and Fall of Brilliant Pebbles."

78. PREM 19/3650: "SDI Programme Status and UK Participation," prepared by SDI Participation Office, October 16, 1990.

79. Slayton, *Arguments that Count*, 174. It is not clear from the archival record currently available if software engineers were involved in the JASONs Brilliant Pebbles study.

80. Baucom, "The Rise and Fall of Brilliant Pebbles," 157.

81. SDIO did conduct a countermeasures evaluation, which also remains classified. According to Baucom, "the general conclusion of this study was that Brilliant Pebbles would be subject to the same countermeasures faced by all space-based elements in the SDI architecture, but faced no special problems in this area. The study's major recommendation was that survivability features should be built into the BP system." See Baucom, "The Rise and Fall of Brilliant Pebbles," 159.

82. FCO 46/7586, "Minute from Clarke to Whitaker, SDI," January 10, 1990, TNA.

83. Stanley Orman interview with the author, July 31, 2020.

84. FCO 46/7586, "Minute from Clarke to Whitaker, SDI," January 10, 1990, TNA.

85. Mackenzie, *Inventing Accuracy*, 8–9.

86. Even in early 1990, British officials were still unsure of Bush's position on the deployment of a strategic defense system. A January 1990 minute said, "In all of this, it is still hard to guage [sic] the intentions of the President. His mild reproach to Congress in response to the first absolute reduction in the SDIO budget in FY90 was rightly characterized as 'hollow' by officials and SDI boosters alike. He says little, if anything, about the program in public." For details, see FCO 46/7586, "Minute from Clarke to Whitaker, SDI," January 10, 1990, TNA. Because many of the relevant documents (e.g., minutes from national security council meetings) from the Bush administration are still classified, we do not have significant insight into the nature of high-level discussions over SDI and the military uses of space during this period.

87. Memorandum for National Space Council, "National Space Policy Review," June 2, 1989, Aerospace Corporation Digital Archive, available at https://csps.aerospace.org/sites/default/files/2021-08/National%20Space%20Policy%20Review%20Jun89.pdf.

88. A point paper for an October 26, 1989, Space Council Meeting stated, "The Space Council process thus far included only minimal interagency participation. More interagency participation is planned but the relatively short time frame may negatively impact on the quality of the information available to support the debate." For details, see "Point Paper for National Space Council Meeting," October 26, 1989, CREST, CIA-RDP91B01306R000300070006-6.

89. Missile Defense Agency, "Historical Funding for MDA."

90. Eliot Brenner, "Cheney: SDI in Danger," *UPI*, October 12, 1989, available at https://www.upi.com/Archives/1989/10/12/Cheney-SDI-in-danger/7068624168000/?spt=su.

91. George H. W. Bush, "Statement on Signing the National Defense Authorization Act for Fiscal Years 1990 and 1991," November 29, 1989, available at https://www.presidency.ucsb.edu/documents/statement-signing-the-national-defense-authorization-act-for-fiscal-years-1990-and-1991.

92. Thomas L. Friedman, "US–Soviet Talks End with Progress on Arms Control," *New York Times*, September 24, 1989, available at https://www.nytimes.com/1989/09/24/world/us-soviet-talks-end-with-progress-on-arms-control.html.

93. Michael Levi, "Strategic Arms Reduction Treaty (START I) Chronology," Federation of American Scientists, undated, available at https://fas.org/nuke/control/start1/chron.htm.

94. "Vzaimosvyaz' mezhdu strategicheskimi nastupatel'nymi i oboronitel'nymi vooruzjenyami, ik vliyanie na strategicheskyu stabil'nost', amerikanskoe predlozhenie o

neyadernykh sistemakh PRO" undated, Box 8, Arms Control, Kataev Archive; "Letter from A. Obukhov to Shevernadze," undated, Box 8, Arms Control, Kataev Archive.

95. "Vzaimosvyaz' mezhdu strategicheskimi nastupatel'nymi i oboronitel'nymi vooruzjenyami, ik vliyanie na strategicheskyu stabil'nost', amerikanskoe predlozhenie o neyadernykh sistemakh PRO."

96. Delo Operativnoi Perepiski No. 2113, "Po Voenno-Strategicheskoi Problematike," 23 Noyabr' 1990, Tom. No. 2, Lithuanian Special Archives.

97. "Po Voenno-Strategicheskoi Problematike."

98. Andrew Rosenthal, "'Star Wars' Funds Cut in the Senate," *New York Times*, September 27, 1989, available at https://www.nytimes.com/1989/09/27/world/star-wars-funds-cut-in-the-senate.html.

99. In January 1990, General Monahan received approval from both the Undersecretary of Defense for Acquisitions and Chairman of the DAB (John Betti) and Cheney to move forward with Brilliant Pebbles outside of the DAB milestone process, at least for the time being. The Pentagon intended to move Brilliant Pebbles through a four-phase process: concept definition, pre-full-scale development, full-scale development, and production. An indication of Brilliant Pebbles being handled more loosely than traditional Department of Defense acquisition efforts was Monahan's invitation for contractors, rather than the SDIO, to come up with detailed specifications for the system. See FCO 46/7586, "Minute from Clarke to Whitaker, SDI," March 7, 1990, TNA. Baucom observes that "on 16 January 1990, General Monahan discussed the SDI program with Secretary of Defense Richard Cheney, who had advised Monahan that he expected the General to proceed with the program. Monahan interpreted these instructions as meaning that a DAB was not required for approval of his acquisition strategy for Brilliant Pebbles." See Baucom, "The Rise and Fall of Brilliant Pebbles," 163. Monahan explained while giving congressional testimony that Brilliant Pebbles had not been entered into the DAB process because the Strategic Systems Committee had conducted a review of it on November 27, 1989, and determined that "a DAB review was . . . unnecessary as the streamlined acquisition approach adopted for Brilliant Pebbles could be executed within existing program funding and schedule constraints." See Department of Defense Appropriations for 1991, Hearings Before a Subcommittee of the Committee on Appropriations, One Hundred First Congress, Second Session, Government Printing Office, Washington, DC, 1990, 657.

100. Baucom, "The Rise and Fall of Brilliant Pebbles," 163.

101. Baucom, "The Rise and Fall of Brilliant Pebbles."

102. FCO 46/7586, "Briefing by General Monahan and Lowell Wood at the Pentagon," February 9, 1990, TNA.

103. FCO 46/7586, "Briefing by General Monahan and Lowell Wood at the Pentagon."

104. John Whitehead, "Propulsion Engineering Study for Small-Scale Mars Missions," Lawrence Livermore National Laboratory, September 12, 1995, available at https://www.osti.gov/servlets/purl/6943660.

105. FCO 46/7586, "Briefing by General Monahan and Lowell Wood at the Pentagon," February 9, 1990, TNA.

106. Roger D. Speed, "ASATs vs. Brilliant Pebbles," Lawrence Livermore National Laboratory, March 1990, 27, available at https://www.osti.gov/servlets/purl/6943660.

107. John Whitehead, "Propulsion Engineering Study for Small-Scale Mars Missions," Lawrence Livermore National Laboratory, September 12, 1995, 27, available at https://www.osti.gov/servlets/purl/6943660.

108. John Whitehead, "Propulsion Engineering Study for Small-Scale Mars Missions."

109. "Dresden, 1989: The Helmut Kohl Speech Showing German Reunification Was the Way," *DW*, undated, available at https://www.dw.com/en/dresden-1989-the-helmut-kohl-speech-showing-german-reunification-was-the-way/a-51735052.

110. Sayle, *Enduring Alliance*, 215.

111. Sayle, *Enduring Alliance*.

112. R. Jeffrey Smith, "Pentagon Increased SDI Push," *Washington Post*, February 18, 1990, available at https://www.washingtonpost.com/archive/politics/1990/02/18/pentagon-increases-sdi-push/3764c083-31c2-4308-9fe2-852d73709f80/.

113. Smith, "Pentagon Increased SDI Push," 1990.

114. FCO 46/7586, "Minute from Clarke to Whitaker, SDI: The Cooper Report," April 4, 1990, TNA.

115. Anne H. Cahn, Martha C. Little, and Stephen Daggett, "Nunn and Contractors Sell ALPS," *Bulletin of the Atomic Scientists* 44, no. 5 (1988).

116. PREM 19/3650, "PM Meeting with Amb. Cooper," October 22, 1990, TNA.

117. PREM 19/3650, "PM meeting with Amb. Cooper."

118. FCO 46/7586, "Transcript of Cheney Speech at the SDI Symposium of the American Defense Preparedness Association," March 19, 1990, TNA.

119. Memorandum for Director SDIO and Chairman, Strategic Systems Committee, Acquisition Decision Memorandum for Strategic Defense Initiative Program, June 19, 1990, available at http://highfrontier.org/wp-content/uploads/2016/08/BP-1990-DAB-Approval.pdf.

120. FCO 46/7586, "Clarke to Whitaker, SDI," July 13, 1990, TNA.

121. Government Accounting Office, "Strategic Defense Initiative: Some Claims Overstated for Early Flight Tests of Interceptors," GAO/NSIAD-92-282, September 1992, 32, available at https://www.gao.gov/assets/nsiad-92-282.pdf.

122. "State of the Union; Transcript of President's State of the Union Message to Nation," January 30, 1991, *New York Times*, available at https://www.nytimes.com/1991/01/30/us/state-union-transcript-president-s-state-union-message-nation.html.

123. "State of the Union."

124. George H. W. Bush, "State of the Union Address," January 29, 1991, available at https://millercenter.org/the-presidency/presidential-speeches/january-29-1991-state-union-address.

125. PREM 19/3650, "GPALS Architecture Annex B," undated, TNA.

126. PREM 19/3650, "Joint MoD/FCO Minute on SDI," July 19, 1991, TNA.

127. Graham Spinardi, "Technical Controversy and Ballistic Missile Defence: Disputing Epistemic Authority in the Development of Hit-to-Kill Technology," *Science as Culture* 23, no. 1 (2013): 27.

128. Spinardi, "Technical Controversy and Ballistic Missile Defence."

129. See "The Laser and Missile Defense Enthusiasts" in chapter 2.

130. See "The Laser and Missile Defense Enthusiasts" in chapter 2.

131. Westwick, "From the Club of Rome to Star Wars," 283.

132. Ronald Reagan, "Remarks at the Johnson Space Center," September 22, 1988, available at https://www.reaganlibrary.gov/archives/speech/remarks-johnson-space-center-houston-texas.

133. General Accounting Office, "Ballistic Missile Defense: Evolution and Current Issues," GAO/NSIAD-93-229, 29, available at https://www.gao.gov/assets/nsiad-93-229.pdf.

134. "Report of the NSC Ad Hoc Panel on Technological Evolution and Vulnerability of Space," October 1976, 1, US National Security Council Institutional Files 1974–1977, Gerald R. Ford Presidential Library.

135. Memorandum for Secretary of Defense, "Air Force Space Policy," December 21, 1988, CREST, CIA-RDP90M00551r002001250013-7.

136. William H. Rohlman, "A Political Strategy for Antisatellite Weaponry" (MA thesis, The Industrial College of the Armed Forces, 1993), 22, available at https://www.hsdl.org/?view&did=3724.

137. Stockholm International Peace Research Institute, *SIPRI Yearbook 1991*, 68.

138. Stockholm International Peace Research Institute, *SIPRI Yearbook 1991*, 58.

139. Stockholm International Peace Research Institute, *SIPRI Yearbook 1991*, 50.

140. R. Jeffrey Smith, "SDI Success Said to Be Overstated," *Washington Post*, September 16, 1992, available at https://www.washingtonpost.com/archive/politics/1992/09/16/sdi-success-said-to-be-overstated/55ae890b-8b81-48d9-8ab0-0212d313e527/.

141. Stephen C. LeSuer, "Battle for SDI Funding Pushed to Background in Debate Over Missile Defenses," *Inside the Pentagon* 7, no. 33 (1991), available at https://www .jstor.org/stable/pdf/43987358.pdf?refreqid=excelsior%3A5d9d0abf9249f5a64ea0546 2648b2347.

142. Smith, "SDI Success Said to Be Overstated."

143. LeSuer, "Battle for SDI Funding."

144. Missile Defense Agency, "Historical Funding for MDA."

145. Baucom, "The Rise and Fall of Brilliant Pebbles," 172.

146. Baucom, "The Rise and Fall of Brilliant Pebbles."

147. Department of Defense, "National Military Strategy of the United States," January 1, 1992, available at https://history.defense.gov/Portals/70/Documents/nms /nms1992.pdf?ver=AsfWYUHa-HtcvnGGAuWXAg%3d%3d.

148. Moltz, *The Politics of Space Security*, 202.

149. Stockholm International Peace Research Institute, *SIPRI Yearbook 1992* (Oxford: Oxford University Press, 1992), 135.

150. PREM 19/3650, "Minute from Wall to Cradock, SDI, December 2, 1991, TNA.

151. PREM 19/3650, "Minute from Prentice to Wall, GPALS," July 22, 1992, TNA.

152. PREM 19/3650, "Minute from Wall to Prime Minister, Anti-Ballistic Missile System," February 14, 1992, TNA.

153. PREM 19/3650, "Anti-Ballistic Missile System."

154. PREM 19/3650, "Draft of Letter from Bush," TNA. It is not clear from the archival record whether Major actually sent the letter. Regardless, it provides insight into the prime minister's mostly negative view of strategic defense deployment.

155. PREM 19/3650, "Minute to Wall, GPALS," May 5, 1992, TNA.

156. PREM 19/3650, "Minute from Prentice to Wall, GPALS," July 22, 1992, TNA.

157. PREM 19/3650, "Cable from UK Embassy in DC to FCO, GPALS," May 13, 1992, TNA.

158. PREM 19/3650, "Cable from UK Embassy in DC to FCO."

159. PREM 19/3650, "Cable from UK Embassy in DC to FCO, President Bush's Proposal for Co-operation on Limited ABM Defences," February 18, 1992, TNA.

160. Patrick E. Tyler, "US Strategy Plan Calls For Insuring No Rival Develop," *New York Times*, March 8, 1992, available at https://www.nytimes.com/1992/03/08/world /us-strategy-plan-calls-for-insuring-no-rivals-develop.html.

161. "Defense Planning Guidance, FY 1994–1999," April 16, 1992, 2, available at https://www.archives.gov/files/declassification/iscap/pdf/2008-003-docs1-12.pdf.

162. "Defense Planning Guidance, FY 1994–1999."

163. PREM 19/3650, "Anti-Ballistic Missile System: Initial Assessment," undated, TNA.

164. Baucom, "The Rise and Fall of Brilliant Pebbles," 174.

165. Baucom, "The Rise and Fall of Brilliant Pebbles," 179.

166. Baucom, "The Rise and Fall of Brilliant Pebbles," 180.

167. Government Accounting Office, "Strategic Defense Initiative: Estimates of Brilliant Pebbles' Effectiveness Are Based on Many Unproven Assumptions," March 1992, GAO/NSIAD-92-91, available at https://www.gao.gov/assets/nsiad-92-91.pdf.

168. Arms Control Association, "Arms Control and the 1992 Election," available at https://www.armscontrol.org/act/1992-09/features/arms-control-1992-election.

169. George H. W. Bush, "Bush Campaign Rally," September 13, 1992, available at https://www.c-span.org/video/?32270-1/bush-campaign-rally.

170. William Clinton, Presidential Review Directive/NSC 31, "US Policy on Ballistic Missile Defenses and the Future of the ABM Treaty," April 26, 1993, available at https://fas.org/irp/offdocs/prd/prd-31.pdf.

171. Clinton, "US Policy on Ballistic Missile Defenses and the Future of the ABM Treaty."

172. Department of Defense, "National Military Strategy of the United States," 1995, available at https://history.defense.gov/Portals/70/Documents/nms/nms1995.pdf?ver=FpT1JOUGguy83LIRFW87Ow%3d%3d.

SDI RECONSIDERED

1. For an overview of Clementine, see Stephanie Roy, "The Origin of the Smaller, Faster, Cheaper Approach in NASA's Solar System Exploration Program," *Space Policy*, vol. 14, no. 3 (1998), 161-162.

2. Lettow, *Ronald Reagan and His Quest*; Brown, *The Human Factor*.

3. Westwick, *Into the Black*, x.

4. Charles, "The Game Changer," 341.

5. For more information on astroculture, see Alexander Geppert, ed., *Imagining Outer Space* (New York: Palgrave Macmillan, 2012).

6. "The SDIO: A Security Threat to NRO Technology," August 14, 1985, CREST, CIA-RDP97M00248R000500220031-8.

7. For a good overview of Israeli involvement in SDI, see Rabinowitz, "'Arrow' Mythology Revisited."

8. Krepon, "Nitze's Strategic Concept."

9. Memorandum of Conversation, "Verification, Krasnoyarsk, NST Instructions, Defense and Space," December 6, 1987, *FRUS*, 1981–1988, Volume XI, START I.

10. Memorandum of Conversation, "The Secretary's Meeting with Gorbachev," April 14, 1987, *FRUS*, 1981–1988, Volume XI, START I.

11. Delo Operativnoi Perepiski No. 2113, "Po Voenno-Strategicheskoi Problematike, 23 Noyabr" 1990, Tom. No. 2, Lithuanian National Archives.

12. Krige, *Sharing Knowledge, Shaping Europe*, 14.

13. Stanley Meisler, "Reagan Recants 'Evil Empire' Description," *Los Angeles Times*, June 1, 1988, available at https://www.latimes.com/archives/la-xpm-1988-06-01-mn -3667-story.html.

14. Minutes of a National Security Planning Group Meeting, "Discussion of Substantive Issues for Geneva," December 17, 1984, *FRUS*, 1981–1988, Volume IV, Soviet Union, January 1983–March 1985.

15. See "A Sense of Déjà Vu" for details.

16. Minutes of NSPG Meeting, "US Options for Arms Control at the Summit," May 23, 1988, *FRUS*, 1981–1988, START I.

A SENSE OF DÉJÀ VU

1. Brian Weeden, "2007 Chinese Anti-Satellite Test Fact Sheet," November 23, 2010, Secure World Foundation, available at https://swfound.org/media/9550/chinese_asat _fact_sheet_updated_2012.pdf.

2. Will Knight, "China Dismisses 'Space Arms Race' Fears," *New Scientist*, January 19, 2007, available at https://www.newscientist.com/article/dn10990-china-dismisses -space-arms-race-fears/.

3. William J. Broad and David Sanger, "China Tests Anti-Satellite Weapon, Unnerving US," *New York Times*, January 18, 2007, available at https://www.nytimes.com/2007 /01/18/world/asia/18cnd-china.html.

4. Benjamin Rojek, "JSpOC Integral to Burnt Frost Success," undated, available at https://www.vandenberg.spaceforce.mil/News/Article-Display/Article/340906/jspoc -intergral-to-burnt-frost-success/.

5. Robert Gates, *Duty: Memoirs of a Secretary at War* (New York: Vintage, 2015), 250.

6. "US Spy Satellite Plan 'A Cover,'" February 17, 2008, available at http://news.bbc.co .uk/2/hi/americas/7248995.stm.

7. Carin Zissis, "China's Anti-Satellite Test," Council on Foreign Relations, February 22, 2007, available at https://www.cfr.org/backgrounder/chinas-anti-satellite -test; "US Leads in Preparing for War in Space," *New York Times*, March 9, 2008,

available at https://www.nytimes.com/2008/03/09/world/americas/09iht-space.4
.10846000.html.

8. "Report of the Commission to Assess United States National Security Space Management and Organization" (Washington, DC: Commission to Assess United States National Security Space, January 11, 2001), available at https://aerospace.csis.org/wp-content/uploads/2018/09/RumsfeldCommission.pdf.

9. "Report of the Commission to Assess United States National Security Space Management and Organization," viii.

10. "Report of the Commission to Assess United States National Security Space Management and Organization," vii–xxxv.

11. George W. Bush, "State of the Union Address," January 29, 2002, available at https://georgewbush-whitehouse.archives.gov/news/releases/2002/01/20020129-11.html.

12. "Remarks by President Bill Clinton on National Missile Defense," September 1, 2000, available at https://www.armscontrol.org/act/2000-09/remarks-president-bill-clinton-national-missile-defense.

13. Ben Shapiro, "Barack Obama's Anti-Military Problem," *Townhall*, May 28, 2008, available at http://townhall.com/columnists/benshapiro/2008/05/28/barack_obamas_anti-military_problem. In 2002, the Ballistic Missile Defense Organization became the Missile Defense Agency.

14. Missile Defense Agency, "Ground-Based Midcourse Defense (GMD)," available at https://www.mda.mil/system/gmd.html.

15. US Government, "National Security Space Strategy," January 2011, available at https://www.hsdl.org/?view&did=10828.

16. James Clapper, "Statement for the Record on the World Threat Assessment of the US Intelligence Community for the Senate Committee on Armed Services," March 10, 2011, available at https://www.dni.gov/files/documents/Newsroom/Testimonies/2011 0310_testimony_clapper.pdf.

17. James Clapper, "Statement for the Record on the World Threat Assessment of the US Intelligence Community for the Senate Committee on Armed Services," February 9, 2016, available at https://www.dni.gov/files/documents/SASC_Unclassified_2016_ATA_SFR_FINAL.pdf.

18. Alexis A. Blanc, Nathan Beauchamp-Mustafaga, Khrystyna Holynska, M. Scott Bond, and Stephen Flanagan, *Chinese and Russian Perceptions of and Responses to U.S. Military Activities in the Space Domain* (Santa Monica, CA: RAND, 2022), 9.

19. Blanc et al., *Chinese and Russian Perceptions*, 11.

20. Blanc et al., *Chinese and Russian Perceptions*, 13.

21. The National Space Council is a body within the Executive Office of the President and advises the president on space policy and strategy. George H. W. Bush established

it in 1989. It was dissolved in 1993, and then Donald Trump reestablished it in 2017. For details, see Sandra Erwin, "Biden administration to continue the National Space Council," *SpaceNews*, March 29, 2021, available at https://spacenews.com/biden-administration-to-continue-the-national-space-council/.

22. Helene Cooper, "Trump Signs Order to Begin Creation of Space Force," *New York Times*, February 19, 2019, available at https://www.nytimes.com/2019/02/19/us/politics/trump-space-force.html.

23. "US Space Force Chief: Space Is 'A Warfighting Domain,'" October 28, 2020, available at https://www.forces.net/news/head-us-space-force-space-warfighting-domain.

24. See "Détente Unravels and a New (Limited) Space Competition Begins" in chapter 1.

25. Carla Babb, "Trump Unveils Space-Based Missile Defense Strategy," January 17, 2019, *VOA News*, available at https://www.voanews.com/a/trump-to-unveil-space-based-missile-defense-strategy/4747188.html.

26. Sandra Erwin, "Trump Unveils Missile Defense Review, Promises Funding of Space Sensors in 2020," *SpaceNews*, January 17, 2019, available at https://spacenews.com/trump-unveils-missile-defense-review-promises-funding-for-space-sensors-in-2020/.

27. Deb Reichmann and Lolita Baldour, "Trump Says US Will Develop Space-Based Missile Defense," *AP*, January 17, 2019, available at https://apnews.com/article/north-america-donald-trump-ap-top-news-north-korea-international-news-33e12cf640a64 08bb11aa3defa864966.

28. Geoff Brumfiel, "Trump's Plan to Zap Incoming Missiles with Lasers is Back to the Future," *NPR*, April 8, 2019, available at https://www.npr.org/2019/04/08/707689746/trumps-plan-to-zap-incoming-missiles-with-lasers-is-back-to-the-future; Reichmann and Baldour, "Trump Says US Will Develop Space-Based Missile Defense."

29. Hank Stuever, "Steve Carell's 'Space Force' Has a Troubled Launch, even with Heroic Efforts from John Malkovich," *Washington Post*, May 28, 2020, available at https://www.washingtonpost.com/entertainment/tv/space-force-review/2020/05/28/4e034738-9f8f-11ea-b5c9-570a91917d8d_story.html.

30. Nathan Strout, "Space Force Chief Says He's Working on a Declassification Strategy, but Offers Scant Details," *C4ISRNet*, March 3, 2021, available at https://www.c4isrnet.com/battlefield-tech/space/2021/03/03/space-force-chief-says-hes-working-on-a-declassification-strategy-but-offers-scant-details/.

31. Sam Sabin, "Nearly Half the Public Wants the US to Maintain Its Space Dominance. Appetite for Space Exploration Is a Different Story," *Morning Consult*, February 25, 2021, available at https://morningconsult.com/2021/02/25/space-force-travel-exploration-poll/.

32. Sabin, "Nearly Half the Public Wants the US to Maintain Its Space Dominance." Director of National Intelligence Avril Haines has publicly stressed that space will

become an increasingly contested domain; see "DNI Haines Speaks at the Activation of the National Space Intelligence Center," June 27, 2022, available at https://www .dni.gov/index.php/newsroom/news-articles/news-articles-2022/item/2304-dni -haines-speaks-at-the-activation-of-the-national-space-intelligence-center.

33. Christopher Miller, Mark Scott, and Bryan Bender, "UkraineX: How Elon Musk's Space Satellites Changed the War on the Ground," *Politico*, June 8, 2022, available at https://www.politico.eu/article/elon-musk-ukraine-starlink/.

34. Barbara Opall-Rome, "Israeli Experts: Arrow-3 Could be Adapted for Anti-Satellite Role," *SpaceNews*, November 9, 2009, available at https://spacenews.com/israeli -experts-arrow-3-could-be-adapted-anti-satellite%E2%80%82role/.

35. Kelsey Davenport, "Indian ASAT Test Raises Space Risks," *Arms Control Today*, May 2019, available at https://www.armscontrol.org/act/2019-05/news/indian-asat -test-raises-space-risks.

36. "UK Space Command Marks One-Year Anniversary," MoD Press Release, April 1, 2022, available at https://www.gov.uk/government/news/uk-space-command-marks -one-year-anniversary; Vivienne Machi, "France Puts Space at Top of National—and European—Security priorities," *Defense News*, March 14, 2022, available at https:// www.defensenews.com/space/2022/03/14/france-puts-space-at-top-of-national-and -european-security-priorities/; Vivienne Machi, "Germany Establishes New Military Space Command," *Defense News*, July 13, 2021, available at https://www.defensenews .com/space/2021/07/13/germany-establishes-new-military-space-command/.

37. "NATO's Approach to Space," NATO, December 2, 2021, available at https://www .nato.int/cps/en/natohq/topics_175419.htm.

38. Theresa Hitchens, "Space Lasers for Satellite Defense Top New French Space Strategy," July 26, 2019, available at https://breakingdefense.com/2019/07/france -envisions-on-orbit-lasers-for-satellite-defense/.

39. Ruth Harrison, "ADF Establishes New Defence Space Command Branch," *Space Australia*, March 29, 2022, available at https://spaceaustralia.com/index.php/news /adf-establishes-new-defence-space-command-branch; Park Si-Soo, "Japan to Launch 2nd Space Defense Unit to Protect Satellites from Electromagnetic Attack," *SpaceNews*, November 15, 2021, available at https://spacenews.com/japan-to-launch-2nd-space -defense-unit-to-protect-satellites-from-electromagnetic-attack/.

40. David C. DeFrieze notes the problems concerning even defining what constitutes a space weapon; see "Defining and Regulating the Weaponization of Space," *Joint Forces Quarterly*, available at https://ndupress.ndu.edu/Portals/68/Documents/jfq/jfq -74/jfq-74_110-115_DeFrieze.pdf.

41. DeFrieze, "Defining and Regulating the Weaponization of Space."

42. Bleddyn Bowen, "Space Oddities: Law, War, and the Proliferation of Spacepower," in *Routledge Handbook of War, Law, and Technology*, ed. James Gow, Ernst Dijxhoorn, Rachel Kerr, and Guglielmo Verdirame (London: Routledge, 2019), 266.

43. Patrick Howell O'Neill, "Russia Hacked an American Satellite Company One Hour before the Ukraine Invasion," *Technology Review*, May 10, 2022, available at https://www.technologyreview.com/2022/05/10/1051973/russia-hack-viasat-satellite-ukraine-invasion/.

44. Valerie Insinna, "SpaceX Beating Russian Jamming Attack was 'Eyewatering': DoD Official," *Breaking Defense*, April 20, 2022, available at https://breakingdefense.com/2022/04/spacex-beating-russian-jamming-attack-was-eyewatering-dod-official/.

45. Elizabeth Howell, "Russia Is Jamming GPS Satellite Signals in Ukraine, US Space Force Says," *Space*, April 12, 2022, available at https://www.space.com/russia-jamming-gps-signals-ukraine.

46. Kari A. Bingen, Kaitlyn Johnson, and Zhanna Malekos Smith, "Russia Threatens to Target Commercial Satellites," *CSIS*, November 10, 2022, available at https://www.csis.org/analysis/russia-threatens-target-commercial-satellites.

47. Rachel Zisk, "The National Defense Space Architecture (NDSA): An Explainer," Space Development Agency, December 5, 2022, available at https://www.sda.mil/the-national-defense-space-architecture-ndsa-an-explainer/.

48. Zisk, "The National Defense Space Architecture (NDSA): An Explainer."

49. A report by the Carnegie Endowment for International Peace notes that the short-term future of space-based interceptors is unclear but that the US "is likely to remain interested in them, while China and Russia are likely to remain concerned about them." See James Acton, Thomas D. MacDonald, and Pranay Vaddi, "Reimagining Nuclear Arms Control," Carnegie Endowment for International Peace, 2021, available at https://carnegieendowment.org/files/Acton_et_al_ReImagining_Arms_Control_fnl_1.pdf.

50. For recent insights into cost of launch, see Thomas G. Roberts, "Implications of Low-Cost Launch," Appendix 1 in "Boost Phase Missile Defense: Interrogating the Assumptions," June 2022, available at https://csis-website-prod.s3.amazonaws.com/s3fs-public/publication/220624_Karako_BoostPhase_MissileDefense.pdf?WjJxlNM58oru1LK21LC9untewoK_UAQD#page=43.

BIBLIOGRAPHY

ARCHIVES

Central Intelligence Agency, Freedom of Information Act Electronic Reading Room
John F. Kennedy Presidential Library, Boston, Massachusetts
Gerald R. Ford Presidential Library, Ann Arbor, Michigan
National Reconnaissance Office, Freedom of Information Act Electronic Reading Room
National Security Archive, Washington, DC
Ronald Reagan Presidential Library, Simi Valley, California
United Kingdom National Archives, Kew, UK
Vitalii Kataev Archive, Hoover Institution, Palo Alto, California
Lithuanian Special Archives, Vilnius, Lithuania
Woodrow Wilson Center Digital Archive

ORAL HISTORIES

Kenneth Adelman (Head of the US Arms Control and Disarmament Agency, 1983–1987)
Michael Griffin (Former member of the Johns Hopkins University Applied Physics Laboratory who later served as the NASA administrator and undersecretary of defense for research and engineering)
Donald Hafner (Arms Control and Disarmament Agency, 1977–1978, and participant in American–Soviet anti-satellite weapons talks)
Richard Matlock (Former senior official with the Missile Defense Agency)
Stanley Orman (Former scientist, UK MoD)
John Poindexter (National Security Advisor, 1985–1986)
Charles Powell (Private secretary for foreign affairs to Prime Minister Margaret Thatcher, 1983–1991)

Colonel Gil Rye, USAF (ret.) (Director of space policy at the National Security Council, 1981–1985)

Horst Teltschik (National Security Advisor to West German Chancellor Helmut Kohl)

TELEVISION SERIES

The Real War in Space, BBC, 1978

NEWSPAPERS AND PERIODICALS

Arms Control Today
Aviation Week and Space Technology
BBC
Bulletin of Atomic Scientists
Chicago Daily Tribune
Christian Science Monitor
Foreign Affairs
Houston Post
Los Angeles Times
Newsweek
New York Times
Scientific American
US News and World Report
Washington Post

BOOKS AND ARTICLES

Abrahamson, James. "SDI and the New Space Renaissance." *Space Policy*, 1, no. 2 (1985).

Adelman, Kenneth. *Reagan at Reykjavik: Forty-Eight Hours that Ended the Cold War* (New York: Broadside Books, 2014).

Afinov, V. V. "Razvitie v SShA vysokotochnogo oruzhiya i perspektivy sozdaniya razvedyvatel'no-udarnykh kompleksov." *Voennaia mysl'*, 4 (1983).

Agar, Jon. *Science Policy under Thatcher* (London: UCL Press, 2020).

Aldous, Richard. *Reagan and Thatcher: The Difficult Relationship* (New York: W. W. Norton, 2012).

Alves, Pericles Gasparini. "Prevention of an Arms Race in Outer Space: A Guide to the Discussions in the Conference on Disarmament." United Nations Institute for Disarmament Research, 1991, available at https://www.unidir.org/files/publications/pdfs/prevention-of-an-arms-race-in-outer-space-a-guide-to-the-discussions-in-the-cd-en-451.pdf.

Amato, Ivan. *Taking Technology Higher: The Naval Center for Space Technology and the Making of the Space Age* (Washington, DC: NRL, 2022).

Ambrose, Matthew J. *The Control Agenda: A History of the Strategic Arms Limitation Talks* (Ithaca, NY: Cornell University Press, 2018).

Anderson, Martin. *Revolution* (San Diego, CA: Harcourt, 1988).

Andrew, Christopher. *For the President's Eyes Only: Secret Intelligence and the American Presidency from Washington to Bush* (New York: Harper, 1996).

Andrew, Christopher, and Gordievsky, Oleg. *KGB: The Inside Story of Its Operations from Lenin to Gorbachev* (New York: Harper, 1991).

Andrew, Christopher, and Mitrokhin, Vasili. *The Sword and the Shield: The Mitrokhin Archive and the Secret History of the KGB* (New York: Basic Books, 1999).

Austerman, Wayne. *Program 437: The Air Force's First Anti-Satellite System* (Colorado Springs: Air Force Space Command History Office, 1991).

Bateman, Aaron. "Intelligence and Alliance Politics: The Strategic Defense Initiative and Anglo-American Relations." *Intelligence and National Security*, 36, no. 7 (2021).

Bateman, Aaron. "Keeping the Technological Edge: The Space Arms Race and Anglo-American Relations in the 1980s." *Diplomacy and Statecraft*, 33, no. 2 (2022).

Bateman, Aaron. "Mutually Assured Surveillance at Risk: Anti-Satellite Weapons and Cold War Arms Control." *Journal of Strategic Studies*, 45, no. 1 (2022).

Bateman, Aaron. "Space Reconnaissance and Anglo-American Relations." *The Space Review*, March 9, 2020, available at https://www.thespacereview.com/article/3896/1.

Bateman, Aaron. "Trust but Verify: Satellite Reconnaissance, Secrecy, and Cold War Arms Control." *Journal of Strategic Studies* (2023). https://www.tandfonline.com/doi/abs/10.1080/01402390.2022.2161522.

Baucom, Donald. *Origins of the Strategic Defense Initiative Organization* (Washington, DC: Pentagon, 1989), 251.

Baucom, Donald. *The Origins of SDI 1944–1983* (Lawrence: University of Kansas Press, 1992).

Baucom, Donald. "The Rise and Fall of Brilliant Pebbles." *The Journal of Social, Political, and Economic Studies*, 29, no. 2 (2004).

Bauer, Martin. "Resistance to New Technology and Its Effects on Nuclear Power, Information Technology and Biotechnology," in *Resistance to New technology: Nuclear Power, Information Technology and Biotechnology*, ed. Martin Bauer (Cambridge: Cambridge University Press, 1995).

Baylis, John. "Exchanging Nuclear Secrets: Laying the Foundations of the Anglo-American Nuclear Relationship." *Diplomatic History*, 25, no. 1 (2001).

Bethe, Hans; Garwin, Richard; Gottfried, Kurt; Kendall, Henry. "Space-Based Missile Defense," A Report by the Union of Concerned Scientists (Cambridge, MA: UCS, 1984).

Bijker, Wiebe. *Of Bicycles, Bakelites, and Bulbs: Toward a Theory of Sociotechnical Change* (Cambridge, MA: MIT Press, 1995).

Blanc, Alexis A.; Beauchamp-Mustafaga, Nathan; Holynska, Khrystyna; Bond, M. Scott; Flanagan, Stephen. *Chinese and Russian Perceptions of and Responses to U.S. Military Activities in the Space Domain* (Santa Monica, CA: RAND, 2022).

Bluth, Christopher. "SDI: The Challenge to West Germany." *International Affairs*, 62, no. 2, (1986).

Bowen, Bleddyn. *Original Sin: Power, Technology, and War in Outer Space* (London: Hurst, 2022).

Bowen, Bleddyn. "Space Oddities: Law, War, and the Proliferation of Spacepower," in *Routledge Handbook of War, Law, and Technology*, ed. James Gow, Ernst Dijxhoorn, Rachel Kerr, and Guglielmo Verdirame (London: Routledge, 2019).

Bowen, Bleddyn. *War in Space: Strategy, Spacepower, Geopolitics* (Edinburgh: University of Edinburgh Press, 2020).

Bozo, Frederico. "The Sanctuary and the Glacis: France, the Federal Republic of German, and Nuclear Weapons in the 1980s (Part 1)." *Journal of Cold War Studies*, 22, no. 3 (2020).

Brands, Hal. *What Good Is Grand Strategy? Power and Purpose in American Statecraft from Harry S. Truman to George W. Bush* (Ithaca, NY: Cornell University Press, 2014).

Brandt, Thomas. "The Military Uses of Space," in *America Plans for Space* (Washington, DC: National Defense University Press, 1986).

Brauch, Hans Gunter. "The Political Debate in the Federal Republic of Germany," in *Star Wars and European Defence*, ed. Hans Gunter Brauch (New York: Palgrave Macmillan, 1987).

Breyman, Steven. "SDI, The Federal Republic of Germany, and NATO: Political, Economic, and Strategic Implications." University of California, San Diego, Institute on Global Conflict and Cooperation, 1987.

Broad, William. "Reagan's Legacy in Space: More Reach than Grasp." *New York Times*, May 8, 1988.

Brown, Archie. *The Human Factor: Gorbachev, Reagan, and Thatcher, and the End of the Cold War* (Oxford: Oxford University Press, 2020).

Brunet, Luc-André. "The Strategic Defence Initiative and the Atlantic Alliance in the 1980s," in *NATO and the Strategic Defence Initiative*, ed. Luc-André Brunet (New York: Routledge, 2023).

Buchholz, Arnold. "The Scientific-Technological Revolution (STR) and Soviet Ideology." *Studies in Soviet Thought*, 30, no. 4 (1985).

Bulkeley, Rip, and Spinardi, Graham. *Space Weapons: Deterrence or Delusion?* (Oxford: TJ Press, 1986).

Buono, Stephen. "Merely a 'Scrap of Paper'? The Outer Space Treaty in Historical Perspective." *Diplomacy and Statecraft*, 31, no. 2 (2020).

Butrica, Andrew. *Single Stage to Orbit: Politics, Space Technology, and the Quest for Reusable Rocketry* (Baltimore: Johns Hopkins University Press, 2006).

Cahn, Anne H., Little, Martha C., and Daggett, Stephen. "Nunn and Contractors Sell ALPS." *Bulletin of the Atomic Scientists*, 44, no. 5 (1988).

Cameron, James. "Soviet–American Strategic Arms Limitation and the Limits of Co-operative Competition." *Diplomacy and Statecraft*, 33, no. 1 (2022).

Cameron, James. *The Double Game: The Demise of America's First Missile Defense System and the Rise of Strategic Arms Limitation* (Oxford: Oxford University Press, 2017).

Carter, Ashton. "Directed Energy Missile Defense in Space." Background paper, April 1984, available at http://www.princeton.edu/~ota/disk3/1984/8410/8410.PDF.

Chagas, Carlos, and Canuto, Vittorio, ed. "The Impact of Space Exploration on Mankind," Vatican City, 1986, available at https://www.pas.va/content/dam/casinapioiv/pas/pdf-volumi/scripta-varia/sv58pas.pdf.

Clark, Ian. *Nuclear Diplomacy and the Special Relationship: Britain's Deterrent and America, 1957–1962* (Oxford: Clarendon Press, 1994).

Codevilla, Angelo. *While Others Build: The Commonsense Approach to the Strategic Defense Initiative* (New York: Free Press, 1988), 87.

Colbourn, Susan. *Euromissiles: The Nuclear Weapons that Nearly Destroyed NATO* (Ithaca, NY: Cornell University Press, 2022).

Colbourn, Susan, and Haeussler, Matthias. "Once More, with Feeling: Transatlantic Relations in the Reagan Years," in *The Reagan Moment: America and the World in the 1980s*, ed. Jonathan R. Hunt and Simon Miles (Ithaca, NY: Cornell University Press, 2021).

Cole, Alistair. *Francois Mitterrand: A Study in Political Leadership* (New York: Routledge, 1997).

Collins, Martin. "The 1970s: Spaceflight and Historically Interpreting the In-Between Decade," in *Limiting Outer Space: Astroculture After Apollo*, ed. Alexander Geppert (New York: Palgrave Macmillan, 2018).

David, James. *Spies and Shuttles: NASA's Secret Relationships with the DoD and CIA* (Gainesville: University of Florida Press, 2015).

Day, Dwayne. "Intersections in Real Time: The Decision to Build The KH-11 KENNEN Reconnaissance Satellite (Part 1)." *The Space Review*, September 9, 2019, available at https://www.thespacereview.com/article/3791/1.

Day, Dwayne, Logsdon, John, and Latell, Brian. "Introduction," in *Eye in the Sky: The Story of the Corona Spy Satellites*, ed. Dwayne Day, John Logsdon, and Brian Latell (Washington, DC: Smithsonian Institute Press, 1997).

Deudney, Daniel. *Dark Skies: Space Expansionism, Planetary Geopolitics and The Ends of Humanity*. Oxford: Oxford University Press, 2020.

Dickey, Robin. "The Rise and Fall of Space Sanctuary in US Space Policy." Aerospace Corporation, September 1, 2020.

Dickson, David. "A European Defense Initiative: The Idea that European Nations Band Together for a Strictly European Version of SDI Is Gaining Support." *Science*, 229, no. 4719 (1985).

Dickson, David. "Europe Tries a Strategic Technology Initiative." *Science*, 229, no. 4709 (1985).

Dietl, Ralph. *The Strategic Defense Initiative: Ronald Reagan, NATO Europe, and the Nuclear and Space Talks, 1981–1988* (London: Lexington Books, 2018).

Dinerman, Taylor. "Missile Defense, RLVs, and the Future of American Spacepower." *The Space Review*, December 1, 2003, available at https://www.thespacereview.com/article/66/1.

Dobrynin, Anatoly, *In Confidence: Moscow's Ambassador to Six Cold War Presidents*. Seattle: University of Washington Press, 2001.

Dobson, Alan. "The Reagan Administration, Economic Warfare, and Starting to Close Down the Cold War." *Diplomatic History*, 29, no. 3 (2005).

Doel, Ronald E. "Scientists, Secrecy, and Scientific Intelligence: The Challenges of International Science in Cold War America," in *Cold War Science and the Transatlantic Circulation of Knowledge*, ed. Jeroen van Dongen (Leiden: Brill, 2015).

Dolman, Everett. *Astropolitik: Classical Geopolitics in the Space Age* (New York: Routledge, 2001.

Doyle, Suzanne. "A Foregone Conclusion? The United States, Britain and the Trident D5 Agreement." *Journal of Strategic Studies*, 40, no. 6 (2017).

Doyle, Suzanne. "The United States Sale of Trident to Britain, 1977–1982: Deal Making in the Anglo-American Nuclear Relationship." *Diplomacy and Statecraft*, 28, no. 3 (2017).

Duric, Mira. *The Strategic Defense Initiative: US Policy and the Soviet Union* (Abingdon, UK: Routledge, 2003).

Eames, Anthony. "A 'Corruption of British Science?' The Strategic Defense Initiative and British Technology Policy." *Technology and Culture*, 62, no. 3 (2021).

Fenske, John. "France and the Strategic Defence Initiative: Speeding up or Putting on the Brakes?" *International Affairs*, 62, no. 2 (1986).

Ferris, John. *Behind the Enigma: The Authorized History of GCHQ, Britain's Secret Cyber-Intelligence Agency* (London: Bloomsbury, 2020).

Fischer, Beth A. "Nuclear Abolitionism, Strategic Defense Initiative, and the 1987 INF Treaty," in *The INF Treaty of 1987*, ed. Philip Gassert, Tim Geiger, and Hermann Wentker (Berlin: Vandenhoeck & Ruprecht, 2020).

FitzGerald, Frances. *Way Out There in the Blue: Reagan, Star Wars, and the End of the Cold War* (New York: Simon and Schuster, 2001).

Flax, Alexander. "Ballistic Missile Defense: Concepts and History." *Daedalus*, 114, no. 2 (1985).

Gaddis, John. "Looking Back: The Long Peace." *The Wilson Quarterly*, 13, no. 1 (1989).

Gaddis, John. *The Long Peace: Inquiries into the History of the Cold War* (Oxford: Oxford University Press, 1987).

Gaddis, John Lewis. *The United States and the End of the Cold War* (New York: Oxford University Press, 1992).

Gates, Robert. *Duty: Memoirs of a Secretary at War* (New York: Vintage, 2015).

Gavin, Francis J. *Nuclear Statecraft: History and Strategy in America's Atomic Age* (Ithaca, NY: Cornell University Press, 2012).

Gayte, Marie. "The Vatican and the Reagan Administration: A Cold War Alliance?" *The Catholic Historical Review*, 97, no. 4 (2011).

Geppert, Alexander. "European Astrofuturism, Cosmic Provincialism: Historicizing the Space Age," in *Imagining Outer Space: European Astroculture in the Twentieth Century*, ed. Alexander Geppert (New York: Palgrave, 2012).

Geppert, Alexander, and Siebeneichner, Tilmann. "Spacewar! The Dark Side of Astroculture," in *Militarizing Outer Space*, ed. Alexander Geppert, Daniel Brandau, and Tilmann Siebeneichner (New York: Palgrave Macmillan, 2021).

Gilpin, Robert. *France in the Age of the Scientific State* (Princeton: Princeton University Press, 1968).

Grabo, Cynthia M. *Anticipating Surprise: Analysis for Strategic Warning* (Washington, DC: Joint Military Intelligence College's Centre for Strategic Intelligence Research, 2002).

Graham, Daniel. *Confessions of a Cold Warrior* (Fairfax, VA: Preview Press, 1995).

Graham, Thomas W., and Kramer, Bernard M. "The Polls: ABM and Star Wars: Attitudes Toward Nuclear Defense, 1945–1985." *The Public Opinion Quarterly*, 50, no. 1 (1986).

Grahn, Sven. "Simulated War in Space—Soviet ASAT Tests." undated, available at http://www.svengrahn.pp.se/histind/ASAT/ASAT.htm#Mark.

Gray, Colin. *American Military Space Policy: Information Systems, Weapon Systems and Arms Control* (Cambridge: Abt Books, 1982).

Gray, Colin. "Strategic Defense, Deterrence, and the Prospects for Peace." *Ethics*, 95, no. 3 (1985).

Gray, Colin. *Weapons Don't Make War: Policy, Strategy, and Military Technology* (Lawrence: University of Kansas, 1993).

Hafner, Donald. "Assessing the President's Vision: The Fletcher, Miller, and Hoffman Panels." *Daedalus*, 114, no. 2 (1985).

Hafner, Donald. "Outer Space Arms Control: Unverified Practices, Unnatural Acts?" *Survival*, 25, no. 6 (1983).

Hall, R. Cargill. "Origins of US Space Policy: Eisenhower, Open Skies, and Freedom of Space," in *Exploring the Unknown: Selected Documents in the History of the US Civil Space Program, Volume I: Organizing for Exploration*, ed. John M. Logsdon, gen. ed. with Linda J. Lear, Jannelle Warren-Findley, Ray A. Williamson, and Dwayne A. Day (Washington, DC: NASA SP-4407, 1995).

Haslam, Jonathan. *Russia's Cold War: From the October Revolution to the Fall of the Wall* (New Haven, CT: Yale University Press, 2013).

Hays, Peter, *United States Military Space: Into the Twenty-First Century* (Maxwell Air Force Base, AL: Air University Press, 2002).

Hecht, Gabrielle. *The Radiance of France: Nuclear Power and National Identity after World War II* (Cambridge, MA: MIT Press, 1998).

Hendricks, Bart, and Day, Dwayne. "Target Moscow: Soviet Suspicions about the American Space Shuttle (Part I)." *The Space Review*, January 27, 2020, available at https://www.thespacereview.com/article/3873/1.

Hennessy, Peter. *The Secret State Whitehall and the Cold War* (London: Allen Lane, 2002).

Henshaw, John H. "The Origins of COCOM: Lessons for Contemporary Proliferation control Regimes." May 1993, available at https://www.stimson.org/wp-content/files/file-attachments/Report7_1.pdf.

Heuer, Richards J., Jr. *Psychology of Intelligence Analysis* (Washington, DC: Center for the Study of Intelligence, Central Intelligence Agency, 1999).

Hoffman, David. *The Dead Hand: The Untold Story of the Cold War Arms Race and its Dangerous Legacy* (New York: Anchor, 2010).

Howe, Geoffrey. "Defence and Security in the Nuclear Age." *The RUSI Journal*, 130, no. 2 (1985).

Hughes, Thomas P. "The Electrification of America: The System Builders." *Technology and Culture*, 20, no. 1 (1979).

Hughes, Thomas P. "The Evolution of Large Technological Systems," in *The Social Construction of Technological Systems*, ed. Wiebe Bijker, Thomas P. Hughes, and Trevor Pinch (Cambridge, MA: MIT Press, 1987).

Jasani, Bhupendra. "Military Activities in Outer Space," in *Outer Space—A New Dimension in the Arms Race*, ed. Bhupendra Jasani (New York: Routledge, 1982).

Jervis, Robert. *Why Intelligence Fails: Lessons from the Iranian Revolution and the Iraq War.* (Ithaca, NY: Cornell University Press, 2011).

Johnson-Freese, Joan. *Space Warfare in the 21st Century: Arming the Heavens* (New York: Routledge, 2016).

Jones, Matthew. *The Official History of the UK Strategic Nuclear Deterrent: Volume II: The Labour Government and the Polaris Programme, 1964–1970* (London: Routledge, 2017).

Kalic, Sean N. "Reagan's SDI Announcement and the European Reaction: Diplomacy in the Last Decade of the Cold War," in *Crisis of Detente in Europe: from Helsinki to Gorbachev, 1975–1985*, ed. Leopoldo Nuti (London: Routledge, 2009).

Kalic, Sean N. *US Presidents and the Militarization of Space, 1946–1967* (College Station: Texas A&M Press, 2012).

Kandiah, Michael D., and Staerck, Gillian. "The British Response to SDI," a conference at King's College London, 2005, available at https://www.kcl.ac.uk/sspp/assets/icbh-witness/sdi.pdf.

Karas, Thomas. *The New High Ground, Strategies and Weapons of Space-Age War* (New York: Simon and Schuster, 1983).

Kilgo, Robert. "The History of the United States Anti-Satellite Program and the Evolution to Space Control and Offensive and Defensive Counterspace." *Quest*, 11, no. 3 (2004).

Kilgore, DeWitt Douglas. *Astrofuturism: Science, Race, and Visions of Utopia in Space* (Philadelphia: University of Pennsylvania Press, 2003).

Kissinger, Henry. *Diplomacy* (New York: Simon and Schuster, 1994).

Kissinger, Henry. *Years of Renewal* (New York: Simon and Schuster, 1999).

Krepinevich, Andrew, and Watts, Barry. *The Last Warrior: Andrew Marshall and the Shaping of Modern Defense Strategy* (New York: Basic, 2015).

Krepon, Michael. *Winning and Losing the Nuclear Peace: The Rise, Demise, and Revival of Arms Control* (Palo Alto, CA: Stanford University Press, 2022).

Krige, John. *American Hegemony and the Postwar Reconstruction of Science in Europe* (Boston: MIT Press, 2006).

Krige, John. "Hybrid Knowledge: The Transnational Co-Production of the Gas Centrifuge for Uranium Enrichment in the 1960s." *The British Journal for the History of Science*, 45, no. 3 (2012).

Krige, John. "Introduction: Writing the Transnational History of Science and Technology," in *How Knowledge Moves: Writing the Transnational History of Science and Technology* (Chicago: University of Chicago Press, 2019).

Krige, John. *Sharing Knowledge, Shaping Europe: US Technological Collaboration and Nonproliferation* (Boston: MIT Press, 2016).

Krige, John. "The European Space System," in *Reflections on Europe in Space*, ed. John Krige and Arturo Russo (Noordwijk: European Space Agency, 1994).

Krige, John, Callahan, Angelina Long, and Maharaj, Ashok. *NASA in the World* (New York: Palgrave, 2013).

Krige, John, Russo, Arturo, and Sebasta, Laurenza. *A History of the European Space Agency, Vol. 1* (Noordwijk: European Space Agency, 2000).

Krige, John, Russo, Arturo, and Sebesta, Laurenza. *A History of the European Space Agency 1958–1987, Vol. II* (Noordwijk: European Space Agency, 2000).

Lakoff, Sanford, and York, Herbert F. *A Shield in Space? Technology, Politics, and the Strategic Defense Initiative* (Berkeley: University of California Press, 1989).

Launius, Roger. *Apollo's Legacy: Perspectives on the Moon Landings* (Washington, DC: Smithsonian Books, 2019).

Law, John. "On the Social Explanation of Technical Change: The Case of the Portuguese Maritime Expansion." *Technology and Culture*, 28, no. 2 (1987).

Leffler, Melvyn P. "Ronald Reagan and the Cold War," in *The Reagan Moment: America and the World in the 1980s*, ed. Johnathan R. Hunt and Simon Miles (Ithaca, NY: Cornell University Press, 2021).

LeSuer, Stephen C. "Battle for SDI Funding Pushed to Background in Debate Over Missile Defenses." *Inside the Pentagon*, 7, no. 33 (1991).

Lettow, Paul. *Ronald Reagan and His Quest to Abolish Nuclear Weapons* (New York: Random House, 2005).

Linenthal, Edward. *Symbolic Defense* (Chicago: University of Illinois Press, 1989).

Lin-Greenberg, Erik. "Allies and Artificial Intelligence: Obstacles to Operations and Decision-Making." *Texas National Security Review*, 3, no. 2 (2020).

Logsdon, John M. *After Apollo?: Richard Nixon and the American Space Program* (London: Palgrave Macmillan, 2015).

Logsdon, John M. *John F. Kennedy and the Race to the Moon* (New York: Palgrave, 2010).

Logsdon, John M. *Ronald Reagan and the Space Frontier* (London: Palgrave Macmillan, 2019).

Logsdon, John M. *Together in Orbit: The Origins of International Participation in the Space Station* (Washington, DC: NASA History Division, 2005).

Lupton, David E. *A Space Power Doctrine* (Maxwell Air Force Base, AL: Air University Press, 1998).

McCray, Patrick. *The Visioneers: How a Group of Elite Scientists Pursued Space Colonies, Nanotechnologies, and a Limitless Future.* Princeton: Princeton University Press, 2012.

McDougall, Walter. *The Heavens and the Earth: A Political History of the Space Age.* Baltimore: Johns Hopkins University Press, 1997.

Mackenzie, Donald. *Inventing Accuracy: A Historical Sociology of Nuclear Missile Guidance* (Cambridge: MIT Press, 1990).

Maher, Neil M. *Apollo in the Age of Aquarius* (Cambridge: Harvard University Press, 2017).

Matlock, Jack. *Reagan and Gorbachev: How the Cold War Ended* (New York: Random House, 2005).

Matovski, Aleksandar. "Strategic Intelligence and International Crisis Behavior." *Security Studies*, 29, no. 5 (2020).

Maurer, John. "The Purposes of Arms Control." *Texas National Security Review*, 2, no. 1 (2018).

Meyer, Paul. "The CD and PAROS: A Short History." April 2011, available at https://www.unidir.org/files/publications/pdfs/the-conference-on-disarmament-and-the-prevention-of-an-arms-race-in-outer-space-370.pdf.

Michaud, Michael. *Reaching For the High Frontier: The American Pro-Space Movement, 1972–84* (Westport, CT: Praeger, 1986).

Mieczkowski, Yanek. *Eisenhower's Sputnik Moment: The Race for Space and World Prestige* (Ithaca, NY: Cornell University Press, 2013).

Miles, Simon. *Engaging the Evil Empire: Washington, Moscow, and the Beginning of the End of the Cold War* (Ithaca, NY: Cornell University Press, 2020).

Mitchell, Gordon R. "Placebo Defense: Operation Desert Mirage? The Rhetoric of Patriot Missile Accuracy in the 1991 Persian Gulf War." *Quarterly Journal of Speech*, 86, no. 2 (2000).

Mitchell, Vance. *Sharing Space—The Secret Interaction Between the National Aeronautics and Space Administration and the National Reconnaissance Office* (Chantilly, VA: National Reconnaissance Office, 2012), 64.

Moakley, John. "Space Weapons and Congress." *Arms Control Today*, 13, no. 11 (1983).

Moltz, James C. *The Politics of Space Security: Strategic Restraint and the Pursuit of National Interests, 3rd Edition* (Stanford, CA: Stanford University Press, 2019).

Monfort, Charles. "ASATs: Star Wars on the Cheap." *Bulletin of the Atomic Scientist*, 45 (1989).

Moore, Charles. *Margaret Thatcher: The Authorized Biography, Vol. 1* (London: Allen Lane, 2013).

Morris, Edmund. *Dutch: A Memoir of Ronald Reagan* (New York: Modern Library, 1999).

Neufeld, Michael. *Von Braun: Dreamer of Space, Engineer of War* (New York: Knopf, 2007).

Neufeld, Michael J. "Cold War—But No War—in Space," in *Militarizing Outer Space*, ed. Alexander Geppert, Daniel Brandau, and Tilmann Siebeneichner (New York: Palgrave Macmillan, 2021).

Nitze, Paul. *From Hiroshima to Glasnost: At the Center of Decision* (New York: Grove Atlantic, 1989).

Nolan, Janne. *Guardians of the Arsenal: The Politics of Nuclear Strategy* (New York: Basic, 1989).

Norman, Colin. "Congress Reins in, Redirects SDI." *Science*, 241, no. 4862 (1988).

Nuti, Leopoldo, Bozo, Frédéric, Rey, Marie-Pierre, and Rother, Bernd, ed. *The Euromissile Crisis and the End of the Cold War* (Washington, DC: Woodrow Wilson Center Press, 2015).

Nye, Joseph S., and Schear, James A., ed. *Seeking Stability in Space: Anti-Satellite Weapons the Evolving Space Regime* (Lanham, MD: University Press of America, 1987).

Oder, Frederick C. E., Fitzpatrick, James C., and Worthman, Paul E. *The Hexagon Story* (Chantilly, VA: National Reconnaissance Office, 1992).

Orman, Stanley. *An Uncivil Civil Servant* (Dinton, UK: TwigBooks, 2013).

Paikowsky, Deganit. *The Power of the Space Club* (Cambridge: Cambridge University Press, 2017).

Paolini, Jerome. "Politique Spatiale Militaire Française et Coopération Europée." *Politique Étrangère*, 52, no. 2 (1987).

Parisi, Ilaria. "France's Reaction Towards the Strategic Defence Initiative (1983–1986): Transforming a Strategic Threat into a Technological Opportunity," in *NATO and the Strategic Defence Initiative*, ed. Luc-André Brunet (New York: Routledge, 2023).

Peebles, Curtis. *High Frontier: The US Air Force and the Military Space Program* (Washington, DC: Air Force Historical Studies Office, 1997).

Peoples, Columba. *Justifying Ballistic Missile Defence: Technology, Security and Culture* (Cambridge: Cambridge University Press, 2010).

Pike, John. "Anti-Satellite Weapons and Arms Control." *Arms Control Today*, 13, no. 11 (1983).

Pike, John. "The Death-Beam Gap: Putting Keegan's Follies in Perspective." October 1992, available at https://fas.org/spp/eprint/keegan.htm#N_5.

Podvig, Pavel. "Did Star Wars Help End the Cold War? Soviet Response to the SDI Program." *Science and Global Security*, 25, no. 1 (2017).

Prados, John. "The Strategic Defense Initiative: Between Strategy, Diplomacy and United States Intelligence Estimates," in *Crisis of Detente in Europe: from Helsinki to Gorbachev, 1975–1985*, ed. Leopoldo Nuti (London: Routledge, 2009).

Pratt, Erik K. *Selling Strategic Defense: Interests, Ideologies, and the Arms Race* (Princeton: Princeton University Press, 1990).

Rabinowitz, Or. "'Arrow' Mythology Revisited: The Curious Case of the Reagan Administration, Israel and SDI Cooperation." *International History Review*, 43, no. 6 (2021).

Rees, Wyn, and Berman, Azriel. "Nuclear Divergence between Britain and the United States: SDI and the Anti-Ballistic Missile Treaty." *Journal of Strategic Studies* (2021). https://doi.org/10.1080/01402390.2021.1907745.

Reinke, Niklas. *The History of German Space Policy: Ideas, Influence, and Interdependence 1932–2002* (Paris: Beauchesne, 2007).

Reiss, Edward. *The Strategic Defense Initiative* (Cambridge: Cambridge University Press, 1992).

Rhodes, Richard. *Arsenals of Folly: The Making of the Nuclear Arms Race* (New York: Vintage Books, 2007).

Richelson, Jeffrey. *America's Space Sentinels: DSP Satellites and National Security* (Lawrence: University of Kansas Press, 2001).

Roy, Stephanie. "The Origin of the Smaller, Faster, Cheaper Approach in NASA's Solar System Exploration Program." *Space Policy*, vol. 14, no. 3 (1998).

Sandholtz, Wayne. *High-Tech Europe: The Politics of International Cooperation* (Berkley: University of California Press, 1992).

Savel'yev, Aleksandr, and Detinov, Nikolay, *The Big Five: Arms Control Decision-Making in the Soviet Union* (New York: Praeger, 1995).

Sayle, Timothy. *Enduring Alliance: A History of NATO and the Postwar Global Order* (Ithaca, NY: Cornell University Press, 2019).

Schelling, Thomas. *Arms and Influence* (New Haven, CT: Yale University Press, 1966).

Schelling, Thomas, and Halperin, Morton. *Strategy and Arms Control* (New York: Twentieth Century Fund, 1961).

Schweizer, Peter. *Reagan's War: The Epic Story of His Forty-Year Struggle and Final Triumph Over Communism* (New York: Anchor, 2003).

Sherr, Alan. "Sound Legal Reasoning or Policy Expedient? The 'New Interpretation' of the ABM Treaty." *International Security*, 11, no. 3 (1986–1987).

Shultz, George. *Turmoil and Triumph: Diplomacy, Power, and the Victory of the American Deal* (New York: Scribner, 1993).

Siddiqi, Asif. "American Space History: Legacies, Questions, and Opportunities for Research," in *Critical Issues in the History of Spaceflight*, ed. Steven J. Dick and Roger D. Launius (Washington, DC: National Aeronautics and Space Administration, 2006).

Siddiqi, Asif. "Competing Technologies, National(ist) Narratives, and Universal Claims: Toward a Global History of Space Exploration." *Technology and Culture*, 51, no. 2 (2010).

Siddiqi, Asif. *The Red Rockets' Glare: Spaceflight and the Russian Imagination, 1857-1957* (Cambridge: Cambridge University Press, 2014).

Siddiqi, Asif. "The Soviet Co-Orbital Anti-Satellite System: A Synopsis." *British Interplanetary Society*, 50, no. 6 (1997).

Siddiqi, Asif. "The Soviet Fractional Orbiting Bombardment System (FOBS): A Short Technical History." *Quest: The History of Spaceflight Quarterly*, 7, no. 4 (2000).

Siebeneichner, Tilmann. "Spacelab: Peace, Progress and European Politics in Outer Space, 1973–85," in *Limiting Outer Space: Astroculture After Apollo*, ed. Alexander Geppert (New York: Palgrave Macmillan, 2018).

Slayton, Rebecca. *Arguments that Count: Physics, Computing, and Missile Defense, 1949–2012* (Boston: MIT Press, 2013).

Slotten, Hugh. *Beyond Sputnik and The Space Race: The Origins of Global Satellite Communications* (Baltimore: Johns Hopkins University Press, 2022).

Smith, Merrit Roe, and Marx, Leo, ed. *Does Technology Drive History? The Dilemma of Technological Determinism* (Cambridge, MA: MIT Press, 1995).

Sofaer, Abraham D. "The ABM Treaty and the Strategic Defense Initiative." *Harvard Law Review*, 99, no. 8 (1986).

Sokolsky, Joel J., and Jockel, Joseph T. "Canada and the Future of Strategic Defense," in *Perspectives on Strategic Defense*, ed. Steven W. Guerrier, Wayne C. Thompson, Zbigniew Brzezinski (New York: Routledge, 1987).

Spinardi, Graham. "Technical Controversy and Ballistic Missile Defence: Disputing Epistemic Authority in the Development of Hit-to-Kill Technology." *Science as Culture*, 23, no. 1 (2013).

Stares, Paul. "Déjà Vu: The ASAT Debate in Historical Context." *Arms Control Today*, 13, no. 11 (1983).

Stares, Paul. "Reagan and the ASAT Issue." *Journal of International Affairs*, 39, no. 1 (1985).

Stares, Paul. *Space Weapons and US Strategy* (New York: Routledge, 1985).

Stares, Paul. *The Militarization of Space, US Policy, 1945–1984* (Ithaca, NY: Cornell University Press, 1985).

Sterling, Michael J. *Soviet Reactions to NATO's Emerging Technologies for Deep Attack* (Santa Monica: RAND, 1985).

Stimmer, Anette. "Star Wars or Strategic Defense: What's in a Name?" *Journal of Global Security Studies*, 4, no. 4 (2019).

Stocker, Jeremy. *Britain's Role in US Missile Defense* (Carlisle, PA: Strategic Studies Institute, 2004).

Stockholm International Peace Research Institute. *SIPRI Yearbook 1990* (Oxford: Oxford University Press, 1990).

Stockholm International Peace Research Institute. *SIPRI Yearbook 1991* (Oxford: Oxford University Press, 1991).

Stockholm International Peace Research Institute. *SIPRI Yearbook 1992* (Oxford: Oxford University Press, 1992).

Stoddart, Kristan. *Facing Down the Soviet Union: Britain, the USA, NATO and Nuclear Weapons, 1976–1983* (Basingstoke, UK: Palgrave Macmillan, 2014).

Stoddart, Kristan. "The British Labour Government and the Development of *Chevaline, 1974–79.*" *Cold War History,* 10, no. 3 (2010).

Streefland, Abel. "Putting a Lid on the Gas Centrifuge: Classification of the Dutch Ultracentrifuge Project, 1960–1961," in *Cold War Science and the Transatlantic Circulation of Knowledge,* ed. Jeroen Van Dongen (Amsterdam: Brill, 2015).

Taubman, William. *Gorbachev: His Life and Times* (New York: W. W. Norton, 2018).

Teller, Edward, and Schoolery, Judith. *Memoirs: A Twentieth-Century Journey in Science and Politics* (New York: Basic Books), 2002.

Thatcher, Margaret. *The Downing Street Years* (New York: Harper Collins, 1993).

"The History of US Antisatellite Weapon Systems," Federation of American Scientists, undated, available at https://fas.org/man/eprint/leitenberg/asat.pdf.

Thomas, Hugh. "Mr. Gorbachev's Own Star Wars." Centre for Policy Studies, March 1986, available at https://www.cps.org.uk/files/reports/original/111028090946 -MrGorbachevsownStarWars1986.pdf.

Turchetti, Simone. "A 'Need-To-Know-More' Criterion? Science and Information Security at NATO during the Cold War," in *Cold War Science and the Transatlantic Circulation of Knowledge,* ed. Jeroen van Dongen (Leiden: Brill 2015).

Urban, Mark. *UK Eyes Alpha: Inside Story of British Intelligence* (Eastbourne, UK: Gardners, 1997).

Vickers, Rhiannon. "Harold Wilson, the British Labour Party, and the War in Vietnam." *Journal of Cold War Studies,* 10, no. 2 (2008).

Vie-publique.fr. "Discours de Jacques Chirac, président du RPR, sur l'attitude de l'Europe et de la France face à l'initiative de défense stratégique." Paris, October 18, 1985.

Viotti, Major Paul, ed. "Military Space Doctrine—The Great Frontier: Final Report for the United States Air Force Academy Military Space Doctrine Symposium," April 1–3, 1981, USAF Academy, 1981.

Walker, James, Bernstein, Lewis, and Lang, Sharon. *Seize the High Ground: The Army in Space and Missile Defense* (Huntsville, AL: Army Space and Missile Defense Command, 2003).

Weigel, George. *Witness to Hope: The Biography of John Paul II* (New York: Harper Collins, 1990).

Weinberger, Caspar. *In the Arena: A Memoir of the 20th Century* (Washington, DC: Regnery, 2001).

Wellerstein, Alex. *Restricted Data: The History of Nuclear Secrecy in the United States* (Chicago: University of Chicago Press, 2021).

Wells, Anthony. *Between the Five Eyes: 50 Years of Intelligence Sharing* (Philadelphia: Casemate, 2020).

Westad, Odd Arne. "The Fall of Détente and the Turning Tides of History," in *The Fall of Détente: Soviet-American Relations during the Carter Years*, ed. Odd Arne Westad (Oslo: Scandinavian University Press, 1977).

Westwick, Peter. "From the Club of Rome to Star Wars: The Era of Limits, Space Colonization and the Origins of SDI," in *Limiting Outer Space: Astroculture After Apollo*, ed. Alexander Geppert (New York: Palgrave Macmillan, 2018).

Westwick, Peter. *Into the Black: JPL and the American Space Program, 1976–2004* (New Haven, CT: Yale University Press, 2007).

Westwick, Peter. "'Space-Strike Weapons' Soviet Response to SDI and the Soviet Response to SDI." *Diplomatic History*, 32, no. 5 (2008).

Westwick, Peter. "The International History of the Strategic Defense Initiative: American Influence and Economic Competition in the Late Cold War." *Centaurus*, 52 (2010).

Wilkinson, John Nicholas. *Secrecy and the Media: The Official History of the United Kingdom's D-Notice System* (Oxford: Routledge, 2015).

Wilson, James G. *The Triumph of Improvisation: Gorbachev's Adaptability, Reagan's Engagement, and the End of the Cold War* (Ithaca, NY: Cornell University Press, 2014).

Yost, David. "Les Inquiétudes Européennes Face aux Systèmes de Défense Anti-Missiles un Point de Vue Américain." *Politique Étrangère*, 49, no. 2 (1984).

Yost, David S. "Western Europe and the US Strategic Defense Initiative." *Journal of International Affairs*, 41, no. 2 (1988).

Zak, Anatoly. "Russian Military Space Program: Yesterday Today and Tomorrow," November 3, 2011, available at https://swfound.org/media/53469/rsw_milspace _web.pdf.

Zelizer, Julian E. "Détente and Domestic Politics." *Diplomatic History*, 33, no. 4 (2009).

Zubok, Vladislav. *A Failed Empire: The Soviet Union in the Cold War from Stalin to Gorbachev* (Chapel Hill: University of North Carolina Press, 2009).

UNPUBLISHED PAPERS, THESES, AND DISSERTATIONS

Andreoni, Edoardo. "Ronald Reagan's Strategic Defense Initiative and Transatlantic Relations, 1983–1986." Diss., Cambridge University, 2017.

Buono, Stephen. "The Province of All Mankind: Outer Space and the Promise of Peace." Diss., Indiana University, 2020.

Charles, Elizabeth. "The Game Changer: Reassessing the Impact of SDI on Gorbachev's Foreign Policy, Arms Control, and US–Soviet Relations." Diss., George Washington University, 2010.

Ellis, Thomas. "Reds in Space: American Perceptions of the Soviet Space Programme from Apollo to Mir 1967–1991." Diss., University of Southampton, 2018.

Hays, Peter. "Struggling Towards Space Doctrine: US Military Space Plans, Programs, and Perspectives during the Cold War." Diss., Tufts, 1994.

Hunter, Roger. "A US ASAT Policy for a Multipolar World." MA thesis, School of Advanced Air Power Studies, 1992.

Kelly, Ricky. "Centralized Control of Space: The Use of Space Forces by a Joint Force Commander." MA thesis, School of Advanced Air Power Studies, 1993.

Mowthorpe, Matthew. "The United States Approach to Military Space During the Cold War," undated paper.

Podvig, Pavel. "Protivoraketnaya oborona kak faktor strategicheskik vzaimootnoshenii SSSR/Rossii i SShA v 1945–2003 gg." Diss., Moskva, 2004.

Rohlman, William. "A Political Strategy for Antisatellite Weaponry." MA thesis, The Industrial College of the Armed Forces, 1993.

Schlickenmaier, William. "Playing the General's Game: Superpowers, Self-Limiting, and Strategic Emerging Technologies." Diss., Georgetown University, 2020.

Skogerboe, David. "The Godfather of Satellites: Arthur C. Clarke and the Battle for Narrative Space in the Popular Culture of Spaceflight, 1945–1995." MA thesis, Utrecht University, 2020.

Wilson, Benjamin T. "Insiders and Outsiders: Nuclear Arms Control Experts in Cold War America." Diss., MIT, 2014.

INDEX

Abrahamson, James
appointment of, 7, 63
and Brilliant Pebbles, 171, 173
and Europe, 111–112, 121
presenting SDI, 87–88, 203
retirement from the Air Force,
173–174, 184
and SDI effectiveness, 134
and SDI skeptics, 145
and SDI system concepts, 158, 161
and space launch, 135, 152–153
and testing in space, 142–143
Accidental Launch Protection, 184
Action-reaction cycle, 8, 25, 199
Adelman, Kenneth, 113, 135, 150, 162,
207
Advanced Launch System, 167
Advanced Research Projects Agency, 78
Afghanistan
Soviet invasion of, 39
US invasion of, 209
Air Force Academy Symposium, 41,
44, 45
Air Force Space Command, 44, 45, 48
Akhromeyev, Sergey, 157
Aldridge, Edward, 187
Allen, Lew, 46, 47
Allen, Richard, 54, 55

American National Security Space Infrastructure, 3, 5, 17, 18, 194, 202
American Physical Society, 80, 92
Anderson, Martin, 50
Andropov, Yuri, 60, 63, 64, 66, 140,
148
Anglo-American intelligence cooperation, 7, 103, 110
Anglo-American space cooperation,
105, 111
Anglo-French space cooperation, 105,
110, 124
Anti-Ballistic Missile Treaty
broad vs. narrow interpretation, 146,
151, 159, 173
establishment of, 4, 11, 17
and Europe, 119, 125, 143–144,
191–192
and ground-based missile defense,
184
and hit-to-kill interceptors, 70
and Nuclear and Space Talks, 154–155,
170, 161–162, 180–181
and post-Cold War missile defense,
193–194
and SDI speech, 59
termination of, 211
terms of, 21

Anti-Ballistic Missile Treaty (cont.)
  and testing in space, 134, 143, 153,
    163
  and US-Soviet disagreements,
    143–145, 155, 160, 180
  verification of, 18, 21, 76
Apollo-Soyuz Test Project, 4, 23, 151
Ariane, 103, 105, 129
Arms Control and Disarmament
    Agency, 31, 32, 36, 37, 38, 67, 113,
    135, 144, 207
Arrow missile defense system, 129, 213
ASATs
  and arms control, 7, 12; 26, 29–32,
    34–35, 37–40, 67, 72, 135–138
  British views of, 108, 109
  Congressional views of, 64–65, 68,
    83–84, 147–148, 161, 174
  connection with missile defense, 1, 6,
    10, 14, 42, 65–66, 135–136, 162
  deployment of, 15, 180
  and deterrence, 20, 25, 29, 45, 81,
    188
  and electronic warfare, 27, 31, 34, 38
  French views of, 104, 107
  low and high-altitude, 31, 36, 67, 82,
    107, 136, 208
  and nonnuclear interceptors, 20, 22,
    26–27, 79, 199
  and nuclear weapons, 13, 20
  rationale for, 4, 24–25, 45, 175,
    188–189
  relationship with nuclear war, 31
  Soviet testing of, 11, 19, 21, 23–24, 64
  and space control, 4, 6, 25, 31, 45, 48,
    79, 187
  and strategic stability, 4, 88–89
  US testing of, 13, 38, 147, 160, 190
  and vulnerability of SDI technologies,
    61, 155, 158–159, 170, 173, 182
ASM-135, 147
Aspin, Les, 184, 188, 194
Astroculture, 83, 98, 199

Astrofuturism, 42
Australia, 113, 214
Aviation Week & Space Technology, 33, 76
Axis of Evil, 210

Ballistic Missile Defense Organization,
    194, 197, 205
BAMBI (Ballistic Missile Boost
    Intercept), 14
Bendetsen, Karl, 51, 52
Bethe, Hans, 89, 112
Biden, Joseph, 213
Bold Orion, 13
Boost-phase defense, 69, 84, 87, 89, 98,
    121, 135, 137, 157, 176, 178, 186,
    199, 216
Brandt, Thomas, 91
Brezhnev, Leonid, 11–12, 27, 29, 39
Brilliant Eyes, 185, 194
Brilliant Pebbles
  cancellation of, 194–195
  and *Clementine* mission, 197
  Congressional opposition to, 189,
    192–193
  European concerns about, 191
  inclusion in first-phase missile
    defense concept, 181, 186
  origins of, 170–171, 173–174,
    177–178
  technical analysis of, 178–179, 183
  technical description of, 182
  testing of, 184–185, 188
  vulnerability of, 182
Brookings Institution, 147
Brown, Harold, 30, 32
Brzezinski, Zbigniew, 30, 33, 84, 142
Buchsbaum panel, 24, 28, 222
Buchsbaum, Solomon, 24–26, 28, 175,
    212
Buckley, William, 169
Bush, George H. W., 123, 166, 169–170,
    174–177, 180–181, 190–195, 205,
    208

Bush, George W., 210
Byeman control system, 75

Camp David, 11–112, 115
Canada, 129, 219
Carlucci, Frank, 161, 167–168
Carter, Ashton, 80, 88–89
Carter, Jimmy, 12, 45, 59, 77, 142
  declassification of satellite reconnais-
    sance, 36
  and new negotiations on space
    militarization, 28–29
  and SALT II, 39–40
  and views on ASATs, 30–31, 33–35, 37
Casey, William, 53–54, 93–94, 114, 136,
    138, 151
Central Intelligence Agency, 11, 18–19,
    25, 28, 50, 53–55, 64, 89, 92–95,
    108, 112–114, 116, 119, 139,
    142
Chagas, Carlos, 85–86, 96
Chernenko, Konstantin, 66–67, 139,
    162
Cheysson, Claude, 107
China, 1, 188, 202, 209–211, 213–214
Chirac, Jacques, 128
Citizens Advisory Council, 91
Clancy, Tom, 1
Clark, William, 58
Clementine, 197
Clements, William, 22
Cline, Ray, 19
Clinton, William, 194, 205, 208
Codevilla, Angelo, 33–34, 49–51
Command and control, 26, 31, 33, 53,
    98, 101, 118, 134, 136, 157, 172,
    178, 202, 207
Commission to Assess National
    Security Space Management and
    Organization, 210
Committee on the Present Danger,
    28, 53
Conference on Disarmament, 78

Congress
  and ASATs, 64, 65, 67, 84, 135–136,
    147–148, 162
  and SDI, 62–63, 87, 106, 114, 152–153,
    165, 168, 173, 177, 180, 183–184,
    186, 188–193, 195
Conservative Space Agenda, 4, 28
Cooper, Henry, 67, 178, 188, 193
Cooper, Robert, 78
Coordinating Committee for Multilateral
    Export Control, 114–115, 151
Coors, Joseph, 51
Cornwall, John, 178
Counter-space capabilities, 211–212,
    215
Craddock, Percy, 108
Craxi, Bettino, 159
Crowe, William, 159
Cruise missiles, 60, 106, 121

D-20 Soviet program, 147
de Cuelar, Javier Perez, 78
Defense Acquisition Board, 171,
    184
Defense Advanced Research Projects
    Agency, 78
Defense Intelligence Agency, 49, 54–55
Defense Science Board, 172, 178–179
Defense Support Program, 76
de Gaulle, Charles, 101
DeLauer, Richard, 47, 52–53, 63–64, 70
Delta-180 experiment, 133, 153, 212
d'Estaing, Valery Giscard, 37, 104
Détente, 4, 9, 11–13, 15, 17–19, 21–23,
    25, 27–28, 42, 78, 102, 139, 198
Dobrynin, Anatoly, 27, 32, 59
Dr. Strangelove, 82
Dual-track strategy, 34

Eagleberger, Lawrence, 191–192
Early warning satellites, 3, 17, 21, 23,
    31, 33, 46, 76, 149
Eisenhower, Dwight, 13–14, 43, 81

Election
    of 1980, 34
    of 1984, 68–69, 86, 111
Eureka, 116–117, 119–120, 122
Euromissiles Crisis, 28
European Defense Initiative, 118
European Space Agency, 99–102, 104,
    110–111, 119–120
Exoatmospheric Reentry-Vehicle Inter-
    ceptor Subsystem, 168
Explorer VI satellite, 13

Falklands War, 105
Federation of German Industry,
    118
First Gulf War, 9, 185, 187, 194, 205,
    211
First space war, 165
Fletcher, James, 60
Fletcher panel, 60–61
Ford, Gerald
    and appointment of ASAT review
        panel, 24
    and arms control, 27
    and ASAT decision, 4, 11, 25, 27, 46,
        59, 79, 199, 212
    Reagan's criticism of, 28
Fractional Orbital Bombardment
    System, 16
France
    and concerns about missile defense,
        99, 191
    and cooperation with Germany in
        space, 103–104
    and cooperation with the UK in
        space, 110, 124, 129
    and domestic politics, 128
    and establishment of space com-
        mand, 214
    and Eureka, 116–117, 119
    and space policy, 101–104, 118–119,
        130
    and nuclear weapons, 99–100, 112

Galosh ABM, 34
Garwin, Richard, 79–80, 88–89
Geneva summit, 93, 124, 138, 146–147,
    149, 151, 153, 156
Genscher, Hans-Dietrich, 107, 117,
    120
German unification, 183
German-US memorandum of under-
    standing, 126–128
Gershwin, Larry, 54, 93–94
Gilpatrick, Roswell, 15
Gingrich, Newt, 4, 103
Glenn, John, 32
Global Positioning System, 165, 215
Global Protection Against Limited
    Strikes, 183–186, 191–192
Goldwater, Barry, 83, 91
Gorbachev, Mikhail
    and concerns about a space arms
        race, 134, 140, 149, 163
    and de-linkage of SDI and INF, 8, 160,
        163
    and negotiations at Geneva, 149
    and negotiations at Reykjavik, 133,
        154, 156
    and opposition to SDI, 147, 150–152,
        155, 158
    and rise to power, 139
Government Accountability Office, 47,
    53, 187–188
Graham, Daniel, 49–53, 57, 171, 177,
    186
Gray, Colin, 79, 150
Griffin, Michael, 212
Gromyko, Andrei, 29, 38, 96–97,
    138–140, 157
Ground-based midcourse defense, 211

Hafner, Donald, 81
Happer, William, 183
Harris, Kamala, 2
Heavy-lift space launch, 135, 167
Heritage Foundation, 51, 90, 184

Herno, Charles, 107
Heseltine, Michael, 106, 108, 115–117,
    121–123, 125–127, 129, 144–145,
    221
High-Endoatmospheric Defense Inter-
    ceptor, 183
High Frontier, 50–51, 83, 91, 176
Hit-to-kill technologies, 6, 66, 70, 136,
    153, 163, 199, 204
Hofdi house, 154
Hoffman, Fred, 61
Hoffman Panel, 61
Holy See, 73, 96–98
Homing Overlay Experiment, 66–67,
    70, 143,
Hostile Acts Agreement, 35–36
House Armed Services Committee,
    183
Howe, Geoffrey, 107–108, 123
Hugh, Thomas, 85
Hume, Jaquelin, 51
Hunter, Maxwell, 33
Hussein, Saddam, 165, 187

Iklé, Fred, 66, 135–136
Incidents in Space Agreement, 67
India, 1, 188, 202, 214
INF Treaty, 160, 163
Inspector satellite, 13, 16
Integrating strategic defense compo-
    nents, 61, 89, 163, 172
Intellectual property rights, 113, 122,
    124, 126
Intelsat, 102
Intercontinental ballistic missiles, 13,
    14–16, 49, 56, 80, 82, 102
International Space Station, 1
Iran, 21, 166, 195, 210, 213
Iraq, 165–166, 185, 187, 205, 209,
    211
Iraq War, 209
Israel, 7, 113, 129, 166, 188, 202, 213
Italy, 129, 136

Jamming, 22, 26–27, 31, 34, 37–38, 88
Japan, 7, 93, 103, 113–114, 129, 141,
    191, 202, 214
JASONs, 178–179, 183
Jastrow, Robert, 84, 88
John Paul II, Pope, 73, 85, 96–97
Johns Hopkins University Applied
    Physics Laboratory, 153
Johnson, Lyndon, 15–18, 76, 104
Johnson Space Center, 186
Joint Chiefs of Staff, 29–32, 57–59, 159,
    171–172, 208
Joint Program Office, 168

Kahn, Herman, 78
Kampelman, Max, 84
Katz, Amrom, 20
Keegan, George, 30, 49
Keeny, Spurgeon, 37
Kennedy, John F., 5, 14–15, 18, 20, 71, 75
Kennedy Space Center, 77
Kennedy, Ted, 63, 82
Kerry, John, 84, 147, 174
KGB (Komitet Gosudarstvennoy Bezo-
    pasnosti), 63, 90, 92, 140, 181
Killer satellites, 32, 36, 38, 77
Killick, John, 37
Kinetic Energy ASAT, 175
Kistiakowsky, George, 13
Kohl, Helmut, 104, 107, 117, 120–121,
    126, 183
Krasnoyarsk radar, 144–145
Kuwait, 165, 185, 187
Kwajalein Missile Range, 66

L5 Society, 186
Labour Party, 126, 186
Laser lobby, 34, 52
Laser weapons, 4, 31, 82, 142, 168, 174
    and ASATs, 24, 26, 37, 46
    and Excalibur X Ray, 70
    and laser advocacy, 48–51
    and Mid-Infrared Chemical, 70, 92

Laser weapons (cont.)
  and missile defense, 30, 34, 47,
    52–55, 58, 64, 133–134, 145–147,
    161
  and satellite interference, 22, 33
  and Skif-DM, 159
Levchenko, Stanislav, 90
Libya, 166, 170, 195
Limited Test Ban Treaty, 31
Linhard, Robert, 94–95, 195

Mahan, Alfred Thayer, 5
Major, John, 190–191
Manned Orbiting Laboratory, 62
Mark, Hans, 43
Marshall, Andrew, 61
Matlock, Jack, 133
Matra, 118, 124
Mauroy, Pierre, 103
McNamara, Robert, 14–16
McPeak, Merrill, 165
Meese, Ed, 65–69
Mid-course defense, 183, 185, 186
Miniature Homing Vehicle, 29, 46–47,
    49, 67, 137, 147–148, 162
Mirrors, 55, 140, 145
Missile Defense Act, 189, 193
Missile Defense Agency, 9, 148, 211–212
Missile defense countermeasures, 66,
    88–89, 91–92, 121, 147, 149, 155,
    158, 160–161, 172, 186, 203
Missile defense testing in space
  ABM Treaty restrictions concerning,
    134, 142–143, 146–147, 155, 173
  and Brilliant Pebbles, 179, 185, 188,
    193, 205
  and Congress, 152, 190
  and Delta 180 test, 133, 153
  Soviet views concerning, 133, 154,
    156–158, 160, 163
  and START, 161–163, 181, 203
Missile proliferation, 166, 176–177
MITRE Corporation, 178

Mitrokhin, Vasili, 63
Moakley, Joe, 83–84
Monahan, George, 173
Mondale, Walter, 68–69, 86
Moon race, 3, 40, 75, 77
Moonraker, 82
Moorman, Thomas, 187
Multilayered missile defense, 51, 60–61
Multiple independently targeted reentry
    vehicles, 56
Murrow, Edward, 15
Musk, Elon, 213, 215
Mutually assured destruction, 41, 50
MX missile, 49, 56–57

Nakasone, Yasuhiro, 306
National Aeronautics and Space Admin-
    istration (NASA), 13, 16, 18–19, 44,
    60, 62–63, 71, 75, 102–103, 111,
    135, 197
National Reconnaissance Office, 2–3,
    14, 19, 30, 43, 45, 76, 89, 153, 200,
    212
National Security Agency, 45
National Security Council, 24, 45, 56,
    59, 66, 87, 139, 151, 194
National Security Strategy, 189
National Space Council, 180, 212
National Space Defense Architecture,
    215
National Space Foundation, 159
National Technical Means, 21–22, 26,
    31, 36
NATO
  and the ABM Treaty, 145
  and nuclear policy coordination, 115
  reliance on space systems, 3, 37, 39,
    101, 105, 208
  and SDI, 7, 100, 112
  and space policy, 208, 214
Netherlands, 122, 129
New world order, 166, 192
Nicholson, Robin, 116–117, 121–122

Nike-X, 14
Nitze Criteria, 96
Nitze, Paul, 95–96, 142, 151, 159, 216
Nixon, Richard, 11, 16–19, 27, 102
Nonproliferation Treaty, 16
North Korea, 170, 195, 205, 210
Nuclear abolitionism, 5, 49, 58, 151,
    169, 197
Nuclear and Space Talks, 68, 138–139,
    151, 178, 198, 204, 207
Nuclear Freeze, 49, 56–57
Nuclear Planning Group, 101, 115,
    144–145, 208
Nunn, Sam, 176, 183, 193

O'Neil, Gerard, 50
Obama, Barrack, 211
Office of the Director of National Intel-
    ligence, 211
Office of Net Assessment, 61
Office of Technology Assessment, 88
Ogarkov, Nikolai, 67, 141
Okean Exercise, 23
Open laboratories, 151
Operation Burnt Frost, 209
Operation Desert Shield, 165
Operation Desert Storm, 165–166, 187
Orbital mechanics, 82
Ostpolitik, 126
Outer Space Treaty, 2, 15–16, 31, 78,
    206

Patriot missiles, 166, 194
Paul VI, Pope, 86
Pax Sovietica in space, 8, 90, 109
Peaceful use of outer space, 8, 12, 15,
    17, 22, 24, 43, 73, 75, 77, 83, 86,
    93, 149, 198
Perle, Richard, 68, 143, 148, 159, 177
Pershing II, 68
Pike, John, 81, 145
Pipes, Richard, 28
Poindexter, John, 59

Polaris, 106
Pontifical Academy of Sciences, 73,
    85, 96
Pournelle, Jerry, 50
Powell, Charles, 109, 116, 122–123,
    126, 145, 170
Powell, Colin, 167, 188
Powers, Francis Gary, 3, 13
Prevention of an Arms Race in Outer
    Space, 36
Program-437, 76
Project Spike, 20
Psychological dimensions of space
    weapons, 15, 33, 40, 141–142

Quayle, Dan, 169–170, 175–176,
    177–180

Radar Ocean Surveillance Satellite, 23, 26
RAF Fylingdales, 144
RAND Corporation, 211
Rassemblement pour la République,
    128
Raymond, Jay, 212
Reagan, Ronald
    decision to pursue SDI, 47, 49–50
    and détente, 28
    formally establishing SDI, 61
    and grand strategy, 43
    and Margaret Thatcher, 111, 112,
        115, 156
    and negotiations at Geneva, 149, 151
    and negotiations at Reykjavik, 128,
        133, 153, 156
    and nuclear abolitionism, 5, 49,
        57–58, 122, 151, 156, 169, 197
    presentation of SDI as a tool of peace,
        6, 42, 65, 69, 72, 87, 137–138, 150,
        162
    rejection of limits on SDI technolo-
        gies, 133, 134, 137–138, 155, 162,
        204
    and SDI speech, 52, 57–60

Reagan, Ronald (cont.)
  and sharing SDI, 7, 69, 149–151, 155,
    200
  and the Soviet space threat, 55–56
  and space policy, 4, 41–42, 44, 46,
    48, 63
  views of the ABM Treaty, 28, 143
  views of space technologies, 9, 69,
    74, 137
Reconnaissance-strike complex, 141
Red Army Faction, 128
Red shield, 90, 161
Revolution in Military Affairs, 24, 107,
    157
Reykjavik summit, 128, 133, 153–154,
    156, 158, 160
Rogue states, 176–177, 184, 194, 205,
    210
Roman Catholic Church, 73, 96
Rome Declaration, 110
Rona, Thomas, 187
Rowney, Ed, 70
Rules of behavior in space, 135, 207, 215
Rumsfeld, Donald, 24, 27, 210
Rush, Kenneth, 22
Russian Federation, 1–2, 190–192, 198,
    208–209, 211–212, 213–215
Rye, Gil, 45, 48, 59, 69

Sagan, Carl, 83–84
Sagdeev, Roald, 90, 156–157
Saigon, Fall of, 27
Sary Shagan, 24, 174
Satellite communications, 101–102,
    105, 110, 124
Satellite Militaire de Reconnaissance
    Optique, 103
Satellite reconnaissance
  and arms control verification, 11,
    17–18, 21, 36, 81
  civilian control over, 3
  and Corona, 13
  declassification of, 36, 78

and Hexagon, 18
interference with, 13, 15, 17, 21–22,
    24, 26, 29, 31–33, 38
and international stability, 84, 86
and Kennen, 19
legal status concerning, 12–13, 15
and NATO, 101, 136
and the NRO, 14
secrecy concerning, 2, 8, 76, 97, 200
and tactical military support, 12,
    30, 35
vulnerability of, 17, 25, 30, 89
and Zaman, 19
Satellite Security Act, 174
Satellites and nuclear command and
    control, 26, 136
Satellites and transparency, 40, 77, 105,
    198
Saudi Arabia, 165–166
Schelling, Thomas, 45
Schriever, Bernard, 13, 41
Scientists and SDI, 58, 73–74, 79–80,
    89, 90, 92, 98, 112, 120, 141, 145,
    156–158, 168, 173, 178, 183, 201,
    207
Scowcroft, Brent, 24, 77, 169, 175, 183,
    191–192
Scud missiles, 166, 195, 205
Seiberling, John, 84
Senate Armed Services Committee, 159,
    193
Senate Select Committee on Intelli-
    gence, 33
Separate Military Space Service, 44, 48
Shcherbitsky, Vladimir, 142
Shevardnadze, Eduard, 140, 148, 180
Shultz, George, 59, 68, 135–136,
    138–139, 145, 154
Significant Technical Achievements in
    Research, 143
SK-1000 Soviet program, 147
Skif-DM, 159–160
Skynet, 101, 105, 124, 129

SM-3 missile, 209
Smart Rocks, 170–171
Smart weapons, 165
Smith, William, 30
Social Democratic Party, 117
Soviet anti-SDI Propaganda, 91–93, 96–97
Soviet arms control violations, 144–145
Soviet ASAT program, 20, 22–23, 25, 32, 46, 64–65, 77, 148, 182
Soviet dependence on space systems, 24, 45
Soviet Military Power, 93–94, 96
Space
    and arms race, 1, 7–8, 11, 15, 17, 34, 36, 64, 68–69, 74, 77–79, 90–91, 100–104, 136, 141, 150, 160, 176, 209–210
    and debris, 1, 148, 190, 215
    as the "high ground," 41, 44, 47–48, 64, 71, 165, 189, 194
    militarization of, 2, 7–8, 10, 13, 15, 28, 40, 64, 66–67, 72, 78, 82–83, 92, 100, 130, 139, 149, 163, 197, 202, 206, 212, 214
    as a sanctuary, 4, 11, 27, 31–33, 47, 85, 206
Space-based interceptors, 84, 87, 89, 92, 133–136, 158, 167–168, 170–171, 176–179, 181–182, 183–186, 188–189, 193, 195, 202, 205, 216
Space control, 6, 44, 48, 50, 91, 158, 160, 168, 184, 187, 203
Space launch, 5, 44, 52, 87, 103, 105, 129, 152–153, 158, 167, 170, 172–173, 178, 181–182, 203, 216
Space logistics, 87, 135, 152
Space mines, 81, 88, 147
Space Pearl Harbor, 210
Spacepower, 9–10, 26, 43, 47, 79, 187
Space shuttle, 4, 44, 102–103
    and *Challenger*, 129, 152–153, 203
    and *Discovery*, 145–146

and launching foreign payloads, 105, 152
    origins of, 18
    and SDI, 5, 65, 89, 145–146, 153, 203
    Soviet fears of, 35, 38
Space-strike weapons, 6, 64, 138, 161
Space superiority, 9, 13–14, 187, 211–212
Space-to-ground attack, 48, 64, 162, 180
Space tug, 102, 117
Space zealots, 41
    Spacelab, 102, 117, 119
Special access program, 74
SS-9, 81
SS-18, 171
SS-20, 27, 29
Stares, Paul, 20, 48, 81
Starlink, 213, 215
Star Wars, 82
Strategic Arms Reduction Treaty, 67, 70, 139, 160–162, 180–183, 198, 203
State Department, 15–16, 19
Strategic Arms Limitation Talks, 11–12, 16–17, 22, 23, 26–27, 29–32, 35–40, 50
Strategic Defense Initiative
    and the ABM Treaty, 4, 134, 143, 146, 151, 153, 159, 163, 173
    and ASATs, 1, 6, 10, 14, 42, 65–66, 135–136, 162
    cost of, 51–52, 87, 95–96, 134–125, 171–174, 176, 181–183, 203
    effectiveness criteria concerning, 61, 134, 171
    European views of, 100, 102, 106–112, 115–118, 191
    feasibility of, 52, 58, 61, 74, 92, 111, 140, 161, 166, 172–173, 178–179, 184, 212
    and the INF Treaty, 8, 160, 163
    international involvement in, 112–114, 120–130
    origins of, 47, 49–50

Strategic Defense Initiative (cont.)
  and morality, 73, 85, 96, 98, 111
  public debate concerning, 73–75,
    79–81, 83–85, 87–90
  relationship with deterrence, 61, 76,
    95, 11, 171, 177, 204, 210
  relationship with space control,
    157–158, 160, 184, 203
  and software concerns, 173, 178, 202
  Soviet propaganda against, 91–93
  Soviet views of, 63, 90, 92, 140–141
  speech announcing, 52, 57–60
  and spin off technologies, 7, 116–117,
    119, 121, 125, 129
  as a tool of peace, 6, 42, 65, 69, 72,
    87, 137–138, 150, 162
Strategic Defense Initiative Organization
  and ASATs, 160
  and cost of SDI, 87, 96, 114, 152, 189,
    193, 203
  establishment of, 62
  and foreign partnerships, 118,
    128–129, 201
  and public diplomacy, 90, 96, 183
  renaming of, 194
  and system development, 95, 112,
    134–136, 170–172, 195, 202, 205
  and testing missile defense, 143,
    145–146, 153, 157–158, 163, 173,
    179, 184–185, 188, 212
Strategic Defense System, 157–158, 171
  deployment timeline concerning, 159
Surveillance, Tracking, and Kill Assess-
    ment, 62
System effectiveness, 7, 61, 79, 95, 134,
    142, 171–172, 181, 195, 204
System of systems, 134
Système Probatoire d'Observation de la
    Terre, 103

Tactical Exploitation of National Capa-
    bilities, 30
Talent Keyhole, 14

Team B, 28, 50, 53
Technological progress, 87–88, 170
Technological surprise, 56
Technology gap, 7, 99, 116, 141
Technology theft, 114
Technology transfer, 7, 100, 102, 114,
    117, 120–121, 124–126, 128–130,
    150–151, 201
Teller, Edward, 51–52, 57–58, 94
Teltschik, Horst, 120–121, 129
Terminal defense, 121
Terrorist attacks on September 11, 2001,
    210
Thatcher, Margaret
  and Camp David Declaration,
    111–112
  and concerns about a space arms
    race, 108, 136
  disagreements with government
    ministers concerning SDI, 109,
    116, 170
  reaction to SDI, 106
  and relationship with James Abraha-
    mson, 62, 158, 173
  support for strategic defense deploy-
    ment, 170, 172, 176
  and UK involvement in SDI,
    121–126
  views of science and technology, 109,
    159
Titan-D, 152
Trident, 106, 110, 125
Trump, Donald, 212–213
Turner, Stansfield, 11

U-2 incident, 3, 13
Ukraine, 213, 215
UK Foreign Office, 37, 90, 108–110,
    125, 143, 168–170, 172, 190–191
UK Ministry of Defence, 46, 90, 103,
    105, 108–109, 110, 118, 121–125,
    158–159, 169–170, 172, 178–179,
    190

Union of Concerned Scientists, 80, 89,
    92, 112
United Kingdom
    and concerns about Soviet missile
        defense, 68
    and establishment of a space com-
        mand, 214
    and intelligence sharing with the US,
        7, 122
    and involvement in SDI, 121–127,
        129
    and nuclear weapons, 99, 106, 112
    and space cooperation with the US,
        101, 105, 111
    and space policy, 101
United Nations, 36, 148, 165
US Air Force, 13–14, 20, 30, 47, 80,
    212
US Air Force Space Command, 48
US Army, 14, 51, 66, 175
US defense contracting laws, 123
US dependence on space systems, 13,
    20, 39, 211
US Information Agency, 15
US National Test Bed, 171
US Navy Space Command, 48
US Space Command, 1, 44–45, 48, 80,
    181, 212
US Space Force, 212–213
US–UK Mutual Defense Agreement, 126

Vance, Cyrus, 29, 32, 37, 38
Vandenberg Air Force (now Space Force)
    Base, 2, 66, 147
Vatican, 96, 207
Velikhov, Evgeny, 90, 160
von Braun, Werner, 13

Wallop, Malcolm, 33–34, 49, 51–52, 53
WarGames, 82
Warner, John, 147, 186
Warnke, Paul, 32, 34
Watergate, 27

Watkins, James, 58, 137
Webb, James, 75
Weinberger, Caspar, 52–53, 70, 97, 162,
    167
    and the ABM Treaty, 144–145
    and the creation of SDIO, 62
    and foreign involvement in SDI, 113,
        115–116, 122–123, 125, 127
    and the SDI speech, 59
    views on ASAT, 68, 135–138, 151,
        157, 159
Weisskopf, Victor, 89
West Germany, 103–104, 113, 117,
    120–122, 127–129, 136, 201
Western European Union, 107
Wolfowitz, Paul, 192
Wood, Lowell, 51, 171, 181–182
Wörner, Manfred, 115, 117, 127